T0289844

Aerodynamics of Wind Turbines

Aerodynamics of Wind Turbines

Editor: Olive Murphy

www.callistoreference.com

Callisto Reference,
118-35 Queens Blvd., Suite 400,
Forest Hills, NY 11375, USA

Visit us on the World Wide Web at:
www.callistoreference.com

ISBN: 978-1-64116-836-6 (Hardback)

Cataloging-in-Publication Data

Aerodynamics of wind turbines / edited by Olive Murphy.
p. cm.
Includes bibliographical references and index.
ISBN 978-1-64116-836-6
1. Wind turbines--Aerodynamics. 2. Wind turbines--Environmental aspects.
3. Aerodynamics. I. Murphy, Olive.
TJ828 .A37 2023
621.312 136--dc23

Table of Contents

Preface

This book has been a concerted effort by a group of academicians, researchers and scientists, who have contributed their research works for the realization of the book. This book has materialized in the wake of emerging advancements and innovations in this field. Therefore, the need of the hour was to compile all the required researches and disseminate the knowledge to a broad spectrum of people comprising of students, researchers and specialists of the field.

A wind turbine converts wind energy into electricity by utilizing the aerodynamic force of the rotor blades. The blades of the wind turbines are developed by using the aerofoil structure, which is also frequently utilized in the construction of airplane wings. There are many distinct varieties of wind turbines, and each one generates electrical energy using a unique idea. Lift and drag are the two significant aerodynamic forces utilized by wind turbines. The wind turbines which use drag forces contain a vertical rotor and are based on the concept of air resistance. The lift propelled wind turbines contain blades which are positioned perpendicular to the wind direction. The maximum speed of air resistance wind turbines cannot be greater than the speed of wind but lift propelled wind turbines can rotate faster than the speed of the wind. This book elucidates the concepts and innovative models around prospective developments with respect to wind turbines and their aerodynamics. It is a resource guide for experts as well as students.

At the end of the preface, I would like to thank the authors for their brilliant chapters and the publisher for guiding us all-through the making of the book till its final stage. Also, I would like to thank my family for providing the support and encouragement throughout my academic career and research projects.

Editor

A Simplified Free Vortex Wake Model of Wind Turbines for Axial Steady Conditions

Bofeng Xu [1,*], Tongguang Wang [2], Yue Yuan [1], Zhenzhou Zhao [1] and Haoming Liu [1]

1 College of Energy and Electrical Engineering, Hohai University, Nanjing 211100, China; yyuan@hhu.edu.cn (Y.Y.); zhaozhzh_2008@hhu.edu.cn (Z.Z.); liuhaom@hhu.edu.cn (H.L.)
2 Jiangsu Key Laboratory of Hi-Tech Research for Wind Turbine Design, Nanjing University of Aeronautics and Astronautics, Nanjing 210016, China; tgwang@nuaa.edu.cn
* Correspondence: bfxu1985@hhu.edu.cn

Abstract: A simplified free vortex wake (FVW) model called the vortex sheet and ring wake (VSRW) model was developed to rapidly calculate the aerodynamic performance of wind turbines under axial steady conditions. The wake in the simplified FVW model is comprised of the vortex sheets in the near wake and the vortex rings, which are used to replace the helical tip vortex filament in the far wake. The position of the vortex ring is obtained by the motion equation of its control point. Analytical formulas of the velocity induced by the vortex ring were introduced to reduce the computational time of the induced velocity calculation. In order to take into account both accuracy and calculation time of the VSRW model, the length of the near wake was cut off at a 120° wake age angle. The simplified FVW model was used to calculate the aerodynamic load of the blade and the wake flow characteristic. The results were compared with measurement results and the results from the full vortex sheet wake model and the tip vortex wake model. The computational speed of the simplified FVW model is at least an order of magnitude faster than other two conventional models. The error of the low-speed shaft torque obtained from the simplified FVW model is no more than 10% relative to the experiment at most of wind speeds. The normal and tangential force coefficients obtained from the three models agree well with each other and with the measurement results at the low wind speed. The comparison indicates that the simplified FVW model predicts the aerodynamic load accurately and greatly reduces the computational time. The axial induction factor field in the near wake agrees well with the other two FVW models and the radial expansion deformation of the wake can be captured.

Keywords: wind turbine; simplified free vortex wake; vortex ring; aerodynamics; axial steady condition

---

1. Introduction

Over the past four decades, free vortex wake (FVW) methods have emerged as robust and versatile tools for modeling the aerodynamic performance of wind turbine rotors. Unlike the blade element momentum (BEM) theory [1,2] where annular average induction is found, the FVW method can determine vortical induction directly at the blade elements from the effect of the modeled wake and the method is more efficient than computational fluid dynamic (CFD) methods. However, the computational cost of the FVW method is much higher than that of the BEM method. Perhaps this is the reason that the FVW method is not commonly used for predicting the aerodynamic loads of wind turbine rotors in the wind energy community.

FVW methods are based on a discrete representation of the rotor vorticity field and a Lagrangian representation of the governing equations for the wake elements. In essence, the wake elements are allowed to convert and deform under the action of the local velocity field to force-free locations.

Clearly, the ability to predict the aerodynamic performance of wind turbines is strongly dependent on the ability to predict the highly complex wake geometry. Free wake analyses are fundamentally better suited to the complex flowfields generated by wind turbines and avoid the difficulty of prescribing a wake geometry but, in so doing, introduce more computational costs. Gohard [3] presented the first full vortex sheet wake (FVSW) model for horizontal axis wind turbines, in which the unconstrained wake was permitted to move freely with the local velocities existing in the wake. The initial wake geometry of the FVSW model is shown in Figure 1a. FVSW methods for the analysis of wind turbine aerodynamics were also used by Arsuffi et al. [4], Garrel [5], Sant et al. [6] and Sebastian et al. [7]. These methods are computationally expensive, making them somewhat impractical as a design tool. To remedy this problem, Rosen et al. [8] divided the wake into two regions, i.e., near and far wakes. The calculations associated with the far wake are speeded up by some approximations. The most popular approximation is the tip vortex wake (TVW) model [9–12] as shown in Figure 1b, in which, the far wake extends beyond the near wake and is comprised of a single helical tip vortex filament, which is appropriate based on the physics of the flow. The strength of the tip vortex is determined by assuming that the sum of the blade bound vorticity outside of the maximum is trailed into the tip vortex. The release point of the tip vortex is usually the tip of the blade. More significant simplifications to the FVW models for wind turbines are attributed to Miller [13], Afjeh et al. [14], and Yu et al. [15]. The vortex wake system was simplified by Miller [13] and Afjeh et al. [14] into three parts: a number of straight semi-infinite vortex lines, a series of vortex rings, and a semi-infinite vortex cylinder. The method has been used for comprehensive rotorcraft and wind turbine analyses. Yu et al. [15] developed a free wake model that combines a vortex ring model with a semi-infinite cylindrical vortex tube, in which the thrust coefficient of the rotor must be known and is used to calculate the strength of the new vortex ring produced in a time step. Except for the approximations of the wake, parallelization techniques have been used to address the computational expense problem of the FVW model. Farrugia et al. [16] modified the FVW code Wake Induced Dynamics Simulator (WInDS) [7] to enable parallel processing. Later, Elgammi et al. [17] used the modified model to investigate the cycle-to-cycle variations in the aerodynamic blade loads experienced by yawed wind turbine rotors operating at a constant speed with a fixed yaw angle. Turkal et al. [18] used a Graphics Processing Unit (GPU), to exploit the computational parallelism involved in the free-wake methodology.

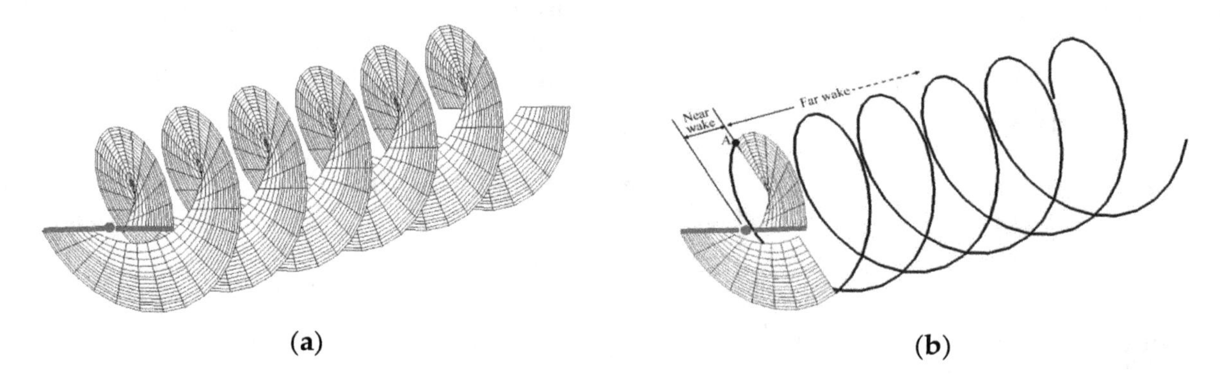

Figure 1. Schematic of the initial wakes of different FVW models. (**a**) FVSW model. (**b**) TVW model.

In FVW methods, the number of discrete elements per vortex filament can be very large, making the tracking process memory-intensive and computationally demanding, although still considerably less demanding than when using CFD methods. The FVSW model has a computation time of about tens of minutes at one steady condition, and the TVW model has a computation time of about several minutes. Although the TVW model requires much less computational time than the FVSW model because it has fewer wake nodes in the far wake, the computational cost makes it somewhat impractical as an iterative optimization design tool. The main objective of this study is to reduce the number of wake nodes that need to be computed and achieve lower computation time in the axial steady condition. The velocity in the plane of rotation of the blade induced by the near wake is remarkable, so the near wake is still modeled by the vortex sheets in order to reflect the actual flow in this study. The effect of the far wake on the plane of rotation is relatively smaller due to the increase in distance, so the far wake is simplified into a series of vortex rings through learning from the studies of Miller [13] and Yu et al. [15]. Therefore, the new simplified FVW model is referred to as the vortex sheet and ring wake (VSRW) model.

In Section 2, the development of the VSRW model is described. Section 3 describes the analysis of the effect of the length of the near wake. The results are presented in Section 4 and include the rate of convergence of the wake iteration, the wake geometry, the induction factor in the wake, and the low-speed shaft torque (LSST). The conclusions are drawn in Section 5.

2. Simplified Free Vortex Wake

The development of the simplified FVW model will be presented in this section. The blade model is first briefly introduced and then the representation of the wake model is provided, including the near wake model and the far wake model. For completeness, the calculation procedure of the simplified FVW model is detailed.

2.1. Blade Model

The FVW model assumes that the flow field is incompressible and potential. Figure 2 shows the rotor body frame coordinates with the z-axis pointing downstream. The blade is modeled by the Weissinger-L model [19], which is a good compromise between the lifting line and the lifting surface models, as a series of straight constant-strength vortex segments lying along the blade quarter chord line. The control points are located at the 3/4-chord line at the center of each vortex segment. The wake vortices extend downstream from the 1/4-chord and form a series of horseshoe filaments. The trailing and shed vortices are modeled by the trailing and shed straight-line vortex filaments as shown in Figure 2.

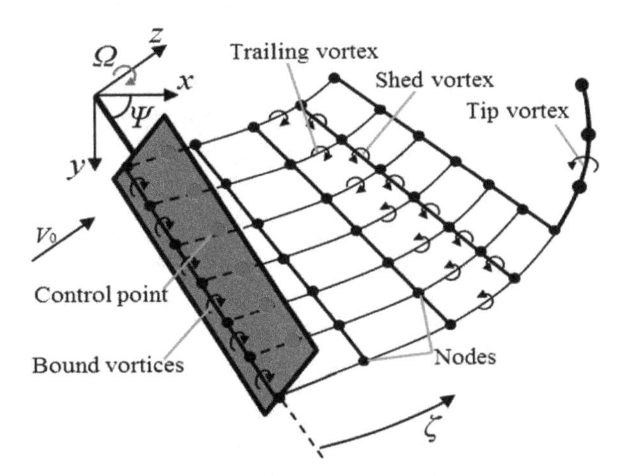

Figure 2. Schematic of the blade model and the vortex wake model.

The blade root section corresponds to the boundary of the first blade element. The remaining boundary distribution along each blade is achieved using the following "arc-cosine" relationship:

$$(\bar{r}_b)_i = \frac{(r_b)_i}{R} = \frac{2}{\pi}\arccos\left(1 - \frac{i-1}{N_E}\right) \tag{1}$$

where N_E is the number of blade elements and i is the element boundary number ($i = 1, \ldots, N_E + 1$). Consequently, there are N_E element control points and $(N_E + 1)$ boundary points.

The strength, Γ_b, of each blade element is evaluated by the application of the Kutta-Joukowski theorem on the basis of airfoil data. The bound vorticity for the ith blade element is given by:

$$(\Gamma_b)_i = \frac{1}{2} W_i C_l c_i \tag{2}$$

where W_i is the resultant velocity at the ith control point and c_i is the chord of the ith blade element. To take into account the three-dimensional rotational effect, the 2D airfoil data is modified by the Du-Selig stall-delay model [20].

2.2. Near Wake Model

The wake that extends downstream from the blade is divided into two parts: the near wake and the far wake (Figure 1b). The near wake is modeled by the vortex sheets, which consist of the trailing and shed vortex filaments. The circulation of each blade element is assumed to be constant. However, the adjacent segments may have unequal loadings and, therefore, have different circulation strengths. The circulation strength of each trailing vortex is equal to the difference between the bound vortex strengths of the two adjacent segments. The circulation strength of each shed vortex is equal to 0 in the axial steady condition. The vortex filaments are allowed to freely distort under the influence of the local velocity field. The convection of these vortex filaments can be described by the Helmholtz equation. For a blade with fixed coordinates, the governing equation of the vortex filaments can be written as the partial differential form:

$$\frac{\partial \boldsymbol{r}_v(\psi,\zeta)}{\partial \psi} + \frac{\partial \boldsymbol{r}_v(\psi,\zeta)}{\partial \zeta} = \frac{1}{\Omega}[\boldsymbol{V}_0 + \boldsymbol{V}_{ind}(\psi,\zeta)] \tag{3}$$

where $\boldsymbol{r}$ is the position vector of the vortex collocation point; $\boldsymbol{V}_{ind}$ equals the mean value of the induced velocities at the surrounding four grid points [9].

To solve the convection equation of the vortex filaments, numerical solutions have been investigated over the last decades, including relaxation methods [21] and time-marching methods [9,11]. The time-marching methods can potentially provide the best level of approximation to the rotor wake problem with the fewest application restrictions. However, these methods have been found to be rather susceptible to numerical instabilities [22]. On the other hand, the relaxation methods improve the ability to control the non-physical wake instabilities by explicit enforcement of the wake periodicity. The relaxation methods can only be used for steady-state problems.

In this study, the axial steady conditions are predicted. It is sufficient to apply a steady relaxation scheme to solve the convection equation of the vortex filaments in the near wake. A five-point central difference approximation is used for both the temporal and spatial derivatives. The detailed information can be found in Reference [21]. A pseudo-implicit technique [23,24] has been introduced to improve the stability of the free-vortex iteration. The effective range of the azimuthal discretization is usually between $\Delta\psi = 5°$ and $\Delta\psi = 20°$ [22,25] and $\Delta\psi = 10°$ was selected for use in this study. The spatial step $\Delta\zeta$ uses the same value as the azimuthal step.

The velocity in the plane of rotation of the blade induced by the near wake is remarkable, so the length of the near wake is significant for the accuracy of the proposed model. The determination of the length of the near wake will be discussed further in Section 3.

2.3. Far Wake Model

In the TVW model, the far wake is modeled by the tip vortex model. In this study, vortex rings are used to replace the helical tip vortex filament as shown in Figure 3. Point A is the release point of the tip vortex and point C is the origin point of the 2nd circular tip vortex filament. The intersection point B of the vortex ring and the replaced helical tip vortex filament is located at the middle of every circular helical filament. The center of the ring is on the rotor's rotation axis. The point 1 opposite the point B on the ring is the control point of the vortex ring. The wake length in this paper is $4R$. The number of the vortex rings is determined by the wake length:

$$N_C = \mathrm{INT}\left(\frac{4R}{V_0} \cdot \frac{\Omega}{2\pi}\right) + 1 \tag{4}$$

where the equation INT is an integer-valued function; therefore, we need to add 1 to consider the missing part.

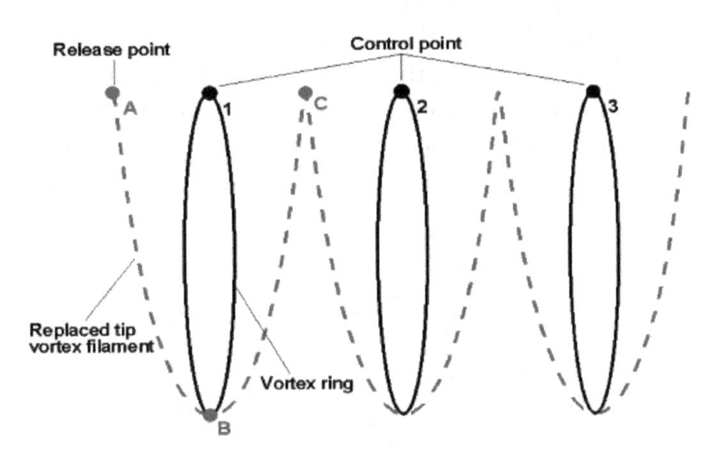

Figure 3. Schematic of the far wake.

As shown in Figure 3, points 1, 2, and 3 are the control points of the 1st, 2nd, and 3rd vortex rings respectively. In the axial steady condition, the vortex rings in the far wake are stationary when the wake geometry is convergent. It is assumed that point A will arrive at the position of point 1 when it moves under the influence of the velocity of point A for one half of the rotational period. Therefore, one period will be experienced between the two control points of the adjacent vortex rings. The position of the nth control point can be represented by its axial position $Z_{cr(n)}$ and its radial position $R_{cr(n)}$. The axial position can be obtained by:

$$\begin{aligned} Z_{cr(1)} &= Z_A + \sum_{k=1}^{N_T/2} \tfrac{\Delta\psi}{\Omega}\left(V_0 + V_{ind,Z}^{A}\right) \\ Z_{cr(n)} &= Z_{cr(n-1)} + \sum_{k=1}^{N_T} \tfrac{\Delta\psi}{\Omega}\left(V_0 + V_{ind,Z}^{n-1}\right), \quad 2 \le n \le N_C \end{aligned} \tag{5}$$

The radial position can be obtained by:

$$\begin{aligned} R_{cr(1)} &= R_A + \sum_{k=1}^{N_T/2} \tfrac{\Delta\psi}{\Omega} V_{ind,R}^{A} \\ R_{cr(n)} &= R_{cr(n-1)} + \sum_{k=1}^{N_T} \tfrac{\Delta\psi}{\Omega} V_{ind,R}^{n-1}, \quad 2 \le n \le N_C \end{aligned} \tag{6}$$

2.4. *Velocity Induced by the Near Wake and the Far Wake*

2.4.1. Velocity Induced by the Near Wake

The vortex filaments of the near wake comprise a series of straight-line vortex elements. The velocity induced by the vortex filaments equals the sum of the velocities induced by the straight-line vortex elements at the control nodes of the vortex elements, which are calculated using the Biot–Savart law as:

$$\boldsymbol{V}_{ind} = \frac{\Gamma}{4\pi h}(\cos\theta_A - \cos\theta_B)\frac{\boldsymbol{r}_A \times \boldsymbol{r}_B}{|\boldsymbol{r}_A \times \boldsymbol{r}_B|} \tag{7}$$

where h, θ_A, θ_B, $\boldsymbol{r}_A$, and $\boldsymbol{r}_B$ are as shown in Figure 4. However, if a collocation point is positioned very close to the vortex-line segment, then $(h \to 0)$ will result in a very high induced velocity. In addition, the self-induced velocity $(h = 0)$ exhibits a logarithmic singularity. These two phenomena cause convergence problems. To avoid these numerical problems, some vortex core models based on the Lamb-Oseen vortex model [26], Ramasamy and Leishman vortex model [27], Vatistas vortex model [28] and β-Vatistas vortex model [29,30] have been applied in a FVW model. In our recent work, we compared the effects of the β-Vatistas model and Lamb-Oseen model in the FVW model [29] and further studied the application [30] of the β-Vatistas model. In this study, a vortex core model based on the Lamb-Oseen vortex model, which is the most widely adopted and simpler model, was used. To account for the effect of viscous diffusion, the vortex core radius growth is used in the Biot-Savart law and is modified by using an empirical viscous growth model [31]. The stretching effect of the vortex filaments is taken into account by the application of a model developed by Ananthan and Leishman [32].

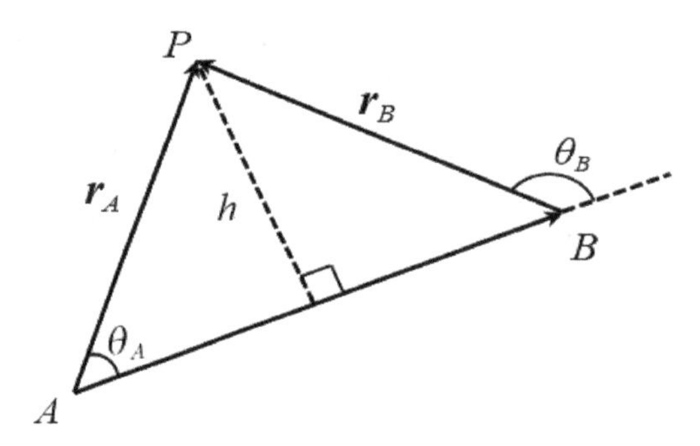

Figure 4. Schematic of the straight-line vortex elements.

2.4.2. Velocity Induced by the Far Wake

The analytical formulas of the velocity induced by a vortex ring are given by Yoon and Heister [33]. Figure 5 shows the Cartesian coordinate system used in this study for the induced velocity calculation. The axial and radial velocities at an arbitrary point P induced by the nth vortex ring are given by:

$$v_{ind,z} = \frac{\Gamma_n}{2\pi a}\left[K(m) - \frac{b^2 + R_p^2 - R_{cr(n)}^2}{a^2 - 4R_pR_{cr(n)}}E(m)\right] \tag{8}$$

$$v_{ind,r} = -\frac{b\Gamma_n}{2\pi R_p a}\left[K(m) - \frac{b^2 + R_p^2 + R_{cr(n)}^2}{a^2 - 4R_pR_{cr(n)}}E(m)\right] \tag{9}$$

where

$$a = \sqrt{\left(Z_p - Z_{cr(n)}\right)^2 + \left(R_p + R_{cr(n)}\right)^2} \tag{10}$$

$$b = Z_p - Z_{cr(n)} \tag{11}$$

$K(m)$ and $E(m)$ are the complete elliptic integrals of the first and second kind, where m is given by:

$$m = \frac{4R_pR_{cr(n)}}{a^2} \tag{12}$$

A fast method [34] is used for evaluating the first and second integrals. When point P is on the vortex ring, the induced velocity is calculated by Equation (7) to avoid the self-induced numerical problem.

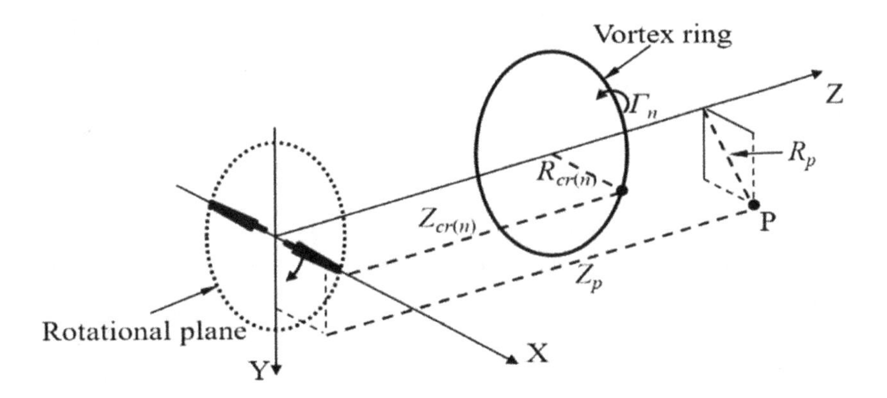

Figure 5. Schematic of the coordinate system for the induced velocity calculation of the vortex ring.

2.5. Calculation Procedure

The calculation process of the proposed simplified FVW model is shown in Figure 6. The following provides detailed explanations for some steps in the flowchart.

- In step 2, the initial wake geometry of the near wake consists of a set of regular helixes. The initial wake geometry of the far wake is calculated by Equations (5) and (6) and the initial induced velocities in the two equations equal 0.
- In step 5 and step 7, the velocity at the node and the control point induced by the vortices is calculated using the methods in Section 2.4.
- In step 6, the five-point central difference approximation is used to solve the convection equation of the vortex filaments in the near wake and Equations (5) and (6) are used to obtain the shape and location of the vortex rings in the far wake.
- In step 8, the root mean square (RMS) change between the new wake geometry and the old wake geometry of the two iteration steps is calculated. If the RMS change is less than a prescribed tolerance of 10^{-4}, convergence is achieved. Otherwise, return to step 3.

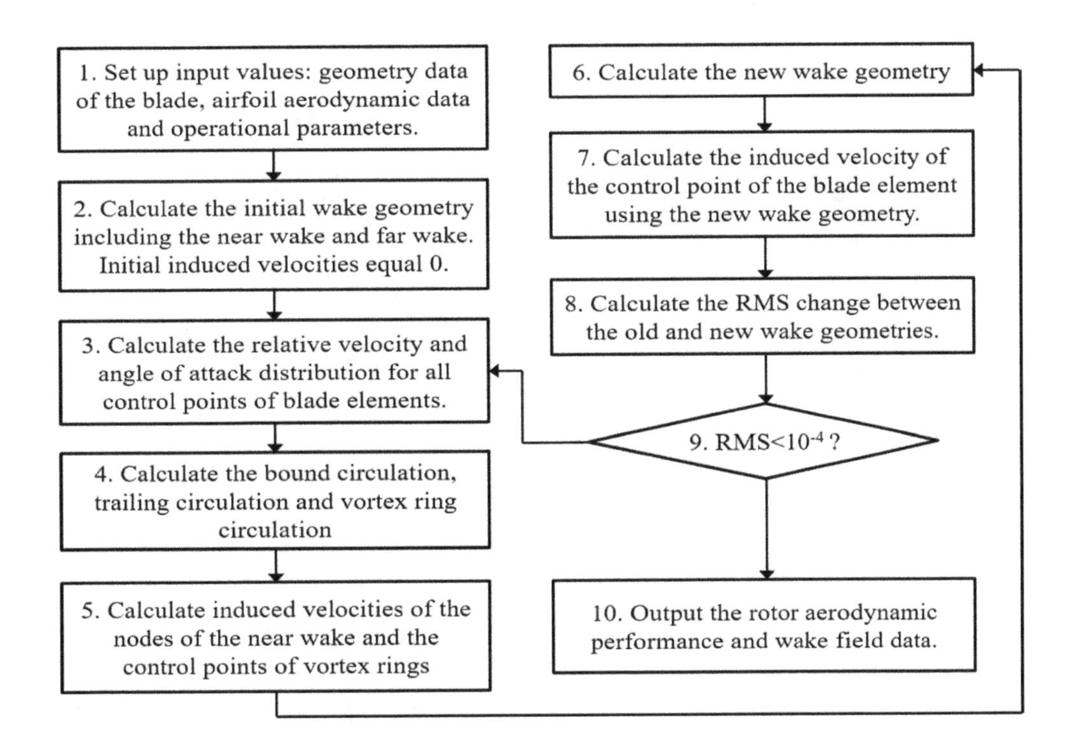

Figure 6. Flowchart of the simplified FVW model.

3. Length of Near Wake

To validate the VSRW model, the National Renewable Energy Laboratory (NREL) Phase VI wind turbine is used in this paper as a test case. The NREL Phase VI is a stall-regulated turbine. This turbine was designed by the NREL. The experiments were performed in the National Aeronautics and Space Administration (NASA) Ames wind tunnel (24.4 × 36.6 m) [35], which is considered a benchmark for the evaluation of wind turbine aerodynamic methods.

In the TVW model, the near wake is truncated after a short azimuthal distance, which is typically an azimuth angle of about 60° [36] in the helicopter field. Beyond this point, the far wake extends and is comprised of a single tip vortex filament. In the proposed VSRW model, one should ensure that the vortex sheet extends sufficiently far downstream. The aerodynamic performance of the rotor should be independent of the length of the vortex sheet. The length of the vortex sheet (or length of the near wake) can be represented by the wake age angle of the near wake. The spatial step equals 10° as mentioned above.

The LSST has been calculated used different lengths of the near wake ranging from 10° to 360°. Figure 7 shows the LSST along the length of the near wake at different wind speeds. As the length of the near wake increases from 10°, the LSST increases at the beginning and reaches the first maximum value at a certain near wake length. When the length of the near wake is longer than this certain length, the LSST is nearly constant at this maximum value. This certain length of the near wake is defined as the dividing length in this study. It is evident that the dividing length of the near wake is about 120° at wind speeds of 13 m/s, 20 m/s and 25 m/s, about 150° at wind speeds of 7 m/s and 10 m/s and about 140° at a wind speed of 15 m/s. At wind speeds of 7 m/s, 10 m/s and 15 m/s, the changes of the LSST between the near wake length of 120° and the dividing length are about 0.73%, 0.08% and 0.04% of the first maximum value. This illustrates that when the length of the near wake is about 120°, the LSSTs achieve nearly constant values at all wind speeds.

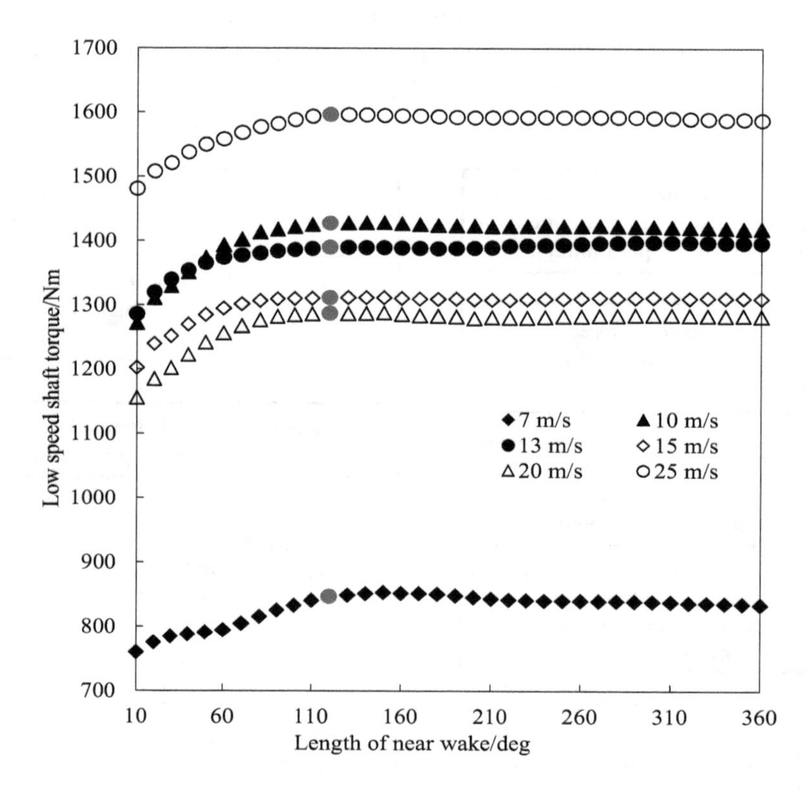

Figure 7. LSST along the length of the near wake at different wind speeds (The red points are the LSST values with a near wake age angle of 120°).

The calculation time of the VSRW model along the length of the near wake was also examined. The central processing unit (CPU) of the computer is an Intel Core i5-4200U and the memory is 8 GB. Figure 8 shows the wake iteration time along the length of the near wake at 7 m/s and 15 m/s. It is apparent that the number of discrete nodes used to approximate the near wake increases as the length of the near wake increases. However, the computational cost for the induced-velocity calculation changes as the square of the number of discrete nodes [25].

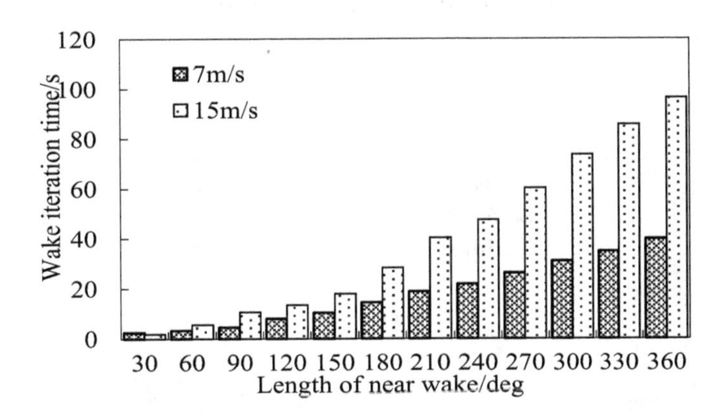

Figure 8. Wake iteration time along the length of the near wake at 7 m/s and 15 m/s.

The vortex sheet of the near wake is cut off at a 120° wake age angle in the VSRW model after considering the accuracy and calculation time. The red points in Figure 7 are the LSST values with a near wake age angle of 120° at every wind speed. It is observed that all LSST values reach the stationary values when the length of the near wake is 120°. It should be noted that this cut-off wake age angle is just an empirical angle rather than a real roll-up angle. In the following section, the aerodynamic analysis is conducted using the VSRW model with the near wake age angle of 120°.

4. Description of TVW and FVSW Models

To compare the calculation time and the accuracy of the proposed VSRW model with other models, the existing and mature models TVW and FVSW are used in this study. The blade models of the TVW and FVSW models are the same as the blade model of the VSRW model. As shown in Figure 2, the wake of the TVW model consists of vortex sheets in the near wake and tip vortex filaments in the far wake; the wake of the FVSW model consists of only vortex sheets in the near wake and far wake. The wake lengths of the two models are $4R$ and the near wake length of the TVW model was also set at 120°. The five-point central difference approximation [21] and the pseudo-implicit technique [23,24] are used to solve the convection equation of the vortex filaments in all the wakes of the TVW and FVSW models. The prescribed tolerances of the RMS change are all set as 10^{-4}.

5. Results and Discussions

5.1. Wake Iteration

In this study, we focus on decreasing the calculation time of the vortex model, which is approximately equal to the computational cost for the wake iteration. Figure 9 shows the wake iteration time of the VSRW model, TVW model, and FVSW model at the wind speeds of 7 m/s, 15 m/s, and 25 m/s. The wake iteration of the VSRW model requires 7.94 s, 13.37 s, and 20.24 s at the three wind speeds respectively; the computation speed is far faster than the two other models. The wake iteration time of the VSRW model at 7 m/s is two orders of magnitude less than for the FVSW model and the other times are one order of magnitude less. The VSRW model can generate results much faster than the two conventional methods mainly for two following reasons.

(1) In the VSRW model, the position of the vortex ring is determined by its control point, so we just have to calculate the induced velocity and the position of the control point of the vortex ring in the far wake. However, in the conventional methods, induced velocities and positions of all nodes of the vortex filaments in the far wake need to be calculated.

(2) The analytical method described in Section 2.4.2 is used to calculate the velocity induced by the far wake in the VSRW model. In the conventional methods, the velocity induced by the far wake is the sum of the velocities induced by the straight-line vortex elements, which are calculated using the Biot–Savart law.

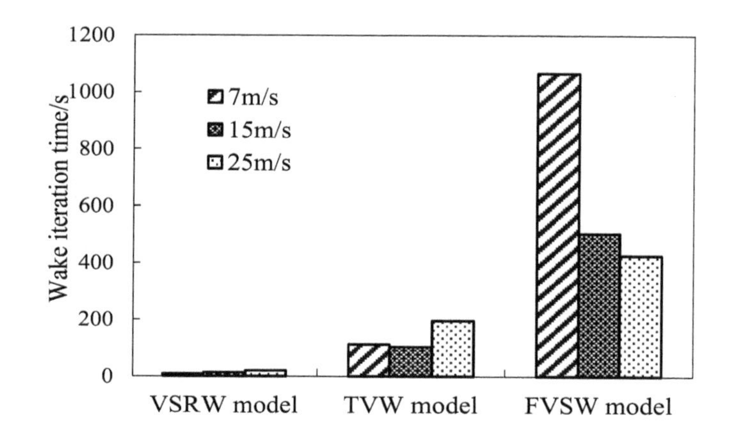

Figure 9. Wake iteration time of the VSRW model, TVW model, and FVSW model at 7 m/s, 15 m/s, and 25 m/s.

Figure 10 shows the convergence histories of the RMS of the error in the wake geometries at wind speeds of 7 m/s, 15 m/s, and 25 m/s using the three wake models. The convergence iteration numbers of the VSRW and TVW models are higher when the wind speed is higher. However, the convergence iteration number of the FVSW model does not change much with increasing wind speed. The wake iteration time depends on the convergence iteration number and the computational time of each step. In the TVW and FVSW models, more resources are needed to calculate the induced velocity field and the positions of the wake nodes; this results in a longer wake iteration time as shown in Figure 9. This is apparent in the FVSW model because of the dense vortex sheets, especially at a wind speed of 7 m/s. The introduction of the vortex ring technology in the far wake of the proposed VSRW model saves a lot of calculation time for determining the induced velocity field of the vortex rings and the positions of the vortex ring control points; as a result, the computational time for each step is greatly reduced. Therefore, the wake iteration time is still very short at a high wind speed although the convergence iteration number increases markedly.

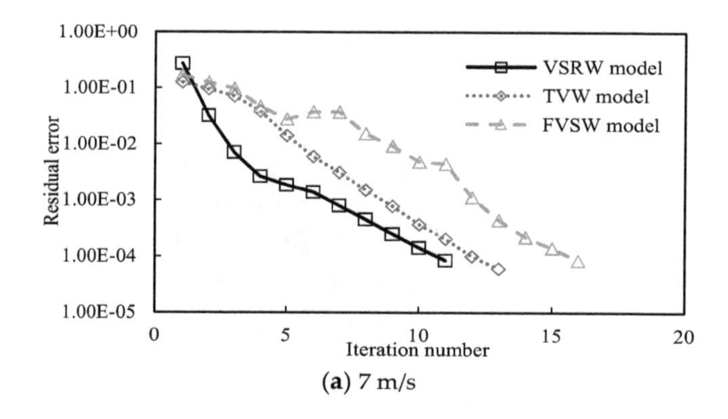

Figure 10. *Cont.*

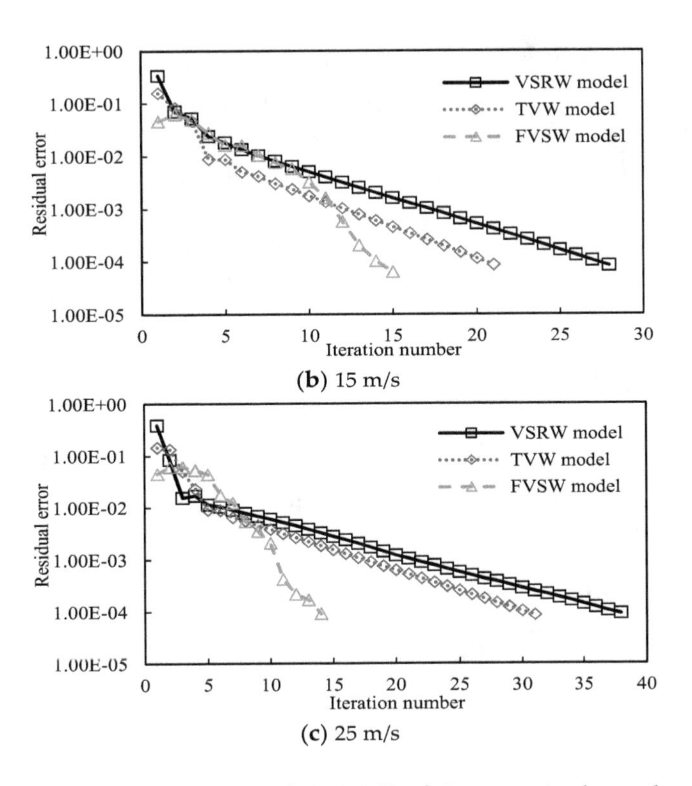

Figure 10. Convergence history of the RMS of the error in the wake geometry.

5.2. *Low Speed Shaft Torque*

The computational time of the proposed VSRW model can be greatly reduced and in the following section, we analyze its accuracy. The aerodynamic load of the LSST is predicted by the three different wake models. Figure 11 shows the aerodynamic estimates from the models compared with the values measured directly at the shaft [35]. The trends of the calculated results are consistent with the measurement results. Above 15 m/s, the estimated values begin to differ from each other and from the value measured at the shaft. The reason is that most of the blade is undergoing a stall flow. Although the 2D airfoil data are modified by the stall-delay model, the accurate prediction of the aerodynamic load in rotational and stall conditions is still challenging. This is also reflected in the studies by Breton et al. [37] who used a prescribed vortex wake technique and by Sant et al. [6] who used a free vortex wake technique. All the results from the VSRW model are slightly higher than those from the TVW and FVSW models. Except for 20 m/s, the results from the VSRW model could have errors no more than 10% relative to the experiment data. At 20 m/s, the error relative to the experiment data is about 16%. However, it is gratifying that the simplification of the far wake in the VSRW model did not significantly affect the accuracy of the aerodynamic load prediction.

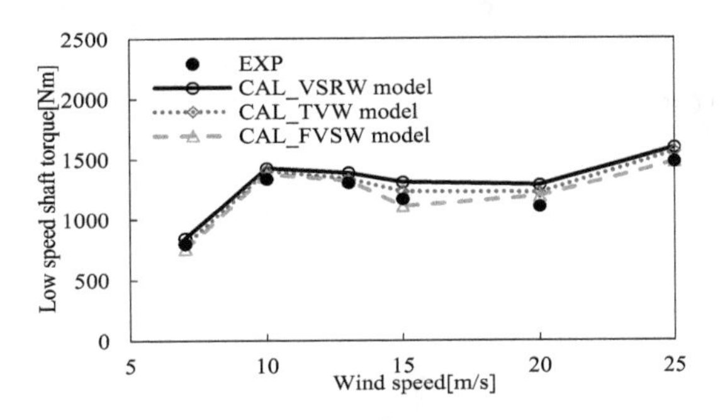

Figure 11. Measured and calculated results of LSST of NREL Phase VI wind turbine for wind speeds ranging from 7 m/s to 25 m/s.

5.3. Radial Distribution of Blade Airloads

The distributions of the normal and tangential force coefficients (C_n and C_t) at different wind speeds are computed by the proposed VSRW model, the TVW model, and the FVSW model. Figures 12 and 13 show the normal force coefficients and the tangential force coefficients respectively at 7 m/s, 15 m/s, and 25 m/s from the models and comparisons with the measurement results [35]. It is evident that the trends of the distributed airloads along the blade span are consistent with the measurement results. At 7 m/s, C_n and C_t derived by the three models agree well with the measurement results; C_n and C_t predicted by the VSRW model could have errors no more than 5% and 9% respectively relative to the experiment data. At high wind speeds of 15 m/s and 25 m/s, the estimated values, especially C_t, begin to differ from each other and from the measurement results because most of the blade is undergoing a stall flow. The comparison results exhibit similarities to the calculation of the LSST and illustrate that the proposed VSRW model predicts the radial distribution of the blade airloads as well as the TVW and FVSW models.

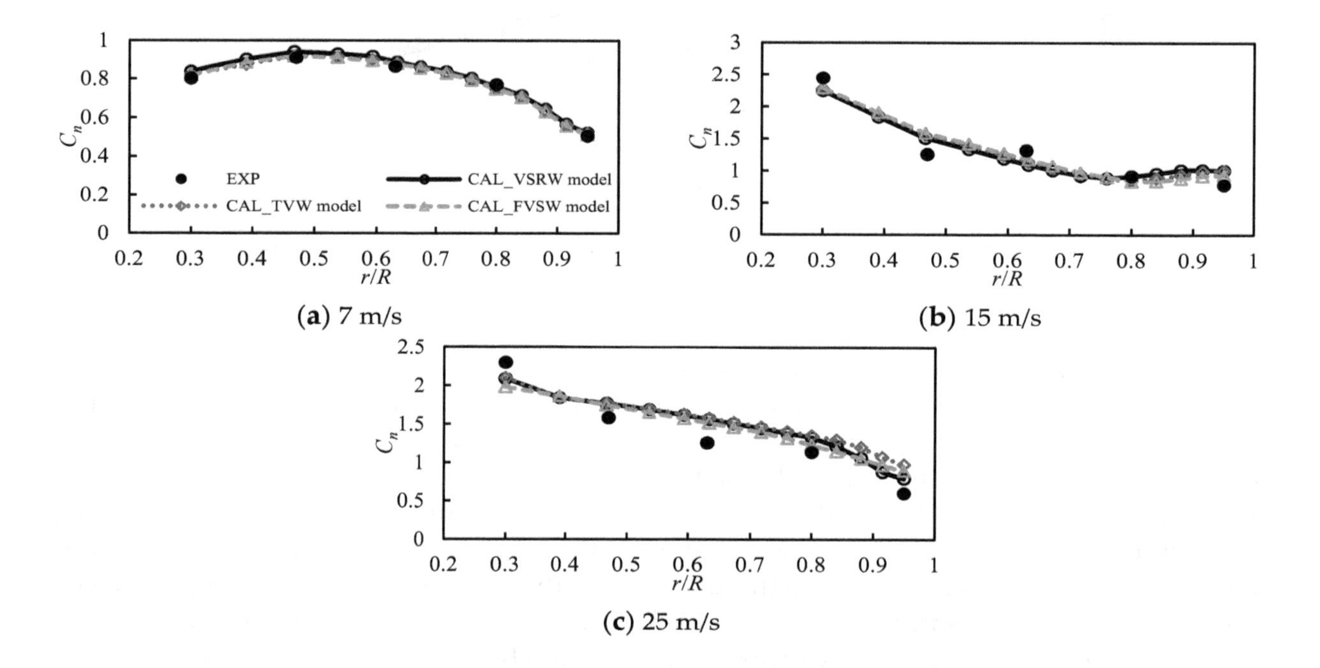

Figure 12. Comparison of the distributions of the computed normal force coefficients at (**a**) 7 m/s, (**b**) 15 m/s, and (**c**) 25 m/s along the blade radial positions for NREL Phase VI in axial steady conditions.

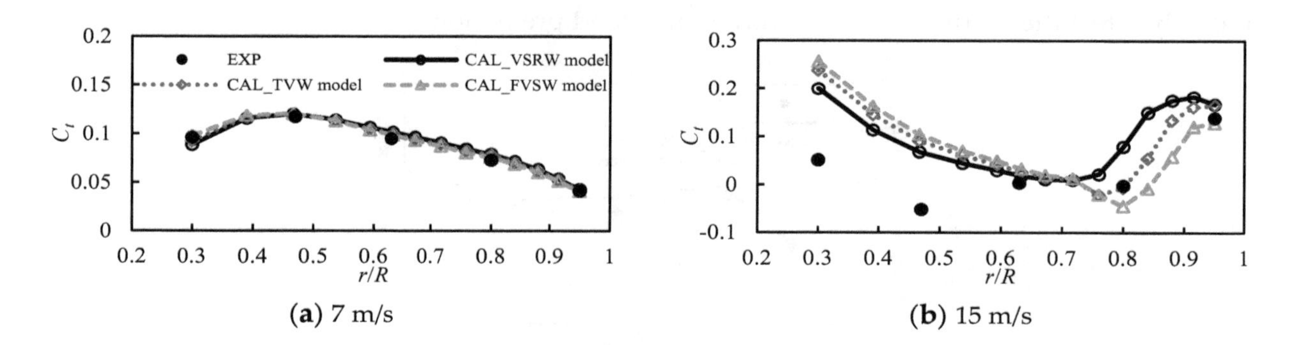

Figure 13. *Cont.*

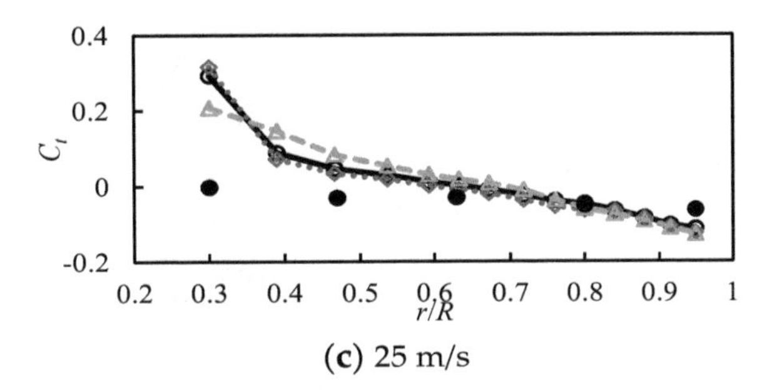

(**c**) 25 m/s

Figure 13. Comparison of the distributions of the computed tangential force coefficients at (**a**) 7 m/s, (**b**) 15 m/s, and (**c**) 25 m/s along the blade radial positions for NREL Phase VI in axial steady conditions.

5.4. Wake Geometry

Figure 14 shows the wake geometries at 10 m/s computed using the three different wake models. The azimuthal angle of the wake geometry is 0°. The wake deformation computed by using the FVSW model differs from that shown in Figure 2a. The wake deformation of the near wake of the VSRW and TVW models is also shown. The number of discrete wake nodes is far higher for the FVSW model than for the VSRW and TVW models; therefore, more computational time is required. It should be noted that the vortex rings from the two blades overlap at the same axial position. Therefore, there are actually six vortex rings in Figure 14a.

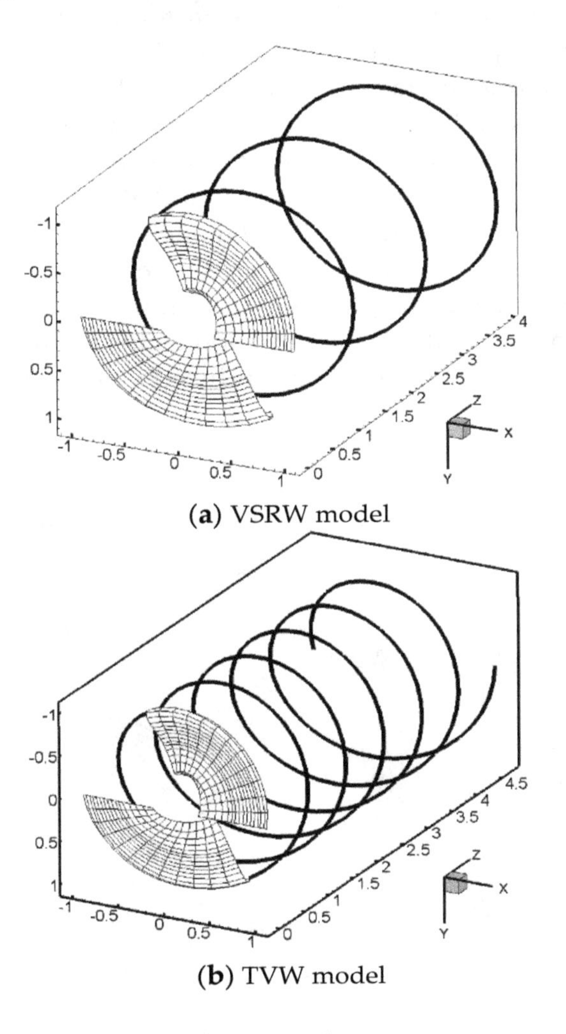

(**a**) VSRW model

(**b**) TVW model

Figure 14. *Cont.*

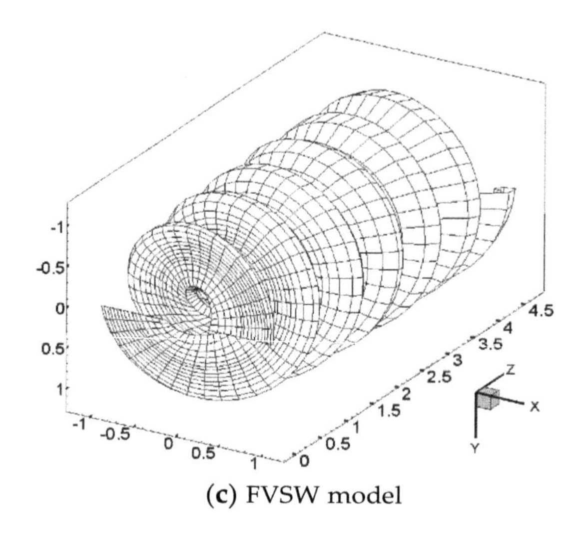

(c) FVSW model

Figure 14. Wake geometries (front view) at four times the radius distance behind the rotor at 10 m/s calculated using the (**a**) VSRW model, (**b**) TVW model, and (**c**) FVSW model.

5.5. Induction Factor in the Wake

Figure 15 shows the contours of the axial induction factor in the XOZ plane at the blade azimuthal angle of 0°. In the front of the plane $Z/R = 0.5$ (on the left of the red dashed line in Figure 14), the distributions of the axial induction factor are similar for the three wake models. On the right of the red dashed line, the patterns of the contours of the axial induction factor are similar for the TVW and FVSW models but a marked difference is observed for the VSRW model. This is because the vortex rings are used to replace the helical tip vortex filament in the far wake. The radial expansion deformation of the wake can be captured by the VSRW model, as well as the other two models. The two overlapping vortex rings at the same axial position are the reason that the maximum axial induction factor around the vortex ring is twice as high as the maximum value in the TVW and FVSW models. On the motion trail of the tip vortex, the axial induction factor remains at about -0.1 in three wake models; therefore, the assumption of the constant induced velocity is reasonable when the control point is moving.

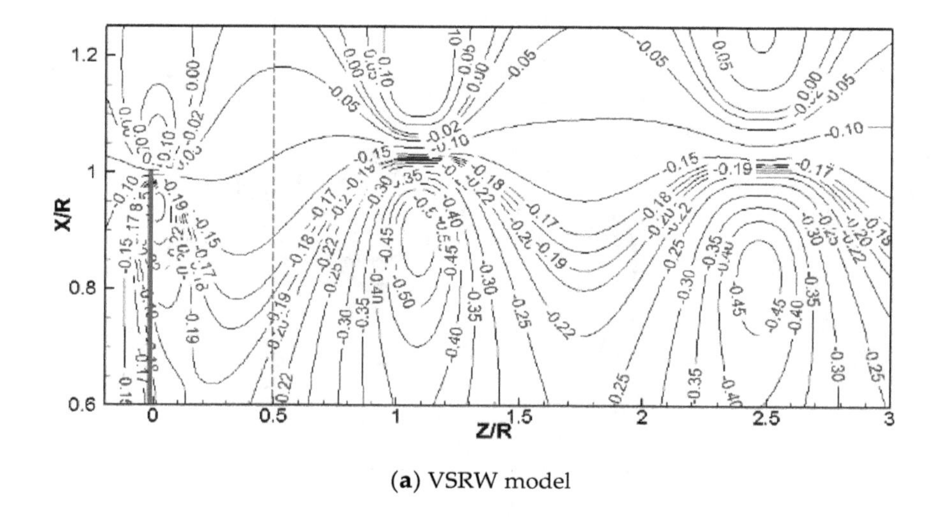

(a) VSRW model

Figure 15. *Cont.*

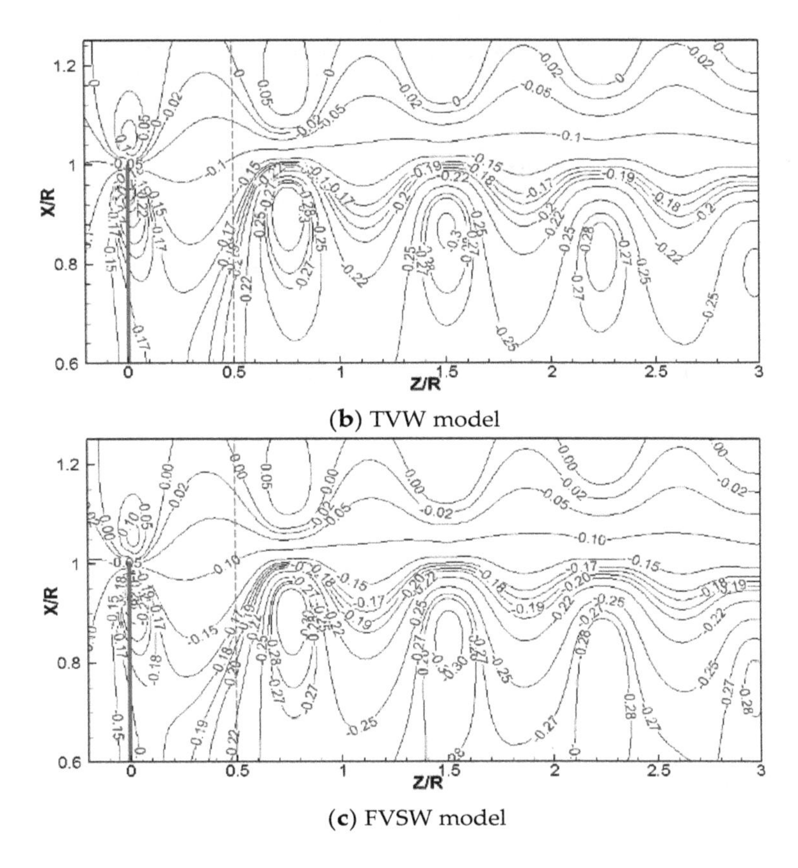

(**b**) TVW model

(**c**) FVSW model

Figure 15. Contours of the axial induction factor.

6. Conclusions

In this study, a simplified FVW model was developed to rapidly calculate the aerodynamic performance of wind turbines in axial steady conditions. The helical tip vortex filament of the far wake in the traditional FVW model is replaced by several vortex rings. The proposed computing method of the vortex ring position and the proposed analytical formulas of the vortex ring-induced velocity greatly reduce the computational time. The length of the near wake was cut off at a 120° wake age angle to take into account both the accuracy and the calculation time of the VSRW model. The computational speed of the simplified FVW model is at least an order of magnitude faster than other two conventional models and the convergence and stability are good.

The simplified FVW model accurately predicts the blade aerodynamic load, including the LSST and the distributions of the span's normal and tangential force coefficients. The error of the low speed shaft torque obtained from the simplified FVW model is no more than 10% relative to the experiment at most wind speeds. The normal and tangential force coefficients obtained from the three models agree well with each other and with the measurement results at the low wind speed. At the high wind speed, the estimated values begin to differ from each other and from the measurement results because most of the blade is undergoing a stall flow

Except for differences in the axial induction factor field in the far wake, the axial induction factor field in the near wake computed by using the simplified FVW model agrees well with the other two FVW models and the radial expansion deformation of the wake is captured. The comparison of the far wake structure illustrates that the helical tip vortex filament, which is used in the TVW model, is more appropriate based on the physics of the flow.

Overall, in axial steady conditions, the computational time of the simplified FVW model is at least an order of magnitude less than that of traditional FVW models and the prediction accuracy is not affected significantly. On this basis, a study of the unsteady simplified FVW model will be conducted in the future.

Author Contributions: B.X. run the FVW codes and prepared this manuscript under the guidance of T.W. and Y.Y. Z.Z. and H.L. supervised the work and contributed in the interpretation of the results. All authors carried out data analysis, discussed the results and contributed to writing the paper.

Acknowledgments: This work was supported by the National Natural Science Foundation of China (grant number 51607058, 11502070); the National Basic Research Program of China (973 Program) (grant number 2014CB046200); and the Fundamental Research Funds for the Central Universities (grant number 2016B01514).

Nomenclature

Variables

a, b	Coefficients in the induced velocity equation (-)
c_i	Chord of the ith blade element (m)
C_l	Lift coefficient (-)
C_n	Normal force coefficient to the rotor disc (-)
C_t	Tangential force coefficients to the rotor disc (-)
h	Distance from the collocation point to the vortex-line segment (m)
i, k, n	Integer variables (-)
N_C	Number of vortex rings (-)
N_E	Number of blade elements (-)
N_T	Number of time steps of a circle (-), $N_T = 2\pi/\Delta\psi$
r	Radial location of the blade (m)
R	Rotor tip radius (m)
$\bar{r}_b$	Dimensionless radial location of the blade element boundary (-)
r_b	Radial location of the blade element boundary (m)
$\boldsymbol{r}_v$	Position vector of the vortex collocation point (m)
$\boldsymbol{r}_A$	Position vector from point A to point P (m)
$\boldsymbol{r}_B$	Position vector from point B to point P (m)
$R_{cr(n)}$	Radial position of the nth vortex ring control point (m)
R_A	Radial position of the tip vortex release (m)
R_P	Radial position of point P (m)
$\boldsymbol{V}_{ind}$	Induced velocity vector (m/s)
$v_{ind,r}$	Radial velocity at point P induced by the nth vortex ring (m/s)
$v_{ind,z}$	Axis velocity at point P induced by the nth vortex ring (m/s)
$V_{ind,R}^{n}$	Radial velocity at the nth vortex ring control point induced by all vortex field (m/s)
$V_{ind,R}^{A}$	Radial velocity at point A induced by all vortex field (m/s)
$V_{ind,Z}^{n}$	Axis velocity at the nth vortex ring control point induced by all vortex field (m/s)
$V_{ind,Z}^{A}$	Axis velocity at point A induced by all vortex fields (m/s)
$\boldsymbol{V}_0$	Free stream velocity vector (m/s)
W_i	Resultant velocity at the control point of the ith blade element (m/s)
X	X axis in the coordinate system pointing right as viewed from the front (m)
Y	Y axis in the coordinate system pointing vertically downwards (m)
Z	Z axis in the coordinate system in the direction of wind flow (m)
$Z_{cr(n)}$	Axis position of the nth vortex ring control point (m)
Z_A	Axis position of the tip vortex release (m)
Z_P	Axis position of point P (m)
Γ	Vortex circulation (m^2/s)
Γ_b	Bound circulation of the blade element (m^2/s)
Γ_n	Vortex circulation of the nth vortex ring (m^2/s)
$\Delta\zeta$	Discretization of the azimuthal angle (rad)
$\Delta\psi$	Discretization of the wake age angle (rad)
ζ	Vortex wake age angle (rad)
θ_A	Angle between vector $\boldsymbol{r}_A$ and vector $\mathbf{AB}$ (rad)
θ_B	Angle between vector $\boldsymbol{r}_B$ and vector $\mathbf{AB}$ (rad)
ψ	Azimuthal angle (rad)

Abbreviations

BEM	blade element momentum
CFD	computational fluid dynamics
CPU	Central Processing Unit
FVSW	full vortex sheet wake
FVW	free vortex wake
GPU	Graphics Processing Unit
LSST	low speed shaft torque
NASA	National Aeronautics and Space Administration
NREL	National Renewable Energy Laboratory
RMS	root mean square
TVW	tip vortex wake
VSRW	vortex sheet and ring wake
WInDS	Wake Induced Dynamics Simulator

References

1. Hansen, M.O.L.; Sørensen, J.N.; Voutsinas, S.; Sørensen, N.; Madsen, H.A. State of the art in wind turbine aerodynamics and aeroelasticity. *Aerosp. Sci.* **2006**, *42*, 285–330. [CrossRef]
2. Fernandez-Gamiz, U.; Zulueta, E.; Boyano, A.; Ansoategui, I.; Uriarte, I. Five megawatt wind turbine power output improvements by passive flow control devices. *Energies* **2017**, *10*, 742. [CrossRef]
3. Gohard, J.C. *Free Wake Analysis of Wind Turbine Aerodynamics*, Wind Energy Conversion, ASRL-TR-184-14; Massachusetts Institute of Technology, Department of Aeronautics and Astronautics, Aeroelastic and Structures Research Laboratory: Cambridge, MA, USA, 1978.
4. Arsuffi, G.; Guj, G.; Morino, L. Boundary element analysis of unsteady aerodynamics aerodynamics of windmill rotors in the presence of yaw. *J. Wind Eng. Ind. Aerodyn.* **1993**, *45*, 153–173. [CrossRef]
5. Garrel, A.V. *Development of a Wind Turbine Aerodynamics Simulation Module*; ECN-C-03-079; Energy Research Centre of The Netherlands: Petten, The Netherlands, 2003.
6. Sant, T.; Kuik, G.V.; Bussel, G. Estimating the Angle of Attack from Blade Pressure Measurements on the NREL Phase VI Rotor Using a Free Wake Vortex Model: Axial Conditions. *Wind Energy* **2006**, *9*, 549–577. [CrossRef]
7. Sebastian, T.; Lackner, M.A. Development of a free vortex wake method code for offshore floating wind turbines. *Renew. Energy* **2012**, *46*, 269–275. [CrossRef]
8. Rosen, A.; Lavie, I.; Seginer, A. A general free-wake efficient analysis of horizontal-axis wind turbines. *Wind Eng.* **1990**, *14*, 362–373.
9. Gupta, S. Development of a Time-Accurate Viscous Lagrangian Vortex Wake Model for Wind Turbine Applications. Ph.D. Dissertation, University of Maryland, College Park, MD, USA, 2006.
10. Qiu, Y.X.; Wang, X.D.; Kang, S.; Zhao, M.; Liang, J.Y. Predictions of unsteady HAWT aerodynamics in yawing and pitching using the free vortex method. *Renew. Energy* **2014**, *70*, 93–105. [CrossRef]
11. Xu, B.F.; Wang, T.G.; Yuan, Y.; Cao, J.F. Unsteady aerodynamic analysis for offshore floating wind turbines under different wind conditions. *Philos. Trans. R. Soc. A* **2015**, *373*, 20140080. [CrossRef] [PubMed]
12. Marten, D.; Lennie, M.; Pechlivanoglou, G.; Nayeri, C.N.; Paschereit, C.O. Implementation, optimization and validation of a nonlinear lifting line free vortex wake module within the wind turbine simulation. In Proceedings of the ASME Turbo Expo 2015: Turbine Technical Conference and Exposition, Montreal, QC, Canada, 15–19 June 2015.
13. Miller, R.H. The aerodynamics and dynamic analysis of horizontal axis wind turbines. *J. Wind Eng. Ind. Aerodyn.* **1983**, *15*, 329–340. [CrossRef]
14. Afjeh, A.A.; Keith, T.G.J. A vortex lifting line method for the analysis of horizontal axis wind turbines. *ASME J. Sol. Energy Eng.* **1986**, *108*, 303–309. [CrossRef]
15. Yu, W.; Ferreira, C.S.; van Kuik, G.; Baldacchino, D. Verifying the blade element momentum method in unsteady, radially varied, axisymmetric loading using a vortex ring model. *Wind Energy* **2017**, *20*, 269–288. [CrossRef]
16. Farrugia, R.; Sant, T.; Micallef, D. Investigating the aerodynamic performance of a model offshore floating wind turbine. *Renew. Energy* **2014**, *70*, 24–30. [CrossRef]

17. Elgammi, M.; Sant, T. Combining unsteady blade pressure measurements and a free-wake vortex model to investigate the cycle-to-cycle variations in wind turbine aerodynamic blade loads in yaw. *Energies* **2016**, *9*, 460. [CrossRef]
18. Turkal, M.; Novikov, Y.; Usenmez, S.; Sezer-Uzol, N.; Uzol, O. GPU based fast free-wake calculations for multiple horizontal zxis wind turbine rotors. *J. Phys. Conf. Ser.* **2014**, *524*, 012100. [CrossRef]
19. Weissinger, J. *The Lift Distribution of Swept-Back Wings*; NACA-TM-1120; National Advisory Committee for Aeronautics: Cleveland, OH, USA, 1947.
20. Du, Z.; Selig, M.S. *A 3-D Stall-Delay Model for Horizontal Axis Wind Turbine Performance Prediction*; AIAA-98–0021; American Institute of Aeronautics and Astronautics: Reston, VA, USA, 1998.
21. Bagai, A.; Leishman, J.G. Rotor free-wake modeling using a relaxation technique—Including comparisons with experimental data. *J. Am. Helicopter Soc.* **1995**, *40*, 29–41. [CrossRef]
22. Bhagwat, M.; Leishman, J.G. Stability, consistency and convergence of time marching free-vortex rotor wake algorithms. *J. Am. Helicopter Soc.* **2001**, *46*, 59–71. [CrossRef]
23. Grouse, G.L.; Leishman, J.G. A new method for improved rotor free vortex convergence. In Proceedings of the 31st AIAA Aerospace Sciences Meeting and Exhibit, Reno, NV, USA, 11–14 January 1993.
24. Wang, T.G.; Wang, L.; Zhong, W.; Xu, B.F.; Chen, L. Large-scale wind turbine blade design and aerodynamic analysis. *Chin. Sci. Bull.* **2012**, *57*, 466–472. [CrossRef]
25. Gupta, S.; Leishman, J.G. Accuracy of the induced velocity from helicoidal vortices using straight-line segmentation. *AIAA J.* **2005**, *43*, 29–40. [CrossRef]
26. Vatistas, G.H.; Kozel, V.; Minh, W. A simpler model for concentrated vortices. *Exp. Fluids* **1991**, *11*, 73–76. [CrossRef]
27. Sant, T.; del Campo, V.; Micallef, D.; Ferreira, C.S. Evaluation of the lifting line vortex model approximation for estimating the local blade flow fields in horizontal-axis wind turbines. *J. Renew. Sustain. Energy* **2016**, *8*, 023302. [CrossRef]
28. Gupta, S.; Leishman, J.G. Validation of a free vortex wake model for wind turbine in yawed flow. In Proceedings of the 44th AIAA Aerospace Sciences Meeting 2006, Reno, NV, USA, 9–12 January 2006; Volume 7, pp. 4529–4543.
29. Xu, B.F.; Yuan, Y.; Wang, T.G.; Zhao, Z.Z. Comparison of two vortex models of wind turbines using a free vortex wake scheme. *J. Phys. Conf. Ser.* **2016**, *753*, 022059. [CrossRef]
30. Xu, B.F.; Feng, J.H.; Wang, T.G.; Yuan, Y.; Zhao, Z. Application of a turbulent vortex core model in the free vortex wake scheme to predict wind turbine aerodynamics. *J. Renew. Sustain. Energy* **2018**, *10*, 023303. [CrossRef]
31. Bhagwat, M.J.; Leishman, J.G. Correlation of helicopter tip vortex measurements. *AIAA J.* **2000**, *38*, 301–308. [CrossRef]
32. Ananthan, S.; Leishman, J.G. The role of filament stretching in the free-vortex modeling of rotor wakes. *J. Am. Helicopter Soc.* **2004**, *49*, 176–191. [CrossRef]
33. Yoon, S.S. Heister, S.D. Analytical formulas for the velocity field induced by an infinitely thin vortex ring. *Int. J. Numer. Methods Fluids* **2004**, *44*, 665–672. [CrossRef]
34. Fukushima, T. Fast computation of complete elliptic integrals and Jacobian elliptic functions. *Celest. Mech. Dyn. Astron.* **2009**, *105*, 305–328. [CrossRef]
35. Hand, M.M.; Simms, D.A.; Fingersh, L.J.; Jager, D.W.; Cotrell, J.R.; Schreck, S.; Larwood, S.M. *Unsteady Aerodynamics Experiment Phase VI: Wind Tunnel Test Configurations and Available Data Campaign*; NREL/TP-500-29955; National Renewable Energy Laboratory: Golden, CO, USA, 2001.
36. Ananthan, S. Analysis of Rotor Wake Aerodynamics during Maneuvering Flight Using a Free-Vortex Wake Methodology. Ph.D. Dissertation, University of Maryland, College Park, MD, USA, 2006.
37. Breton, S.P.; Coton, F.N.; Moe, G. A study on rotational effects and different stall delay models using a prescribed wake vortex scheme and NREL Phase VI experiment data. *Wind Energy* **2008**, *11*, 459–482. [CrossRef]

2

Condition Monitoring of Wind Turbine Blades using Active and Passive Thermography

Hadi Sanati, David Wood * and Qiao Sun

Department of Mechanical and Manufacturing Engineering, University of Calgary, 2500 University Drive N.W., Calgary, AB T2N 1N4, Canada; hadi.sanati@ucalgary.ca (H.S.); qsun@ucalgary.ca (Q.S.)
* Correspondence: dhwood@ucalgary.ca

Abstract: The failure of wind turbine blades is a major concern in the wind power industry due to the resulting high cost. It is, therefore, crucial to develop methods to monitor the integrity of wind turbine blades. Different methods are available to detect subsurface damage but most require close proximity between the sensor and the blade. Thermography, as a non-contact method, may avoid this problem. Both passive and active pulsed and step heating and cooling thermography techniques were investigated for different purposes. A section of a severely damaged blade and a small "plate" cut from the undamaged laminate section of the blade with holes of varying diameter and depth drilled from the rear to provide "known" defects were monitored. The raw thermal images captured by both active and passive thermography demonstrated that image processing was required to improve the quality of the thermal data. Different image processing algorithms were used to increase the thermal contrasts of subsurface defects in thermal images obtained by active thermography. A method called "Step Phase and Amplitude Thermography", which applies a transform-based algorithm to step heating and cooling data was used. This method was also applied, for the first time, to the passive thermography results. The outcomes of the image processing on both active and passive thermography indicated that the techniques employed could considerably increase the quality of the images and the visibility of internal defects. The signal-to-noise ratio of raw and processed images was calculated to quantitatively show that image processing methods considerably improve the ratios.

Keywords: thermography; wind turbine blades; defects; image processing; condition monitoring

---

1. Introduction

The most crucial components of wind turbines, the blades, are susceptible to different types of damage during their operation. The failure of one blade may damage nearby blades and wind turbines, increasing the total damage cost. Most blades consist of two halves made of a fiberglass composite and shear webs, which are glued together with strong adhesive materials [1]. The main function of the shear webs is to increase the strength of the structure. These bonded zones are potential sites for damage initiation and propagation [2]. Different surface and subsurface defects, including delamination, cracks, air inclusion, fiber-matrix debonding, and others, may be introduced to the blade during manufacturing or operation [3]. Harsh environmental conditions and airborne particles such as hail, snow, rain, ice, and dirt, expose wind turbine blades to more potential harm. Defective blades are rarely replaced because of the high cost of manufacturing. To prevent failure, blades need to be continuously monitored through Non-Destructive Testing (NDT) methods [4]. Different NDT techniques such as Ultrasonic Testing (UT), Acoustic Emission (AE), Fiber Bragg Grating (FBG) strain sensors, Vibration Analysis, and Tap Tests have been employed to inspect the integrity of wind turbine blades [5–9]. Conventional NDT techniques generally require close proximity between the sensor and the blade [10]. Since access to a blade is difficult and requires an industrial climber or crane, which can be dangerous and/or time-consuming, the practical implementation of conventional methods sometimes requires blade removal. Developing new NDT techniques that are capable of detecting faults in the blades from larger distances is essential.

Infrared (IR) thermography is a non-contact, long-distance NDT technique that can inspect extensive areas quickly by capturing thermal images of the object's surface. In general, defective areas alter the temperature distributions on the surface that are measured by IR cameras. Thermographic inspection is typically divided into two categories: active and passive. In active thermography, different heating sources such as flash and halogen lamps are employed for heating the object making the technique less usual for operating wind turbines. It is used here largely to allow comparison with passive thermography which utilizes solar radiation [2] to heat a blade (usually around sunrise) or to cool it at sunset. This method has been widely used to detect subsurface defects of different materials including metals [11], composites [12,13], and concrete [14].

Different studies have used thermography to detect faults in wind turbine blades. Meinlischmidt and Aderhold [2] employed passive thermography to detect internal structural features and subsurface defects such as poor bonding and delamination. Beattie and Rumsey [15] employed thermography to inspect blades during fatigue tests of a 13.1 m blade made from wood-epoxy-composite and a 4.25 m fiberglass blade. This experiment identified the root region of the blade as a defective area. Shi-bin [16] employed infrared thermal wave testing to detect subsurface faults such as foreign matter and air inclusions at various depths of a blade section. Galleguillos [17] conducted a new experiment to inspect an installed wind turbine blade. They mounted an IR camera on an unmanned aerial vehicle (UAV) and captured thermal images of installed blades while the blades were stationary, demonstrating the capability for fast data acquisition and inspection with this setup. Other research evaluated the suitability of different weather conditions for revealing the internal features of a blade section with thermal imaging [18,19]. Doroshtnasir [20] proposed a new passive thermography technique that can inspect operating blades from the ground. This experiment developed a new image processing technique to improve the thermal contrast quality by removing the effect of disruptive factors such as environmental reflections. Active thermography has been used by different researchers to quantitatively evaluate the presence of defects in different materials including composite and metallic samples. Lahiri [21] employed phase information obtained from the processing of active thermography data to determine quantitative information associated with flat-bottomed holes (FBHs) embedded in different materials including glass fiber reinforced polymer, high-density rubber and low-density rubber. Shin [22] used the Pulsed Phase Thermography (PPT) method to inspect the subsurface fatigue damage in adhesively bonded joints between fiber reinforcement polymer components. Maierhofer [23] compared the phase values obtained from pulsed and lock-in thermography applied on steel and Carbon Fiber Reinforced Plastic (CFRP) materials. They also compared the spatial resolution calculated from data captured through flash and lock-in thermography at different frequencies. In another study, Almond [24] used long pulsed thermography to detect the FBHs of different sizes and depths created in different materials including aluminum alloy, mild steel and stainless steel, and a CFRP composite plate.

Different image processing methods have been developed to improve the contrast of subsurface defects in images. Maldague and Marinetti [25] proposed PPT, which has the advantages of both pulsed and lock-in thermography. Lock-in thermography can detect deeper defects by continuously heating the surface using a periodic heat source [26] such as a modulated halogen lamp [27] but it may take a long time to detect a fault while pulsed thermography is fast. PPT uses a transform-based algorithm such as Fast Fourier Transform (FFT) to convert time domain data to frequency. Shin [24] employed this method to detect the initiation and propagation of defects in adhesively bonded joints under fatigue loading. This method has been widely used by different researchers to quantitatively and qualitatively evaluate the subsurface damages in different materials. Pawar [28], for example, inspected barely visible impact damage from low-velocity impacts. For this purpose, he first calibrated the defect depth with a blind frequency, the limited frequency at which the subsurface defect at a certain depth is visible in the recorded thermal data, for carbon epoxy laminate, and then applied the findings of depth and the blind frequency relationship to the specimen with barely visible impact damage on it. Castanedo [29] proposed an interactive methodology in a PPT experiment that connected acquisition parameters such as time and frequency resolution and storage capacity to each other in order to inspect defects at different depths with a single test. He used a combination of phase contrast and blind frequency and applied his proposed interactive methodology to a CFRP specimen with artificial defects at different depths to quantify the depth of the defects. Thermographic Signal Reconstruction [30] increases the quality of the thermal signatures associated with internal defects based on the known behavior of simple forms of the heat conduction equation. It improves the signal-to-noise ratio (SNR), while at the same time reducing image blurring and increasing the sensitivity [31]. This method was recently applied to the thermal image sequence of step heating thermography and gave reliable results [32]. Matched Filters (MF) have also been proposed to improve image contrast of subsurface defects by increasing the contrast of defective areas and reducing the signals from sound areas [31,33,34].

The present investigation developed passive and active thermographic inspection of wind turbine blades. During the active experiment, both pulsed and step heating thermography were employed and their results compared with each other (images were obtained and processed in some cases after the heating had finished and so strictly should be called "step cooling". We will use the term "step heating" in a purely generic manner to cover both cases). Several image processing techniques were applied to the raw thermal images to increase the contrast associated with internal defects. Once the most appropriate technique was determined, a passive thermography experiment was designed and the image processing technique was applied to the thermograms and the maps of surface temperature, to improve their quality. Passive thermography was also performed at different times of the day to assess the most favorable times for the best results. This paper is arranged as follows: Section 1 reviews the literature regarding NDT techniques and thermal imaging methods. The experimental procedures and materials are outlined in Section 2. Section 3 provides the theory of quantitative evaluation. Section 4 contains the results and discussion regarding experimental thermography for the inspection of subsurface defects. Finally, Section 5 provides a summary and conclusion.

2. Experimental Procedure

2.1. Materials

All samples used in this experiment came from a 50 m long wind turbine blade made of fiberglass composite obtained after it had been damaged in transit to a wind farm. The blade was never installed or operated. Figure 1a shows the 3 m long blade section with significant surface damage that was used for the passive thermography experiments. The laminate thickness in the damaged section was 14 mm. The chord length was approximately 1 m. The yellow/orange regions in Figure 1a are the exposed sandwich core on the rear section of the suction surface where there is very little laminate. Some patches of glue on the suction side resulted in different effusivity than the background and therefore generated spots with different brightnesses on the thermograms. The upper side of the blade section contained a crack which was visible to the naked eye.

A "defect plate" with dimensions of 170 mm $\times$ 195 mm $\times$ 8 mm was cut from the laminate skin of another section of the blade closer to the tip where the laminate skin was thinner. Flat-bottomed holes with different diameters and depths were drilled from the rear to produce a range of known "defects". The holes had diameters ranging from 4 mm to 20 mm with depths between 0.5 mm and 3 mm. Figure 1c is a schematic of the plate that illustrates the geometry and pattern of the defects. No holes penetrated the outer surface of the plate which corresponded to the outer surface of a blade and all thermograms were of the outer surface. The defect plate was attached to the surface of the damaged blade section during passive thermography experimentation. It was also the only blade material tested with active thermography. The holes were used to evaluate the minimum defect that can be detected using passive and active thermography.

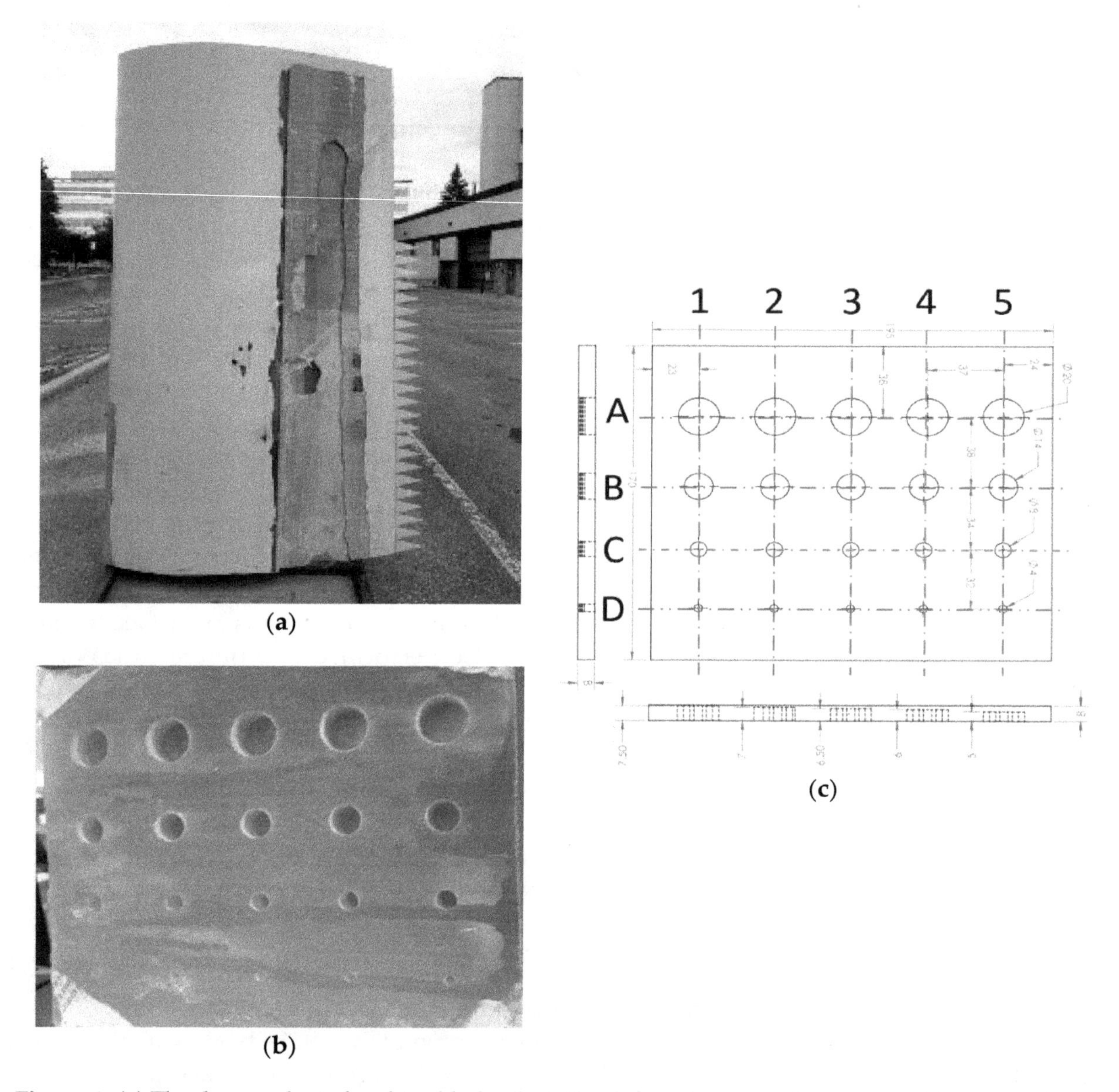

Figure 1. (a) The damaged wind turbine blade. (b,c) The defect plate with flat-bottomed holes. All holes were drilled from the rear and did not penetrate the outer surface.

2.2. Passive Thermography

In the passive thermography experiment, the suction side of the blade was monitored outdoors during a sunny day from morning until the afternoon. The blade's position was not changed relative to the IR camera during this time. This experiment, whose setup is depicted in Figure 2, sought to determine the most favorable conditions to reveal the most defects and to evaluate the fault detection capability of thermography when the blade is heated by the sun.

The experiment was conducted on a sunny day in July 2017 from 9:00 a.m. to 7:30 p.m. Sunrise and sunset on this day were 5:53 a.m. and 9:30 p.m., respectively. During the experiment, the sky was clear, the humidity was almost 36%, and the temperature varied between 16 °C and 26 °C. A T1030Sc IR camera made by FLIR Systems was located 4 m from the blade section and equipped with a 21.2 mm lens, resulting in a spatial resolution of 4 mm per pixel. ResearchIR, a software package developed by FLIR Systems that provides high-speed data recording and image analyzing capabilities was used to record thermograms at a frequency of 1 Hz.

Figure 2. Passive thermography experiment.

2.3. Pulsed and Step Heating Thermography

These techniques were used only on the defect plate mounted 1.5 m from the IR camera. Despite being mounted near windows, the defect plate was always in a shadow during the experiments. During pulsed thermography, the defect plate was heated by a 2400 W flash lamp and thermal images were recorded at a frequency of 15 Hz immediately after flashing the sample. To heat the surface uniformly, the flash lamp was around 0.3 m from the object with the angle of around 15° with respect to the normal of the defect plate. The pulsed thermography experiment is depicted in Figure 3. The spatial resolution of thermal images obtained by active thermography was almost 0.5 mm per pixel.

Figure 3. Pulsed thermography experiment.

The step heating thermography experiment, shown in Figure 4, used two 500 W halogen lamps to continuously heat the surface. The sample was heated between 10 and 75 s and the thermal evolution on the surface of the specimen was recorded at a frequency of 15 Hz. Once heating was finished, thermal decay was recorded with the same frequency. The room temperature was 23 ± 2 °C during the experiment. The thermal contrast associated with the defects could be observed after a few seconds of heating.

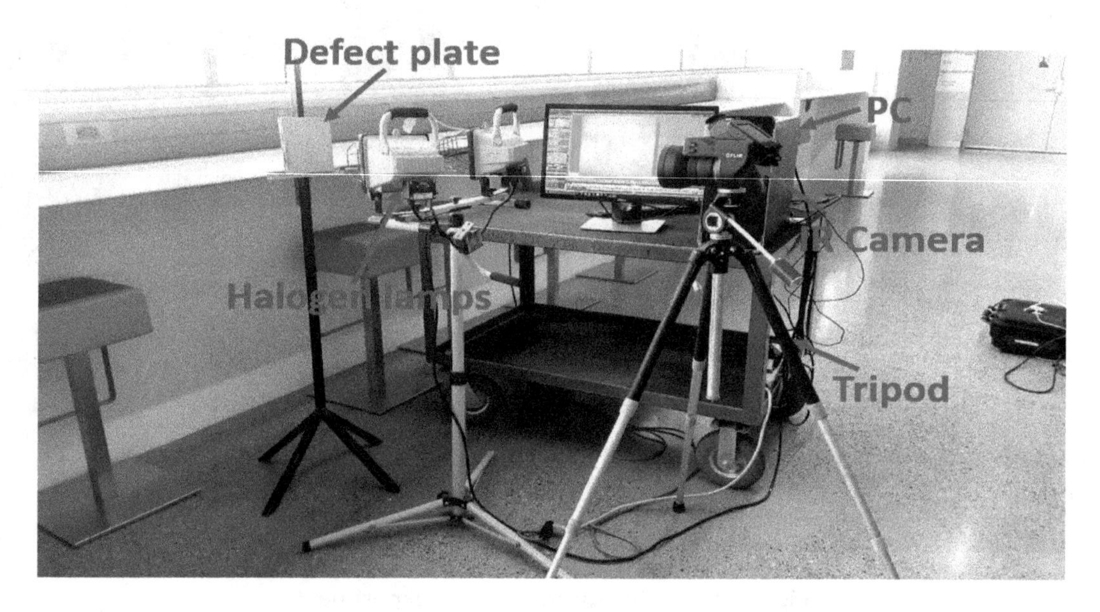

Figure 4. Step heating thermography experiment.

3. MF Algorithms

MF is a method to improve image contrast of subsurface defects by increasing the contrast of defective areas and reducing the signals from sound areas [31,33,34]. Different types of MF algorithms have been developed, all of which are based on the assumption that [31,34]:

$$T_{\text{obs}} = T_{\text{ref l}} = T_{\text{ideal}} \tag{1}$$

at any time. T_{obs} is the temperature recorded by the IR camera, T_{refl} is the temperature response reflected from the defective area, and T_{ideal} is the ideal temperature response of the sound area. Equation (1) can be represented in vector form:

$$X = S + W \tag{2}$$

where these vectors collect all recordings over time and $X = T_{\text{obs}}$, $S = T_{\text{refl}}$, and $W = T_{\text{ideal}}$. Equation (2) is multiplied by a vector q that maximizes the visibility of the reflected temperature from the defective area and minimizes the response of non-defective areas. The vector q can, therefore, be calculated by [31,34]:

$$\max_{q} \|q^T S\|_2 \text{ subject to } \min_{q} \|qW\|_2 \tag{3}$$

where q^T is the transpose of q. Different methods for finding q result in the different types of MF algorithms. Each MF algorithm considers a certain q vector to increase the contrast of defective areas and decrease the signals from sound areas. The q vector for the spectral angle map (SAM) is

$$SAM = \frac{S^T X_{ij}}{\sqrt{S^T S}\sqrt{X_{ij}^T X_{ij}}} \tag{4}$$

where i and j imply that the calculation is repeated for all pixels of all thermal images to provide a single correlation image and S^T. The adaptive coherence estimator (ACE) uses as q:

$$ACE = \frac{S^T C^{-1} X_{ij}}{\sqrt{S^T C^{-1} S}\sqrt{X_{ij}^T C^{-1} X_{ij}}} \tag{5}$$

where C is covariance matrix of the ideal temperature vector defined as:

$$C = \frac{1}{m}\sum_{i=1}^{m} WW^T. \tag{6}$$

The t- and F-statistics use the different q vectors to improve the contrast associated with defective areas:

$$F_{stat} = \frac{\left(S^T R^{-1} x_{ij}\right)^2}{X_{ij}^T R^{-1} X_{ij} - \rho^2 \left(S^T R^{-1} X_{ij}\right)^2} \rho^2 (d-1) \tag{7}$$

and

$$t_{stat} = \frac{S^T R^{-1} x_{ij}}{\sqrt{X_{ij}^T R^{-1} X_{ij} - \rho^2 \left(S^T R^{-1} X_{ij}\right)^2}} \rho\sqrt{d-1} \tag{8}$$

where $\rho = \left(S^T R^{-1} S\right)^{-1/2}$.

4. Quantitative Evaluation

The SNR of the images is a measure of the quality of data. The traditional definition of SNR is the ratio of the average magnitude of a signal to the magnitude of the background noise, determined as described in Section 5. If a distorted signal, $y(n)$, is considered as:

$$y(n) = x(n) + u(n), \tag{9}$$

where $x(n)$ is a signal and $u(n)$ is the background noise, the SNR can be written for N samples as [35]:

$$\text{SNR} = \frac{\sum_{\text{n}=0}^{N-1} |[x(\text{n})]|^2}{N\sigma^2}, \tag{10}$$

where $|[x(\text{n})]|$ is the magnitude of a signal and σ^2 is the variance of the background noise. Since most signals have wide dynamic ranges, the SNR is usually given in decibels (dB), [36]:

$$\text{SNR} = 10\log\left(\frac{\Delta^2}{\sigma^2}\right), \tag{11}$$

where Δ is the peak value of the signal. Equation (11) can be simplified as:

$$\text{SNR} = 20\log\left(\frac{\Delta}{\sigma}\right). \tag{12}$$

5. Results and Discussion

The results of both the passive and active thermography are presented in this section. All temperatures were measured using the IR camera. The different image processing algorithms, employed to increase the quality of the thermal images, are described and their results are discussed in detail.

5.1. *Active Thermography*

The temperature distribution profiles depicted in Figure 5b are plotted along with the rows identified in Figure 5a. The most visible contrasts of the defects revealed, as expected, that the deeper the defect, in terms of the distance from the bottom of the hole to the surface, the less detectable it was. Moreover, the size of the defect was important. The raw thermograms do not show smaller defects located deep within the plate. Defects with a diameter of 4 mm were barely detected. It can also be seen that the defects in the middle part of the plate are clearer than those towards the plate edges. This was primarily caused by non-uniform heating where the middle part of the sample received more thermal energy than the boundaries. The temperature distribution profiles during the step heating thermography for all pixels along the lines shown in Figure 5a are illustrated in Figure 5c. Figure 5c was recorded after heating for 75 s. The signals generated by the defects located in the last row, which have the smallest diameters, were not strong enough to be easily detected. Contrary to pulse thermography, Figure 5c shows that the surface of the object has been uniformly heated during the step heating thermography, which increased the efficiency of this method in detecting internal defects.

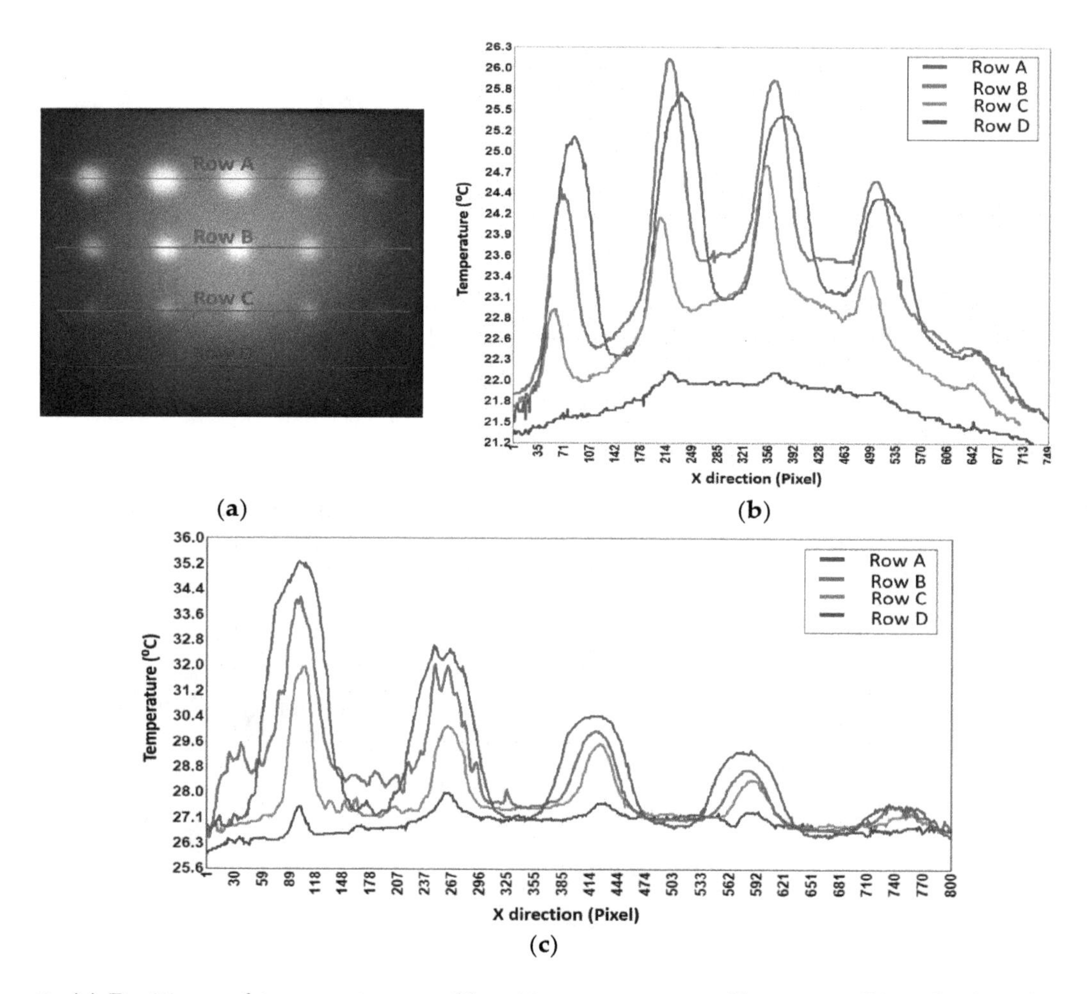

Figure 5. (**a**) Positions of temperature profiles. Temperature profiles using (**b**) pulsed and (**c**) step heating thermography. The rows are identified in (**a**).

Temperatures near the defects were measured and analyzed to better evaluate the effects of the defect's depth and size on the temperature variation. The temperatures measured at the locations of the flat-bottomed holes with various depths are illustrated in Figure 6. The sample was heated by two halogen lamps for 75 s, during which time data were recorded. It can be concluded from this figure that an increase in a defect's depth reduces its temperature variation, which makes detection more difficult because there is not a significant change in the surrounding temperature.

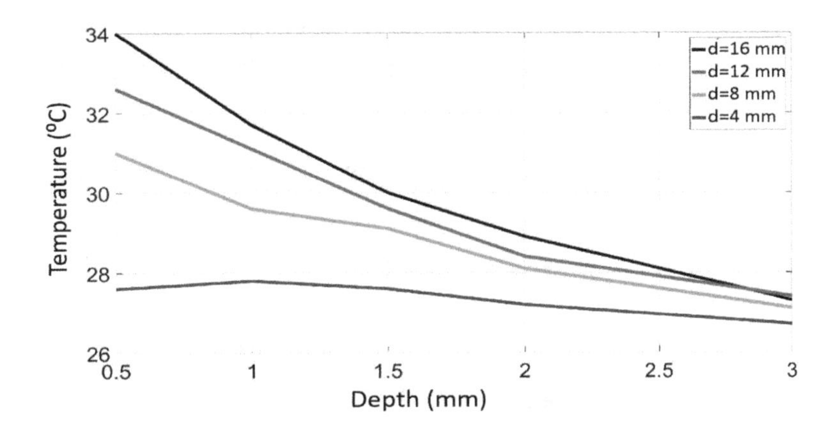

Figure 6. Effect of depth and size of the defect on the temperature distribution in a sample heated up by a halogen lamp for 75 s.

By increasing the size of the defect, the slope of the temperature variation also increased, which implied that the effect of the depth variation on the temperature was more significant for larger defects.

The SNRs for the defect plate profiles shown in Figure 5a were determined to provide a quantitative evaluation of the detection. The signal values of the defects were measured first. The difference between the signal values of the defects and the background signals was identified as the peak value signal for the SNR calculation. Then, by calculating the standard deviation of the background noise and using Equation (12), the SNR of each defect was calculated. By comparing the visual results obtained from image processing and the corresponded SNR tables, defects with a SNR of more than almost three times the background noise was considered as a defect that can be identified by this method. The SNRs, related to raw thermal data captured by flash and step heating (heating and cooling periods) thermography, are listed in Tables 1–3. Defect position names used in these tables are defined in Figure 1c. It can be observed from these tables that bigger defects closer to the surface generated high SNRs. It can also be concluded from the results presented in these tables that raw thermograms of flash thermography can provide more details than step heating thermography. It should be noted that the SNRs of raw thermograms are compared later with values of the processed thermal images to show the strength of each of the image processing algorithms.

Table 1. Signal to Noise Ratio (SNR) (dB) related to raw data captured by step heating thermography (heating period).

	1	**2**	**3**	**4**	**5**
A	18.4	18.2	17.8	14.4	3
B	18.1	18	17.7	12.9	6
C	17.9	17.5	17	11	NA
D	12.3	12	8.6	NA	NA

Table 2. SNR related to raw data captured by step heating thermography (cooling period).

	1	**2**	**3**	**4**	**5**
A	25.7	24.9	22.8	19.4	12.3
B	24.4	20	20.4	17.5	11.9
C	7.4	10.2	15.2	11.5	5.9
D	NA	2.58	2.5	NA	NA

Table 3. SNR related to raw data captured by flash thermography.

	1	2	3	4	5
A	16.3	18.5	18.5	13.7	3.4
B	10.2	15.8	17	11.9	2.9
C	2.5	4.2	13.1	4.9	2.6
D	NA	2.1	2.3	2.1	NA

5.2. Image Processing and Quantitative Evaluation

The results for the fourth row in Figure 5 show that some subsurface defects cannot generate visible contrasts. This suggests the need for image processing to improve the resolution. Different algorithms, including MF, and a combination of PPT as a transform-based technique were employed to increase the SNR and the visibility of the defects. We will use the term "Step Phase and Amplitude Thermography" (SPAT) for the analysis of thermograms after heating and successfully used for passive thermal data processing.

5.2.1. Matched Filters (MF)

Applying MF to the step heating thermograms improved the quality of the results, as shown in Figure 7. All four versions of MF increased the visibility and contrast of the defects. Three of these filters, including the SAM, the ACE, and the *F*-statistic, showed very promising results where all defects could be detected. However, the *t*-statistic highly improved the quality of the raw thermal data and compared to other MF methods provided less details regarding subsurface defects.

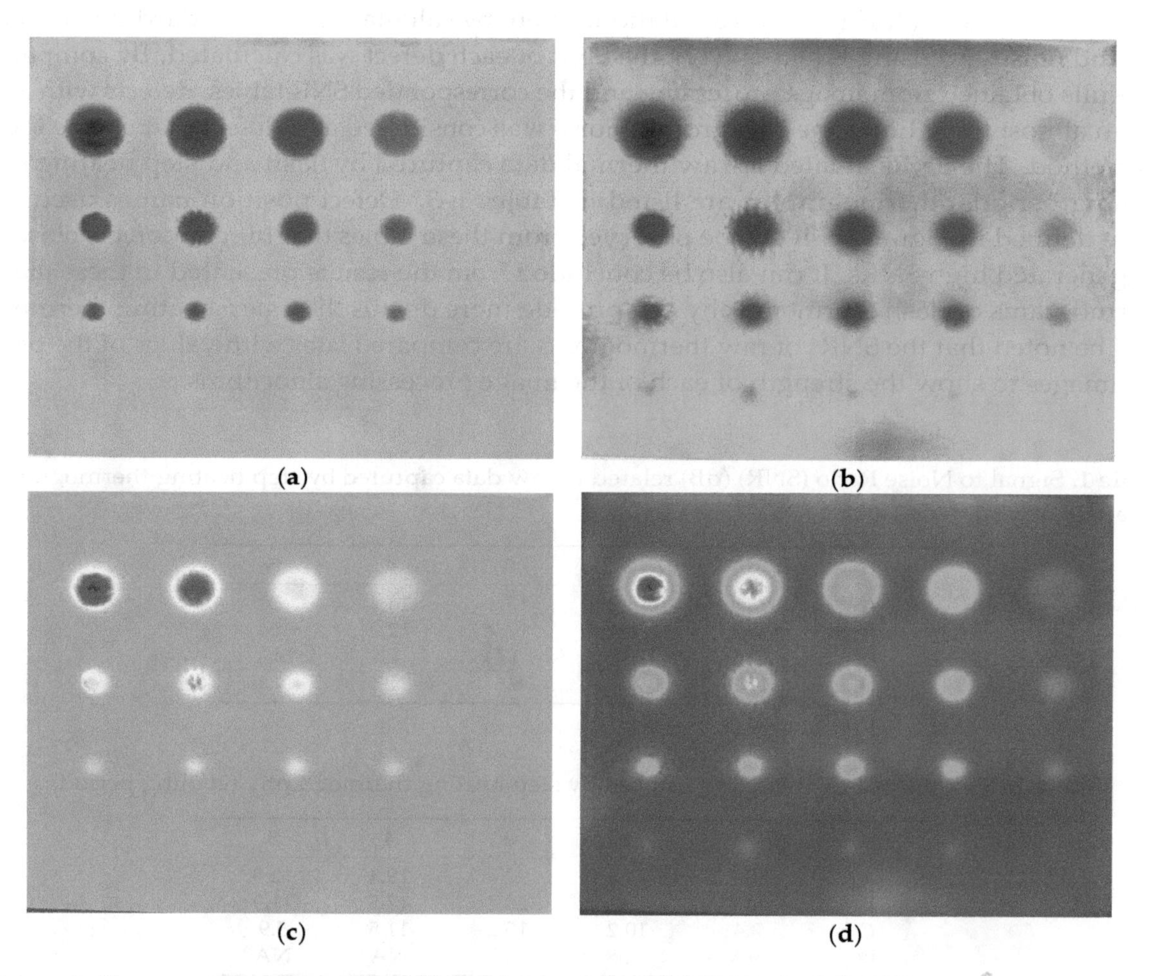

Figure 7. Four matched filters including (**a**) Spectral Angle Map (SAM), (**b**) Adaptive Coherence Estimator (ACE), (**c**) *t*-statistic and (**d**) *F*-statistic when the specimen was being step heated.

By using the above results, it can be seen that the diameter-to-depth ratio of the minimum detectable defect for each of SAM, ACE and *F*-statistic was 1.33, while this value for *t*-statistic was 2. To quantitatively evaluate the MF results, the SNRs for defective areas were determined. Figure 8a illustrates the signals of the *F*-statistic results obtained from the step heating data. The background noise around A2 is also shown in Figure 8b. These results were averaged to get the representative noise. It should be noted that all of the processed images have been normalized by dividing the intensity of each pixel over the difference between the maximum and minimum intensity available on the image.

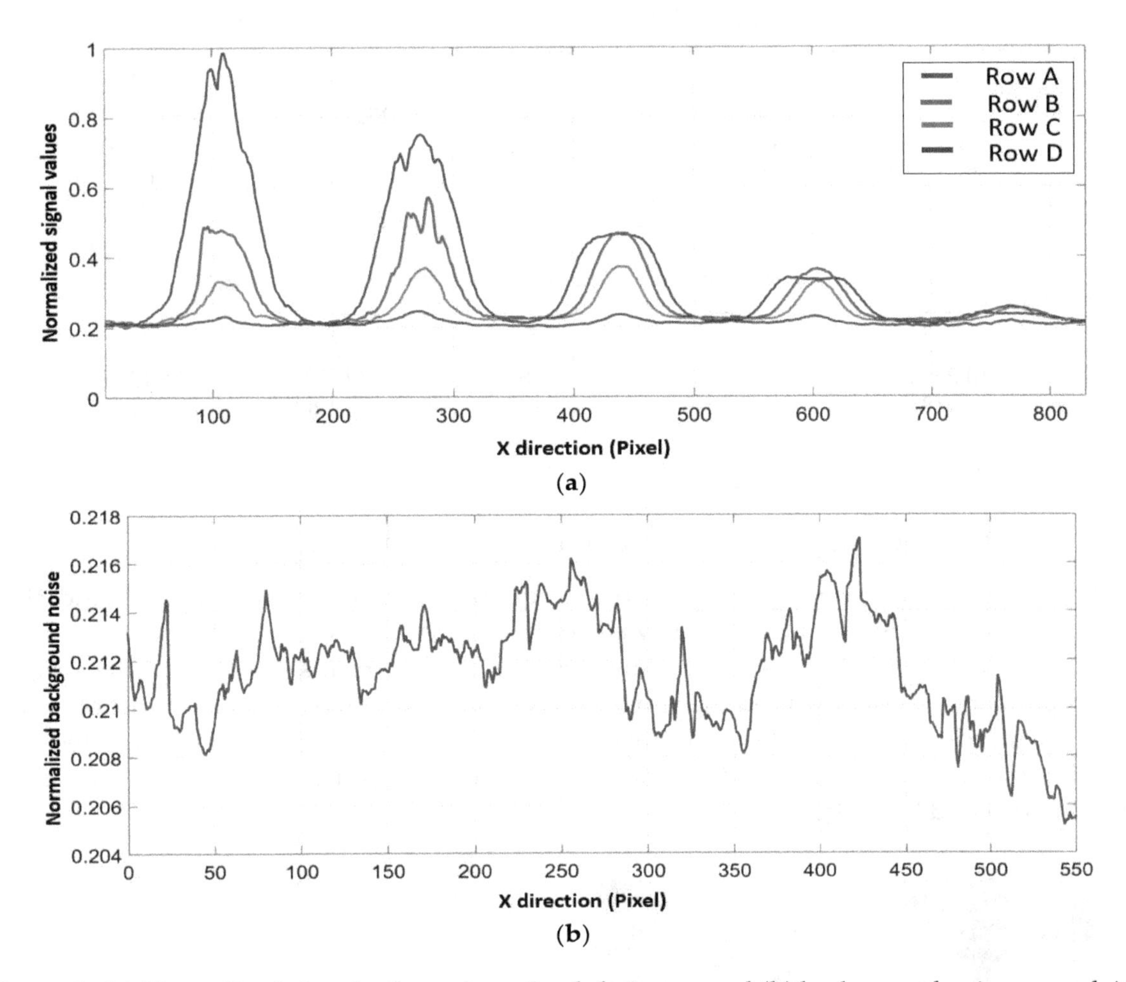

Figure 8. (**a**) Normalized signal values along the defect rows and (**b**) background noise around A2.

The SNRs are presented in Tables 4–7. The percentage improvement in the SNRs compared to the raw values is shown in parenthesis for each defect. These tables show that higher SNR values were obtained when MFs were applied to the step heating data. Therefore, the application of MFs to step heating data improved the resolution of the internal defects.

Tables 4–7 also reveal that larger defects closer to the surface generated higher SNR values. There were some exceptions to this general conclusion, however, primarily due to the non-uniformity of the heating.

Table 4. SNR (dB) for SAM on step heating thermal data.

	1	2	3	4	5
A	36.8 (99%)	35.3 (94%)	31.8 (78%)	26.9 (87%)	15.1 (396%)
B	33.5 (84%)	33.7 (87%)	31.2 (76%)	27.9 (116%)	17 (185%)
C	30.7 (72%)	29.3 (67%)	29.1 (70%)	26.3 (139%)	18.8
D	16.1 (31%)	18.7 (55%)	18.6 (116%)	13.9	5.5

Table 5. SNR (dB) for ACE on step heating thermal data.

	1	2	3	4	5
A	33.2 (79%)	31.9 (76%)	29.1 (63%)	24.3 (68%)	11.6 (282%)
B	30.6 (68%)	29.9 (66%)	28.5 (60%)	25.4 (97%)	15.3 (185%)
C	26.4 (48%)	27.1 (54%)	26.5 (55%)	24 (118%)	16.7
D	14.7 (19%)	16.9 (40%)	16.9 (96%)	11.7	8.5

Table 6. SNR (dB) for the *t*-statistic on step heating thermal data.

	1	2	3	4	5
A	44.3 (139%)	41.6 (129%)	34.9 (95%)	28.5 (90%)	15.7 (416%)
B	35.6 (95%)	37 (105%)	34.5 (95%)	38.8 (201%)	22.7 (280%)
C	28.2 (58%)	30.2 (72%)	30.3 (77%)	27.6 (151%)	19.4
D	13.5 (10%)	18.4 (53%)	16 (85%)	11.3	6.6

Table 7. SNR (dB) for the *F*-statistic on step heating thermal data.

	1	2	3	4	5
A	40.2 (117%)	37.6 (107%)	32.7 (83%)	27 (87%)	13.8 (354%)
B	33 (82%)	33.8 (88%)	32.2 (82%)	28.3 (120%)	17.8 (198%)
C	27.2 (52%)	28.9 (65%)	28.6 (68%)	26.1 (137%)	18.5
D	13.5 (10%)	17.2 (44%)	16.5 (92%)	11.6	2.7

5.2.2. Transform-Based Techniques: Step Phase and Amplitude Thermography

Pulsed thermograms were converted to a frequency domain using the FFT and the phase images were analyzed. Since the surface was excited uniformly during step heating, both phase and amplitude images of the transformed step heating thermograms were evaluated. Figure 9a,b shows a comparison of the raw pulsed thermograms and the normalized phase contrast obtained using the FFT. PPT significantly increased the contrast so that all defects, except D5, were detected. The same normalization method as the one used for the MF method was applied to the results obtained using the transformed-based techniques.

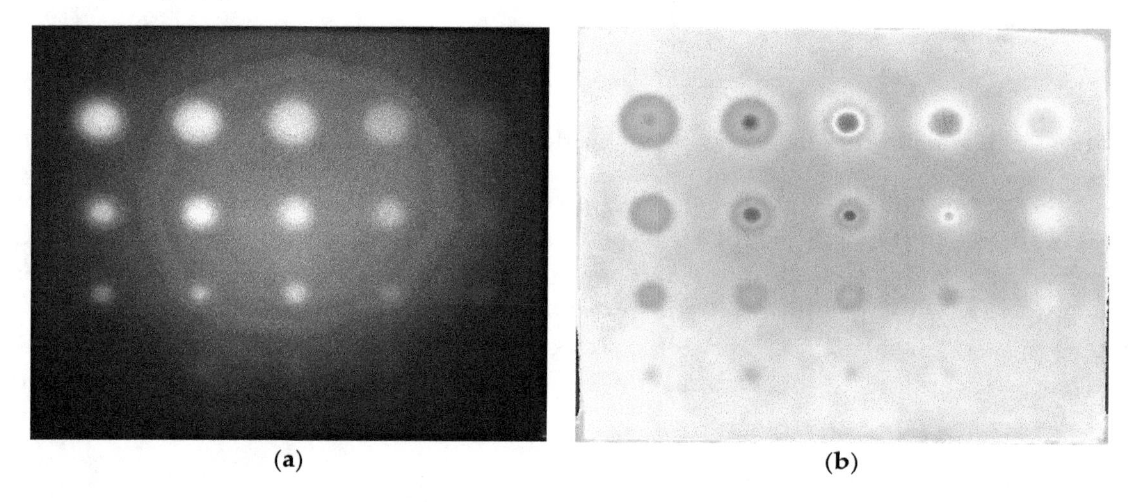

Figure 9. (**a**) Raw thermogram and (**b**) phase image (acquisition time = 53.2 s) obtained from the thermal image sequence recorded during cooling after flashing the surface.

The SNR values for the phase images depicted in Figure 9 are presented in Table 8, which shows that larger defects that were closer to the surface had higher SNRs. The first column of defects is an exception to this observation, mainly due to non-uniform heating of the surface.

Table 8. SNR (dB) for Pulsed Phase Thermography (PPT) data.

	1	2	3	4	5
A	25.5 (56%)	26.8 (45%)	24.9 (35%)	23.2 (69%)	20.7 (511%)
B	23.1 (127%)	26.5 (67%)	26.9 (58%)	23.4 (96%)	19.9 (591%)
C	19.6 (677%)	20.8 (397%)	19.8 (52%)	14.2 (190%)	12.9 (401%)
D	13.9	16.4	12.8	7 (234%)	NA

Figure 10a,b shows the phase and amplitude images at the minimum frequency where the FFT was applied to the step heating thermograms captured during cooling after heating for 75 s.

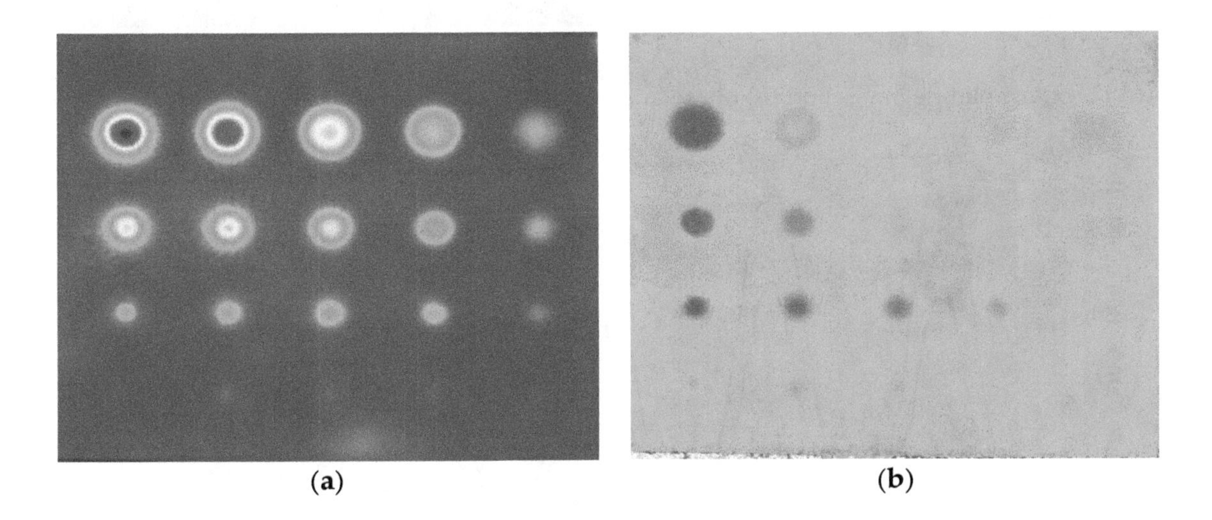

(a) (b)

Figure 10. (**a**) Amplitude image and (**b**) phase image of the thermograms captured during cooling after 75 s of heating.

The diameter-to-depth ratio of the minimum detectable defect in each of phase images of PPT and SPAT was 2, while the amplitude of SPAT provided a better value of 1.33. It can be concluded from these results that the application of the FFT transform to step heating thermography was more effective than it was for flash thermography. This was especially true for the amplitude data, where clearer results with higher contrasts were achieved. Phase images revealed that defects with a better contrast compared to the amplitude images, in some cases. This demonstrated that a reliable inspection could be achieved by evaluating both results. Figure 11 shows the results of an FFT transform applied to thermograms captured during heating for 75 s. The phase image captured all defects—assumed to be indicated by the signals that were more than three times the background level—demonstrating that phase images extracted from the heating data provided more visibility compared to the data obtained from cooling. The amplitude images obtained during cooling were less noisy and contained more detail, which allowed the shape of defects to be determined.

Figure 12a presents a further analysis of the normalized amplitude contrast distribution of the thermograms captured during cooling after 75 s of heating. The curves in this figure are the amplitude variation along the lines shown in Figure 5a. A significant change occurred between the sound and defective areas in the amplitudes, leading to a sharp boundary around the defects. The normalized phase contrast distribution of thermograms obtained after heating for 75 s is depicted in Figure 12b. Bigger defects closer to the surface generated higher phase contrasts and were subsequently more visible in the phase image. By analyzing these plots, all defects, except D5, on the defect plate were detected.

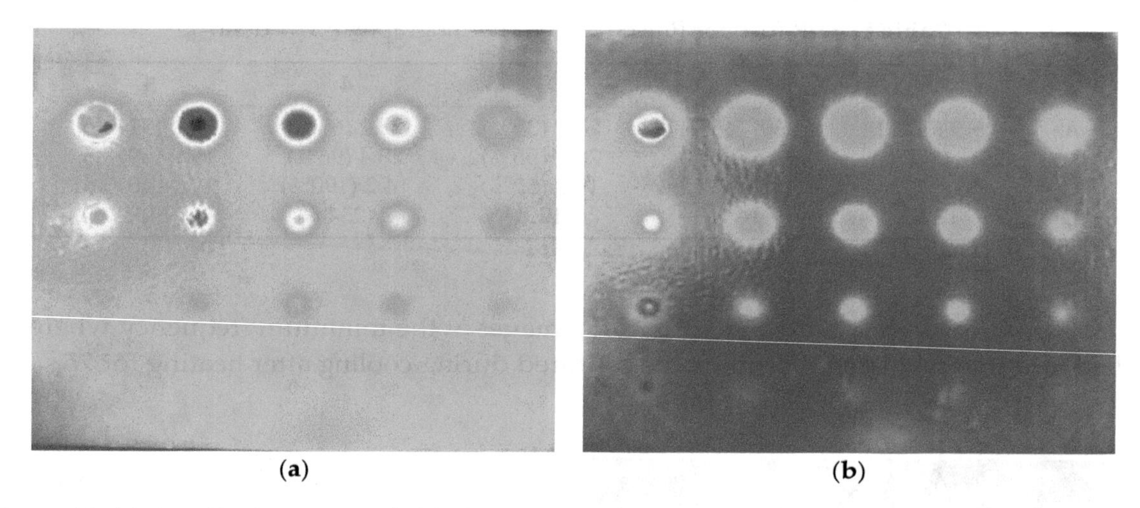

(a) (b)

Figure 11. (**a**) Amplitude image and (**b**) phase image of thermograms captured during heating for 75 s.

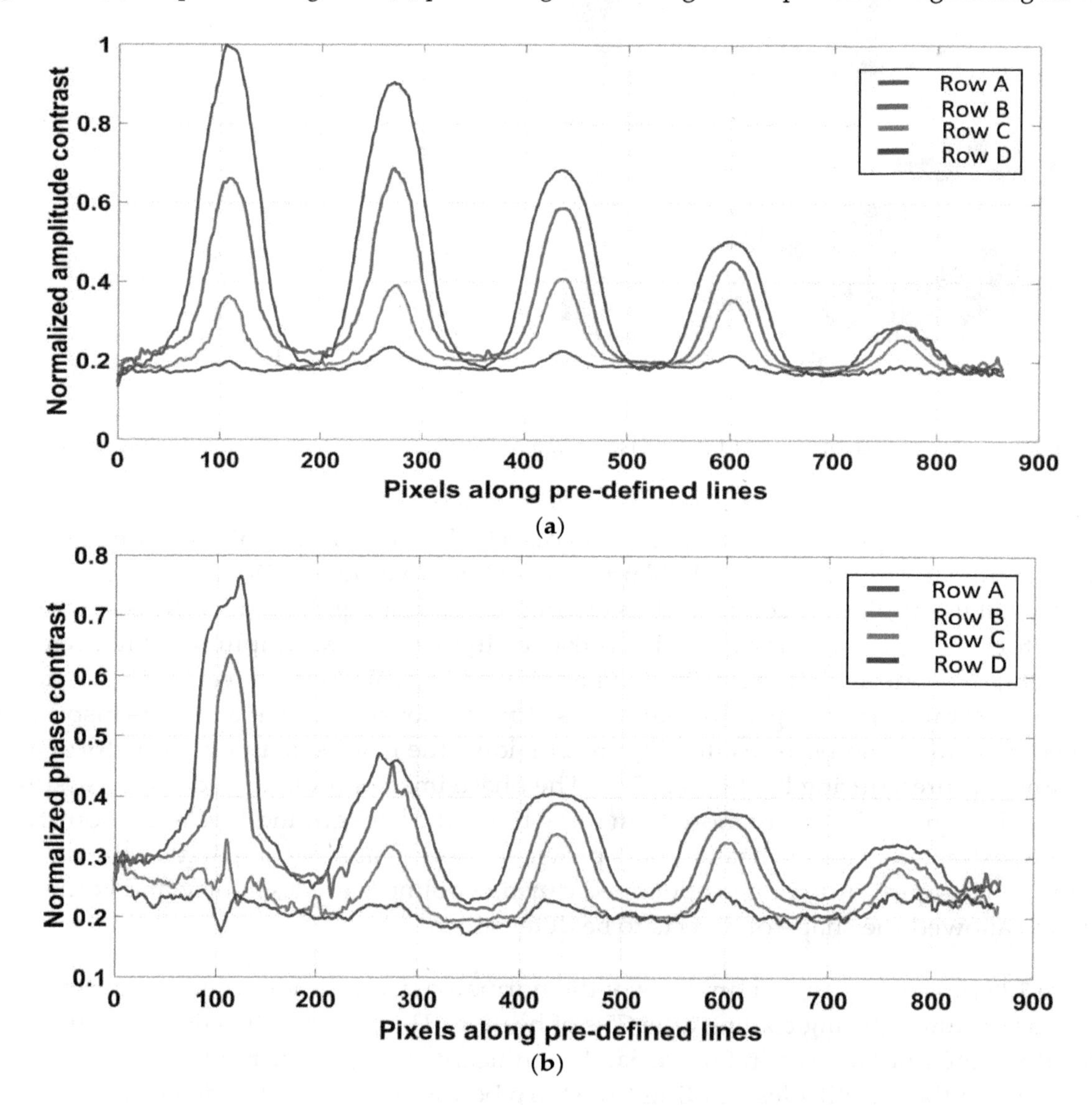

Figure 12. (**a**) Normalized amplitude value and (**b**) normalized phase value distributions of the defects where the thermograms were obtained during cooling and heating, respectively.

The SNR values for the phase and amplitude images captured by the application of FFT on the step heating data are listed in Tables 9–11. These results demonstrate that amplitude images extracted from thermal data captured during cooling had higher SNR values and revealed more details of subsurface defects. The SNRs of both the phase and amplitude images show that larger defects closer to the surface provided higher values.

Table 9. SNR (dB) of amplitude image captured during cooling after 75 s heating.

	1	2	3	4	5
A	46.2 (80%)	45.1 (81%)	42.2 (85%)	38.4 (98%)	29.7 (141%)
B	41 (68%)	41.3 (106%)	39.6 (94%)	36.7 (109%)	28.7 (140%)
C	33.1 (349%)	33.7 (231%)	34.6 (128%)	33.4 (190%)	25.9 (338%)
D	14.9	23.5 (809%)	21.2 (763%)	17	13.6

Table 10. SNR (dB) of phase image captured during 75 s heating.

	1	2	3	4	5
A	38.5 (183%)	30.9 (144%)	29.2 (127%)	27.6 (180%)	23.6 (774%)
B	36 (302%)	32 (160%)	29.2 (132%)	27.5 (208%)	23.2 (896%)
C	21.5 (1211%)	26.3 (707%)	27.8 (164%)	26.6 (580%)	22.9 (907%)
D	19	13.3 (1044%)	18.3 (827%)	17.7 (710%)	12.7

Table 11. SNR (dB) of amplitude image captured during 75 s heating.

	1	2	3	4	5
A	33.6 (106%)	41.9 (127%)	40.8 (120%)	38.3 (179%)	31.6 (834%)
B	26.9 (163%)	38.5 (142%)	37.6 (121%)	34.7 (191%)	29.2 (915%)
C	20.3 (706%)	26.2 (527%)	30.1 (130%)	28.7 (484%)	24.2 (842%)
D	8	13.1 (537%)	12.7 (455%)	10.8 (416%)	9.4

By comparing the results presented in Tables 8–11, it can be concluded from the results presented in Tables 9–11 that the application of the FFT transform to step heating thermography was more effective than its application to flash thermography. This was especially true for the amplitude data, where higher SNRs were achieved. The amplitude images extracted from step heating thermography were an effective means of revealing subsurface damage, as they detected even small and deep defects.

By comparing the results presented in Tables 1–9 related to the SNRs of raw and processed thermal data, it is concluded that the image processing algorithms significantly increased the visibility of subsurface defects especially those of smaller sizes, which are embedded in deeper areas.

5.3. *Passive Thermography*

Passive thermal imaging of the damaged blade section was conducted at different times of the day. Typical results are shown in Figure 13. The results demonstrated that passive thermography was capable of capturing cracks, delamination, and internal features of the blade section.

Early morning experiments provided a visible contrast of the defects on the defect plate attached to the damaged blade section primarily due to the considerable temperature change on the blade during this period [2,18,19]. All defects in the plate, except the smallest 4 mm ones, were detected during this period. Less useful information about the defects was obtained when the experiment was performed around noon. None of the defects were visible during the evening (around 6 pm), which was mainly due to the balanced temperature on the blade surface after several hours of heating.

Thermal contrasts associated with dirt (glue on the surface), within the blue box in Figure 13, were most pronounced at noon. These contrasts faded during the afternoon. The blade's internal features such as shear webs were detected as cold regions during the morning and at noon but became hot signatures during the evening after several hours of heating.

Cracks and delamination on the suction side were apparent near the blade's trailing edge. At noon, with peak sunlight, cracks and delamination were detected. Delamination in the upper area, identified by the green box in Figure 13, were not detected clearly during the morning. The evening thermograms did not provide any information regarding cracks and delamination, so noon was the best time for crack and delamination monitoring.

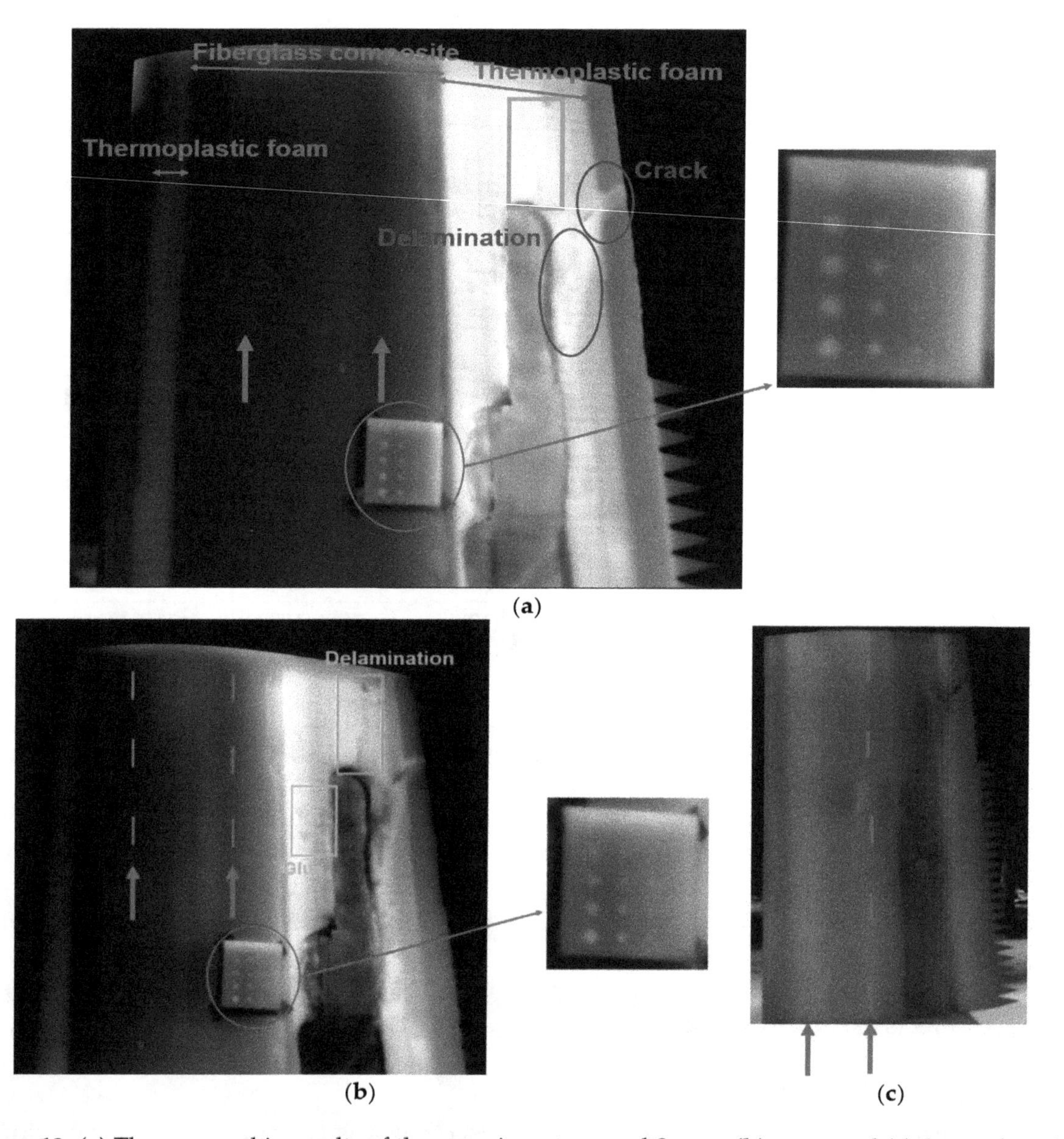

Figure 13. (**a**) Thermographic results of the experiment around 9 a.m., (**b**) noon and (**c**) 6 p.m. (sunrise and sunset were around 5.53 a.m. and 9.30 p.m., respectively). The vertical arrows and dashed lines indicate the shear webs.

The transform-based technique, discussed in Section 5.2.2, SPAT, was employed for the first time to increase the quality of the passive thermography results. The phase images captured using this method were not sensitive to non-uniform heating. The FFT was applied to passive thermograms obtained at morning. The results are shown in Figure 14. Part (b) illustrates that the amplitude images considerably increased the quality and visibility of the visualized subsurface defects, as the visibility of cracks, delamination, and a large portion of the flat-bottomed hole defects was improved. Phase images have noticeably increased the contrast of shear webs signatures, marked by red arrows in Figure 14a.

Cracks and delamination are marked by white boxes in Figure 14b. The thermoplastic foam near the leading edge was detected in the amplitude results and is highlighted by the red box in Figure 14b. This method not only increased the quality of the thermal images and improved the detectability of thermography but also eliminated the false indications associated with environmental reflections, dirt, and dust on the surface.

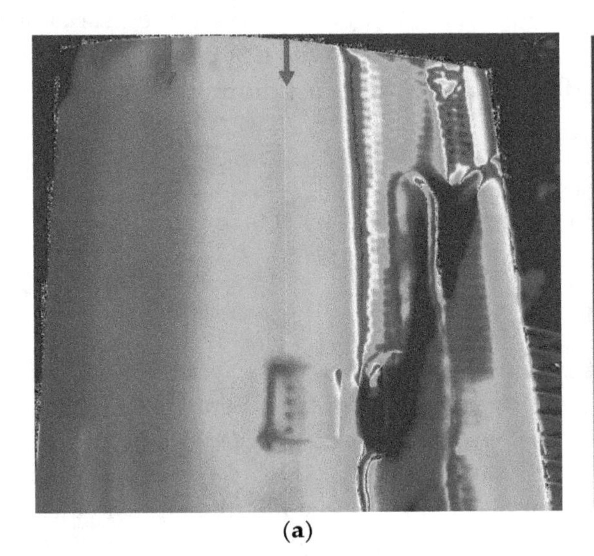

(a)

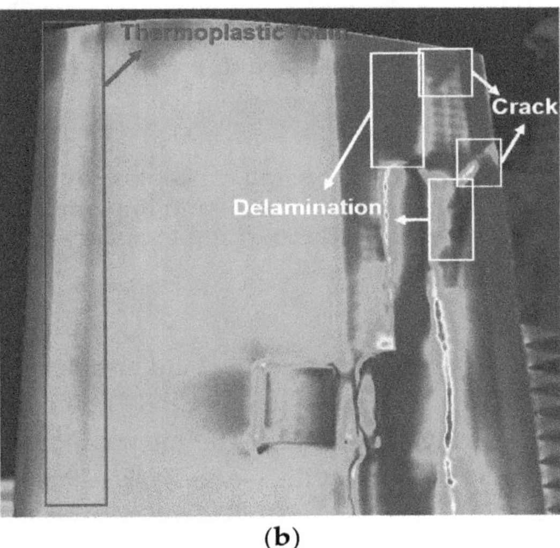

(b)

Figure 14. (**a**) Phase images of the passive thermograms captured during the morning at a frequency of 0.00184 Hz and (**b**) amplitude image of the passive thermograms recorded during the morning at a frequency of 0.0165 Hz.

6. Conclusions

The blades, the most critical components of wind turbines, are susceptible to failure due to initiation and propagation of subsurface damage in a number of forms. This study investigated thermography techniques for monitoring the condition of wind turbine blades. Active thermography, using pulsed and step heating, was conducted on a specially-constructed defect plate to evaluate this method's potential for detecting subsurface defects. The 170 × 195 mm plate was cut from the laminate skin of a wind turbine blade. Flat-bottomed holes were drilled from the inside to produce "defects" with known diameters and penetrations. The results demonstrated that active thermography is a powerful method for the monitoring and fault detection of wind turbine blades but that the signals generated by some small defects could not be detected.

Passive thermography was conducted on a damaged blade section and attached defect plate. This experiment was conducted at different times of day to determine the most favorable time of a day for maximum defect detection. The results showed that early morning and noon were best for detecting certain types of defects. Cracks, delamination, and surface dirt generated the most visible signatures around noon, while defects such as flat-bottomed holes in the defect plate were more visible in the morning as the sun heated the targets. This conclusion applies to the day time experiments as there were not any overnight tests.

The raw thermograms obtained by both passive and active thermography could not reveal the small defects located deep in the laminate skin. Different image processing algorithms including Matched Filters, Thermal Signal Reconstruction, and Pulsed Phase Thermography were used to increase the quality of active thermograms. "Step Phase and Amplitude Thermography", was used to analyze the step heating data. This technique gave the best detection of defects as measured by the diameter-to-depth ratio of 1.33. All the image processing algorithms improved the contrast in the active thermograms but some drawbacks can be considered for the Matched Filters method. The Matched Filters method is not fully automated and requires manual selection of points in a sound area of the damaged blade, which is time-consuming and affects the quality of results. In order to quantitatively evaluate the results, the signal-to-noise ratios of the raw and processed images were calculated. Higher ratios can be obtained when image processing algorithms are applied to the raw thermal data. As an obvious observation, larger defects closer to the surface generated higher ratios.

Step Phase and Amplitude Thermography, as a successful method for improving active thermography results, was applied to passive thermal data. The quality of passive thermograms was increased as a result. This method could also eliminate the false indications associated with environmental reflections and dirt on the surface. Nevertheless, it was not possible to resolve the smallest defects of 4 mm diameter whatever the depth. The minimum diameter-to-depth ratio for these defects was thus 1.25.

Author Contributions: The overall conception of this project was by D.W. and Q.S. who obtained the damaged wind turbine blade. The specific aims and implementation of them was done by H.S. as were all the experiments. All three authors contributed to writing the paper.

References

1. Jüngert, A. Damage Detection in wind turbine blades using two different acoustic techniques. In Proceedings of the 7th fib PhD Symposium in Civil Engineering, Stuttgart, Germany, 10–13 September 2008.
2. Meinlschmidt, P.; Aderhold, J. Thermographic inspection of rotor blades. In Proceedings of the 9th European Conference on NDT, Berlin, Germany, 25–29 September 2006.
3. Tao, N.; Zeng, Z.; Feng, L.; Li, X.; Li, Y.; Zhang, C. The application of pulsed thermography in the inspection of wind turbine blades. In *International Symposium on Photoelectronic Detection and Imaging 2011: Terahertz Wave Technologies and Applications*; SPIE: Washington, DC, USA, 2011; p. 819319.
4. Habali, S.M.; Saleh, I.A. Local design, testing and manufacturing of small mixed airfoil wind turbine blades of glass fiber reinforced plastics: Part II: Manufacturing of the blade and rotor. *Energy Convers. Manag.* **2000**, *41*, 281–298. [CrossRef]
5. Lading, L.; Mcgugan, M.; Sendrup, P.; Rheinländer, J.; Rusborg, J. *Fundamentals for Remote Structural Health Monitoring of Wind Turbine Blades—A Preproject Annex B—Sensors and Non-Destructive Testing Methods for Damage Detection in Wind Turbine Blades*; Risø National Laboratory: Roskilde, Denmark, 2002.
6. Rumsey, M.; Paquette, J. Structural health monitoring of wind turbine blades. *Proc. SPIE* **2008**, *6933*, 69330E.
7. Schroeder, K.; Ecke, W.; Apitz, J.; Lembke, E. A fibre Bragg grating sensor system monitors operational load in a wind turbine rotor blade. *Meas. Sci. Technol.* **2006**, *17*, 1167.
8. Liu, W.; Tang, B.; Jian, Y. Status and problems of wind turbine structural health monitoring techniques in China. *Renew. Energy* **2010**, *35*, 1414–1418. [CrossRef]
9. Larsen, G.; Hansen, M.; Baumgart, A.; Carlén, I. *Modal Analysis of Wind Turbine Blades*; Risø National Laboratory: Roskilde, Denmark, 2002.
10. Borum, K.; McGugan, M. Condition monitoring of wind turbine blades. In Proceedings of the 27th Riso International Symposium on Materials Science: Polymer Composite Materials for Wind Power Turbines, Roskilde, Denmark, 4–7 September 2006.
11. Li, T.; Almond, D.P.; Rees, D.A.S. Crack imaging by scanning laser-line thermography and laser-spot thermography. *Meas. Sci. Technol. Mar.* **2011**, *22*, 35701.
12. Pawar, S.S.; Peters, K. Through-the-thickness identification of impact damage in composite laminates through pulsed phase thermography. *Meas. Sci. Technol.* **2013**, *24*, 115601. [CrossRef]
13. Maierhofer, C.; Myrach, P.; Reischel, M.; Steinfurth, H.; Röllig, M.; Kunert, M. Characterizing damage in CFRP structures using flash thermography in reflection and transmission configurations. *Compos. Part B Eng.* **2014**, *57*, 35–46.
14. Maierhofer, C.; Arndt, R.; Röllig, M.; Rieck, C.; Walther, A.; Scheel, H.; Hillemeier, B. Application of impulse-thermography for non-destructive assessment of concrete structures. *Cem. Concr. Compos.* **2006**, *28*, 393–401. [CrossRef]
15. Beattie, A.; Rumsey, M. Non-destructive evaluation of wind turbine blades using an infrared camera. In Proceedings of the 37th Aerospace Sciences Meeting and Exhibit, Reno, NV, USA, 11–14 January 1999.
16. bin Zhao, S.; Zhang, C.; Wu, N. Infrared thermal wave nondestructive testing for rotor blades in wind turbine generators non-destructive evaluation and damage monitoring. In *International Symposium on Photoelectronic Detection and Imaging 2009: Advances in Infrared Imaging and Applications*; SPIE: Washington, DC, USA, 2009.
17. Galleguillos, C.; Zorrilla, A.; Jimenez, A.; Diaz, L.; Montiano, Á.L.; Barroso, M.; Viguria, A.; Lasagni, F. Thermographic non-destructive inspection of wind turbine blades using unmanned aerial systems. *Plast. Rubber Compos.* **2015**, *44*, 98–103. [CrossRef]

18. Worzewski, T.; Krankenhagen, R.; Doroshtnasir, M. Thermographic inspection of a wind turbine rotor blade segment utilizing natural conditions as excitation source, Part I: Solar excitation for detecting deep structures in GFRP. *Infrared Phys. Technol.* **2016**, *76*, 756–766. [CrossRef]
19. Worzewski, T.; Krankenhagen, R. Thermographic inspection of wind turbine rotor blade segment utilizing natural conditions as excitation source, Part II: The effect of climatic conditions on thermographic inspections–A long term outdoor experiment. *Infrared Phys. Technol.* **2016**, *76*, 767–776. [CrossRef]
20. Doroshtnasir, M.; Worzewski, T.; Krankenhagen, R.; Röllig, M. On-site inspection of potential defects in wind turbine rotor blades with thermography. *Wind Energy* **2016**, *19*, 1407–1422. [CrossRef]
21. Lahiri, B.B.; Bagavathiappan, S.; Reshmi, P.R.; Philip, J.; Jayakumar, T.; Raj, B. Quantification of defects in composites and rubber materials using active thermography. *Infrared Phys. Technol.* **2012**, *55*, 191–199. [CrossRef]
22. Shin, P.H.; Webb, S.C.; Peters, K.J. Pulsed phase thermography imaging of fatigue-loaded composite adhesively bonded joints. *NDT E Int.* **2016**, *79*, 7–16. [CrossRef]
23. Maierhofer, C.; Röllig, M.; Krankenhagen, R.; Myrach, P. Comparison of quantitative defect characterization using pulse-phase and lock-in thermography. *Appl. Opt.* **2016**, *55*, D76–D86. [CrossRef] [PubMed]
24. Almond, D.P.; Angioni, S.L.; Pickering, S.G. Long pulse excitation thermographic non-destructive evaluation. *NDT E Int.* **2017**, *87*, 7–14. [CrossRef]
25. Maldague, X.; Marinetti, S. Pulse phase infrared thermography. *J. Appl. Phys.* **1996**, *79*, 2694–2698. [CrossRef]
26. Chatterjee, K.; Tuli, S.; Pickering, S.; Almond, D. A comparison of the pulsed, lock-in and frequency modulated thermography nondestructive evaluation techniques. *NDT E Int.* **2011**, *44*, 655–667. [CrossRef]
27. Montanini, M. Quantitative determination of subsurface defects in a reference specimen made of Plexiglas by means of lock-in and pulse phase infrared thermography. *Infrared Phys. Technol.* **2010**, *53*, 363–371. [CrossRef]
28. Pawar, S.S. Identification of Impact Damage in Composite Laminates through Integrated Pulsed Phase Thermography and Embedded Thermal Sensors. Ph.D. Thesis, North Carolina State University, Raleigh, NC, USA, 2013.
29. Ibarra, C.C. Quantitative Subsurface Defect Evaluation by Pulsed Phase Thermography: Depth Retrieval with the Phase. Ph.D. Thesis, Laval University, Quebec, QC, Canada, 2005.
30. Shepard, S.M.; Lhota, J.R.; Rubadeux, B.A.; Wang, D.; Ahmed, T. Reconstruction and enhancement of active thermographic image sequences. *Opt. Eng.* **2003**, *42*, 1337. [CrossRef]
31. Larsen, C. Document Flash Thermography. Master's Thesis, Utah State University, Logan, UT, USA, 2011.
32. Roche, J.M.; Balageas, D.L. Common tools for quantitative pulse and step-heating thermography—Part II: Experimental investigation. *Quant. Infrared Thermogr. J.* **2015**, *12*, 1–23. [CrossRef]
33. Foy, B.R. *Overview of Target Detection Algorithms for Hyperspectral Data; Rep. to NNSA, Rep. No. LU-UR-09-00593;* Los Alamos Natl. Lab.: Los Alamos, NM, USA, 2009.
34. Kretzmann, J. Evaluating the industrial application of non-destructive inspection of composites using transient thermography. Master's Thesis, Stellenbosch Stellenbosch University, Stellenbosch, South Africa, 2016.
35. Xia, X.G. A Quantitative Analysis of SNR in the Short-Time Fourier Transform Domain for Multicomponent Signals. *IEEE Trans. Signal Process.* **1998**, *46*, 200–203.
36. Jain, J.; Jain, A. Displacement Measurement and Its Application in Interframe Image Coding. *IEEE Trans. Commun.* **1981**, *29*, 1799–1808. [CrossRef]

3

Design and Testing of a LUT Airfoil for Straight-Bladed Vertical Axis Wind Turbines

Shoutu Li [1,2,3], Ye Li [4,5,6,7,*], Congxin Yang [1,2,3], Xuyao Zhang [1,2,3], Qing Wang [1,2,3], Deshun Li [1,2,3], Wei Zhong [8] and Tongguang Wang [8]

1 School of Energy and Power Engineering, Lanzhou University of Technology, Lanzhou 730050, China; lishoutu@lut.edu.cn (S.L.); ycxwind@163.com (C.Y.); zxy0932@163.com (X.Z.); wangqing_lut@foxmail.com (Q.W.); lideshun_8510@sina.com (D.L.)
2 Gansu Provincial Technology Centre for Wind Turbines, Lanzhou 730050, China
3 Key Laboratory of Fluid machinery and Systems, Lanzhou 730050, China
4 School of Naval Architecture, Ocean and Civil Engineering, Shanghai Jiao Tong University, Shanghai 200240, China
5 State Key Laboratory of Ocean Engineering, School of Naval Architecture, Ocean and Civil Engineering, Shanghai Jiao Tong University, Shanghai 200240, China
6 Collaborative Innovation Center for Advanced Ship and Deep-Sea Exploration, Shanghai Jiao Tong University, Shanghai 200240, China
7 Key Laboratory of Hydrodynamics (Ministry of Education), Shanghai Jiao Tong University, Shanghai 200240, China
8 Jiangsu Key Laboratory of Hi-Tech Research for Wind Turbine Design, Nanjing University of Aeronautics and Astronautics, Nanjing 210016, China; zhongwei@nuaa.edu.cn (W.Z.); tgwang@nuaa.edu.cn (T.W.)
* Correspondence: ye.li@sjtu.edu.cn

Abstract: The airfoil plays an important role in improving the performance of wind turbines. However, there is less research dedicated to the airfoils for Vertical Axis Wind Turbines (VAWTs) compared to the research on Horizontal Axis Wind Turbines (HAWTs). With the objective of maximizing the aerodynamic performance of the airfoil by optimizing its geometrical parameters and by considering the law of motion of VAWTs, a new airfoil, designated the LUT airfoil (Lanzhou University of Technology), was designed for lift-driven VAWTs by employing the sequential quadratic programming optimization method. Afterwards, the pressure on the surface of the airfoil and the flow velocity were measured in steady conditions by employing wind tunnel experiments and particle image velocimetry technology. Then, the distribution of the pressure coefficient and aerodynamic loads were analyzed for the LUT airfoil under free transition. The results show that the LUT airfoil has a moderate thickness (20.77%) and moderate camber (1.11%). Moreover, compared to the airfoils commonly used for VAWTs, the LUT airfoil, with a wide drag bucket and gentle stall performance, achieves a higher maximum lift coefficient and lift–drag ratios at the Reynolds numbers 3×10^5 and 5×10^5.

Keywords: airfoil design; aerodynamic; wind tunnel experiment; VAWTs (Vertical axis wind turbines)

1. Introduction

In an urban or offshore environment, Vertical Axis Wind Turbines (VAWTs), especially the lift-driven type, have unique advantages due to their lower cost, lower noise, simpler structure, and ability to actively accept wind energy from different directions [1–3]. Moreover, with the development of distributed wind energy, VAWTs function more efficiently in certain areas, particularly areas with large fluctuations in wind direction [4,5]. However, compared to Horizontal Axis Wind Turbines (HAWTs), for a long time, VAWTs have experienced slow development and received very little financial support. One reason is the lack of completed theoretical research on VAWTs. For example, although the multiple flow tube theory, which is based on blade element momentum (BEM), is applied to the VAWT design, the results are still far from satisfactory because of the lack of adequate theoretical correction [6]. Another possible reason is that HAWTs have thousands of special airfoils, such as the FFA (FLYGTEKNISKA FORSOKSANSTALTEN) series, DU (Delft University) series, RISØ series, and S series. However, little research has been performed on the airfoils for VAWTs, resulting in the lack of good blades for VAWTs, which further affects VAWT development. Meanwhile, the airfoil plays an essential role in the design of a wind turbine and greatly affects the wind turbine's performance. Therefore, the above factors have led to not only the slow development of VAWTs but also to a heavily debated topic, i.e., the types of airfoil to use in a VAWT [7].

Many studies that have been conducted on the aerodynamic performance of VAWTs have been carried out with a focus on commonly used aircraft airfoils. In particular, in previous research, the symmetrical airfoils of the NACA 00XX series, such as NACA 0012, NACA 0015, and NACA 0018, are widely applied to VAWTs because they help to improve the power coefficient of VAWTs [8]; nevertheless, symmetrical airfoils are unable to properly self-start in VAWTs [9]. Other studies [10,11] have pointed out that VAWTs with a symmetrical airfoil result in poor performance at low Reynolds numbers (R_e), because the self-start difficulty is worsened under such conditions. Studying the effects of the airfoil on the aerodynamic performance of VAWTs, Mohamed et al. [12] focused on 25 kinds of airfoils, and the results indicate that the cambered airfoil LS(1)-0413 increases the power coefficient compared to NACA 0018. In addition, the cambered airfoil NACA 63-415 has a wider operating range. The thickness and camber of the airfoil influence VAWTs according to studies of the NACA-4 digit modified airfoil family or the NACA four-series airfoils [13–15]. However, when the VAWTs work at a higher tip-speed ratio, the aerodynamic performance of the VAWTs decreases with the constant increase in both the camber and thickness of the airfoil [11,13]. Therefore, based on these previous results, it is necessary to consider a moderate camber and moderate thickness in the design of an airfoil for use in a VAWT.

Additionally, according to previous research, some specially purposed airfoils are developed for VAWTs. Achieving great success, the Sandia National Laboratory and Natural Resources Canada carried out a huge volume of work on airfoils for VAWTs through wind tunnel and field experiments [16,17]. The studies of the Sandia National Laboratory show that the natural laminar flow airfoil is suitable for VAWTs. As a result, the Somers airfoils are designed based on their research results; however, they are applied to the curved-bladed Darrieus turbines primarily [7]. Meanwhile, Islam performed meaningful work on VAWTs [9,18,19] by demonstrating the proper camber and thickness most suitable for VAWTs. Moreover, the MI-VAWT1 airfoil is designed to solve the problems of the overall cost, the aerodynamic performance, and blade strength. At the same time, the aerodynamic performance of the MI-VAWT1 airfoil is superior to that of NACA 0015. In addition, the Technical University of Denmark (DTU) has been dedicated to the research of airfoils for VAWTs. Its studies [20,21] depict some very thick airfoils that were designed by considering the structural stiffness of the blade and the aerodynamic characteristics. For instance, the AIR001 airfoil (max. thickness of 25%) can improve the aerodynamic performance in the case of a larger tip-speed ratio, while the DU12W262 airfoil (max. thickness of 26.2%) can increase blade strength. Other research [9,22,23] has explored specially purposed airfoils for VAWTs, such as the DU06-W-200, EN0005, WUP1615, and "TWT series" airfoils from Tokai University. In brief, there are different design aims for different special-purpose airfoils.

As mentioned above, a great deal of investigation has been done to understand the aerodynamic performance of VAWTs and the characteristics of the airfoil. However, compared to HAWTs, the research on special-purpose VAWT airfoils, which affect the aerodynamic performance of VAWTs, is limited. In addition, there exists room for improvement of many aspects, such as the maximum lift coefficient, maximum lift–drag coefficient, wide drag bucket, and stall performance, in the aerodynamic performance of the existing airfoil. Additionally, it is very important that the airfoil has good performance at a low R_e due to the operating environment of VAWTs. Therefore, the objective of this paper is to design an airfoil—named the LUT airfoil, a reference to Lanzhou University of Technology—that shows good aerodynamic characteristics for VAWTs and to investigate the characteristics of the airfoil at low R_e values by wind tunnel experiments. This paper is divided into three main parts: illustrating the design basis and optimization method in Section 2, describing the experimental equipment and procedures in Section 3, and reporting the profile and the aerodynamic performance of the LUT model in Section 4.

2. Basis and Method of Design for the LUT Airfoil

2.1. Design Basis

Figure 1 shows the structure and operation of an H-type VAWT (H-VAWT). The rotational axis of H-VAWTs is always perpendicular to the incoming airflow V_{in}, which results in a constantly changing local angle of attack and a readily occurring dynamic stall [24], especially with a rotor at low speed. Moreover, the VAWTs typically operate in low-R_e environments. Nevertheless, reasonable design of special airfoil for VAWTs is one of the effective methods to solve the above problems. Therefore, the characteristics of an airfoil at a low R_e and a wide range of angles of attack (α) were also considered factors in the design of the LUT airfoil.

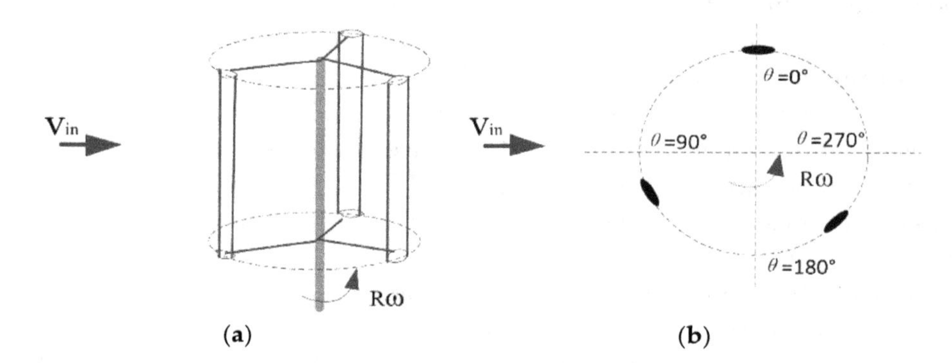

Figure 1. Schematic of an H-type Vertical Axis Wind Turbines (H-VAWT): (**a**) 3D schematic; and (**b**) 2D cross-section. The rotor of the H-VAWT turns counterclockwise; the azimuth angle θ is defined to help describe the active conditions of an H-VAWT. R is the radius of the rotor in units of mm. ω is the rotation angular velocity of the rotor in units of rad/s.

2.2. LUT Airfoil Optimization and Design Method

Optimization algorithm and numerical simulation are widely used in design of the airfoil for VAWTs. Therefore, in this study, the sequential quadratic programming (SQP) method [25,26] was adopted to design the LUT airfoil, because it can properly solve the problem of constrained optimization, maintain the approximate second-order drop speed, and effectively handle constraint conditions. The optimization process is detailed in the following.

The optimization problem can essentially be boiled down to a nonlinear programming problem, as follows:

$$\begin{cases} Min & f(X) \\ s.t. & \begin{matrix} c_i(X) = 0 & i = 1, 2, \cdots, m_e \\ c_i(X) \geq 0 & i = m_e + 1, \cdots, m \end{matrix} \end{cases} \quad (1)$$

where $f(X)$ is the objective function, $c_i(X)$ is the constraint condition, m_e is the number of constraint conditions of the equation, and m is the number of total constraint conditions.

For the convenience of derivation, the Lagrange function of Equation (1) is:

$$L(X, \lambda) = f(X) - \sum_{i=1}^{m_e} \lambda_i \cdot c_i(X) \quad (2)$$

The gradient vector of this function of x and the Hesse matrix are denoted, respectively, by:

$$\nabla_x L(X,\lambda) = \nabla f(X) - \sum_{i=1}^{m_e} \lambda_i \nabla c_i(X) \tag{3}$$

$$\nabla_x^2 L(X,\lambda) = \nabla^2 f(X) - \sum_{i=1}^{m_e} \lambda_i \nabla^2 c_i(X) \tag{4}$$

For $\lambda = (\lambda_1, \lambda_2, \cdots, \lambda_{m},)$, the gradient vector and Hesse matrix of the Lagrange function of X and λ, corresponding to $\nabla L(X,\lambda)$ and $\nabla^2 L(X,\lambda)$, respectively, are defined as follows:

$$\nabla L(X,\lambda) = \begin{bmatrix} \nabla f(X) - \nabla c(X)\lambda \\ -c(X) \end{bmatrix} \tag{5}$$

$$\nabla^2 L(X,\lambda) = \begin{bmatrix} \nabla_x^2 L(X,\lambda) & -\nabla c(X) \\ -\nabla c(X)^T & 0 \end{bmatrix} \tag{6}$$

where $\nabla c(X)$ is the Jacobian matrix for the vector function at X, that is, the matrix of $n \times m$ with the partial derivative $\frac{\partial c_j(X)}{\partial x_i}$ of the (i, j) element. The first-order Taylor of the gradient vector is expanded to:

$$\nabla L(X,\lambda) = \nabla L\left(X^k, \lambda^k\right) + \nabla^2 L\left(X^k, \lambda^k\right) \begin{bmatrix} X - X^k \\ \lambda - \lambda^k \end{bmatrix} \tag{7}$$

where k represents the current iteration. It is a necessary condition that the gradient is equal to zero for the optimal solution of the Lagrange function, that is, $\nabla L(X,\lambda) = 0$. Substituting Equations (5) and (6) into the first-order Taylor expansion of the gradient vector produces:

$$\begin{bmatrix} \nabla_x^2 L\left(X^k, \lambda^k\right) & -\nabla c\left(X^k\right) \\ -\nabla c\left(X^k\right)^T & 0 \end{bmatrix} \begin{bmatrix} X - X^k \\ \lambda - \lambda^k \end{bmatrix} = \begin{bmatrix} -\nabla f\left(X^k\right) + \nabla c\left(X^k\right)\lambda^k \\ c\left(X^k\right) \end{bmatrix} \tag{8}$$

By defining $d = X - X^k$ and replacing $\nabla_x^2 L\left(X^k, \lambda^k\right)$ with the Hessen matrix B^k, we can obtain:

$$\begin{bmatrix} B^k & -\nabla c\left(X^k\right) \\ -\nabla c\left(X^k\right)^T & 0 \end{bmatrix} \begin{bmatrix} d \\ \lambda \end{bmatrix} = \begin{bmatrix} -\nabla f\left(X^k\right) \\ c\left(X^k\right) \end{bmatrix} \tag{9}$$

This equation is the exact K-T condition of the quadratic programming problem:

$$\begin{cases} Min & \frac{1}{2} d^T B^k d + \nabla f\left(X^k\right)^T d \\ s.t. & \nabla c\left(X^k\right)^T d + c\left(X^k\right) = 0 \end{cases} \tag{10}$$

where the vector d represents the search direction, and B^k is the approximation of the Hessen matrix. In this way, a common constraint problem is transformed into a quadratic programming problem. Moreover, on this basis, it can be naturally extended to inequality constraints:

$$\begin{cases} Min & \frac{1}{2} d^T B^k d + \nabla f\left(X^k\right)^T d \\ s.t. & \nabla c\left(X^k\right)^T d + c\left(X^k\right) = 0 \\ & \nabla c\left(X^k\right)^T d + c\left(X^k\right) \geq 0 \end{cases} \tag{11}$$

The above SQP method is only locally convergent in theory. To converge to the optimal solution, a one-dimensional search is required to ensure its overall convergence. The search direction d is obtained by Equation (11), which is the downward direction of many penalty functions. For example:

$$F_\sigma(X) = f(X) + \sigma\left[\sum_{i=1}^{m_e}|c_i(X)| + \sum_{i=m_e+1}^{m}|min\{c_i(X),0\}|\right] \quad (12)$$

where σ is the penalty factor that satisfies $\sigma > \max\{|\lambda_i|\, i = 1,2,\cdots,m\}$. A one-dimensional search based on $F_\sigma{}'\left(X^k,d\right) < 0$ is carried out along with the d direction, and then the step length t^k that satisfies $F_\sigma\left(X^k + t^k d\right) < F_\sigma\left(X^k\right)$ is obtained. Meanwhile, the next iteration point can be obtained:

$$X^{k+1} = X^k + t^k d^k \quad (13)$$

The flowchart describing the optimization of the airfoil's aerodynamic shape based on the SQP method is shown in Appendix A.

Additionally, the aerodynamic performance of the LUT airfoil is obtained under free transition by employing ANSYS Fluent code during the process of optimization; the numerical solution of the Reynolds-averaged Navier–Stokes (RANS) equation is used, and the Shear–Stress–Transport (*SST*) *k-ω* turbulence model is employed [27]. Pressure–velocity coupling is achieved by the SIMPLEC scheme, where the second-order upwind discretization scheme is utilized to calculate the pressure, and the second-order implicit formula is adopted to test temporal discretization. Figure 2a demonstrates the geometric scheme and the boundary conditions of the numerical simulation. The walls of the airfoil are set as non-slip walls, while the other walls are set as free-slip walls. The O-grids are positioned near the airfoil, and the first grid's spacing is less than 1×10^{-5} to ensure that the Y^+ is less than 1.

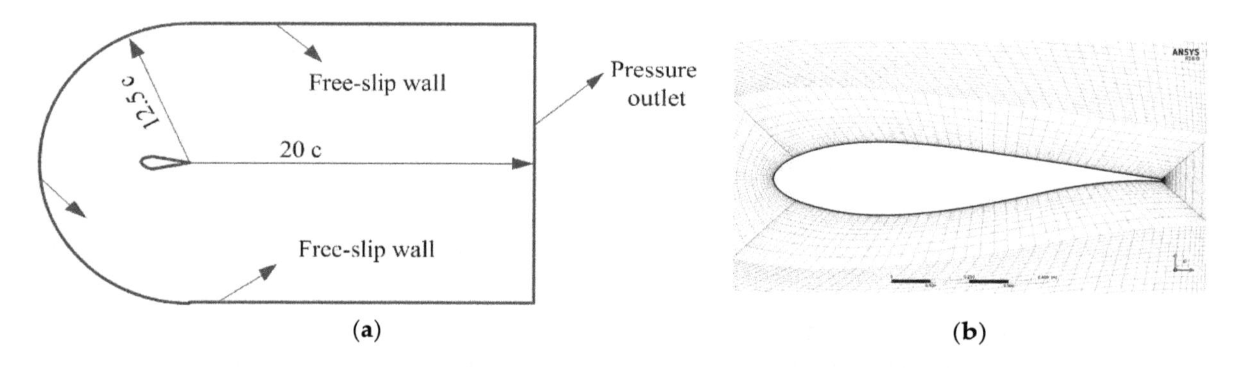

Figure 2. Scheme of the geometry and boundary conditions for the numerical simulation: (**a**) The semicircular far field is set as a velocity inlet located at 12.5 c (c is chord length) from the trailing edge of the airfoil, while the outlet condition is the pressure outlet located at 20 c from the trailing edge of the airfoil. (**b**) The boundary layer consists of 40 cell layers: the height of first layer is 1×10^{-6}; the growth rate of the grid is 1.05 in the boundary layer.

The design results are introduced in the results and discussion section.

This study required a LUT airfoil with a good aerodynamic performance; meanwhile, to ensure the structural strength of the blade for the LUT airfoil, the corresponding airfoil thickness and bending constraints were considered. Therefore, the objective function and constraint conditions can be summarized as:

$$\begin{cases} \quad Min \quad C_d/C_L \\ s.t. \quad \max(thickness) > \max(thickness0) \\ \qquad \max(curve) > \max(curve0) \end{cases} \quad (14)$$

where the thickness0 and curve0 represent the optimized benchmark corresponding to the maximum thickness and the maximum camber of the NACA 0018 airfoil, respectively; the design Reynolds number is 5×10^5; and the design angle of attack is $\alpha = 9°$.

Figure 3 exhibits the results of optimization for $R_e = 5 \times 10^5$. In Figure 3a, the profile of the LUT airfoil was obtained through the SQP method. Figure 3b shows the checked result of the mesh independence in the numerical simulation, where the maximum lift–drag ratio (C_L/C_d) of the NACA 0018 airfoil was investigated at an angle of attack of $\alpha = 9.25°$ [28]. When the total number of cells was higher than 5.8×10^5, the maximum C_L/C_d value of NACA 0018 was converged. Therefore, in this study, more than 6×10^5 cells were employed in the numerical simulation method.

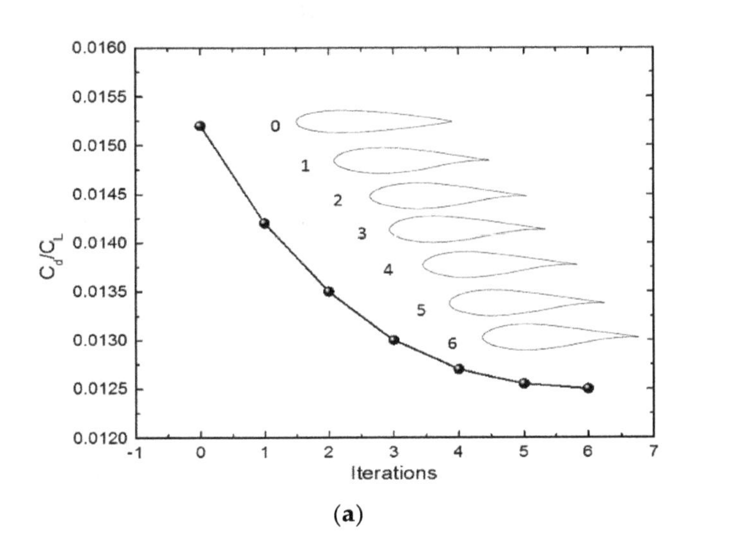

(a)

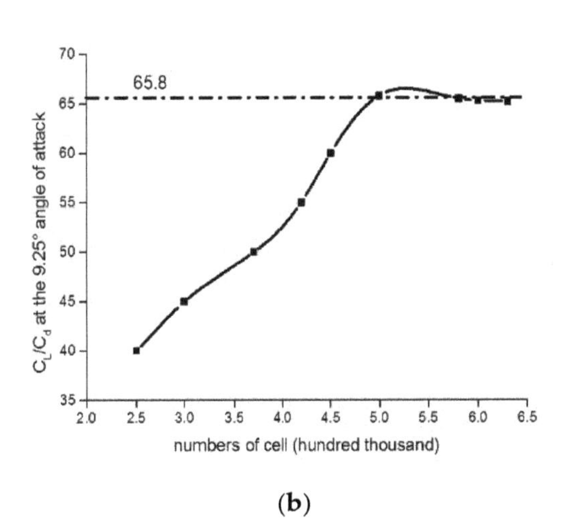

(b)

Figure 3. Airfoil optimization and convergence process. The process of the optimization was from 0 to 6, where the 0 and 6 represent the NACA 0018 airfoil and Lanzhou University of Technology (LUT) airfoil, respectively. (**a**) The constraint conditions were the minimum C_d/C_L, thickness, and camber of the airfoil for the LUT airfoil. (**b**) The maximum lift–drag ratio of the NACA 0018 is 65.8 at an angle of attack of 9.25° [28]. In this Figure, the R_e is 5×10^5.

Figure 4 presents the profile of the LUT airfoil. The LUT airfoil has a moderately maximum camber of 1.11% that is positioned at 78.9% of the chord; a maximum thickness of 20.77% is located at 28.3% of the chord; and the leading-edge radius (r_0) is at 3.84% of the chord. Compared with the DU06-W-200 and NACA 0018 airfoils, the position of the maximum camber is close to the trailing edge for the LUT, which can form a trailing-edge loading to get effective lift [29,30]. Moreover, the moderately maximum thickness not only ensures the strength of the blade for the LUT airfoil, but also helps improve the performance of the VAWTs with the LUT airfoil at high tip speed ratio.

Meanwhile, in the design, we tried to maintain a constant curvature of the upper surface to avoid an increase in flow velocity on the upper surface. Thus, the maximum thickness of the LUT was achieved by adding the thickness of the lower surface, which helps to improve the resistance of the leading-edge roughness. Additionally, the LUT airfoil has a larger leading-edge radius, helping to improve the resistance of the leading-edge roughness. Theoretically, as a result, the LUT has a low sensitivity to roughness. The sensitivity to the roughness of the LUT airfoil will be investigated in a future wind tunnel experiment. However, compared with the DU06-W-200 and NACA 0018, the position of the maximum thickness is slightly closer to the leading edge.

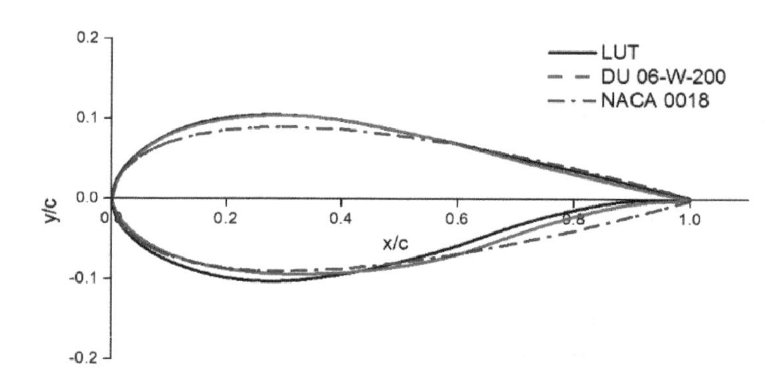

Figure 4. The profiles of the three airfoils. The profiles of the NACA 0018 and DU06-W-200 models come from the material specified on the website [28]. As shown in this Figure, the marks in black, red, and blue represent the profiles of the LUT, DU06-W-200, and NACA 0018 airfoils, respectively.

3. Experimental Equipment and Procedures

In this work, the aerodynamic performance of the LUT model was investigated by measuring the surface pressure of the LUT model at different values of R_e. A description of the main equipment used in the experiment is as follows.

3.1. Wind Tunnel

Wind tunnel tests were carried out in an open-circuit, closed-test-section, low-speed, and low-turbulence wind tunnel at the Northwestern Polytechnical University, as shown in Figure 5. The rectangular test section is 400 mm × 1000 mm, the length is 2800 mm, and the maximum wind speed is 70 m/s. The turbulence intensity is adjustable between 0.02% and 0.3% by changing the screen numbers. For the case of the empty test section in the stream direction with 12 layers of screens and a contraction ratio of 22.6, the turbulence intensity is 0.03%, 0.025%, and 0.02% at 12, 15, and 30 m/s freestream velocities, respectively [31], and the accuracy of the wind speed is less than ±3%.

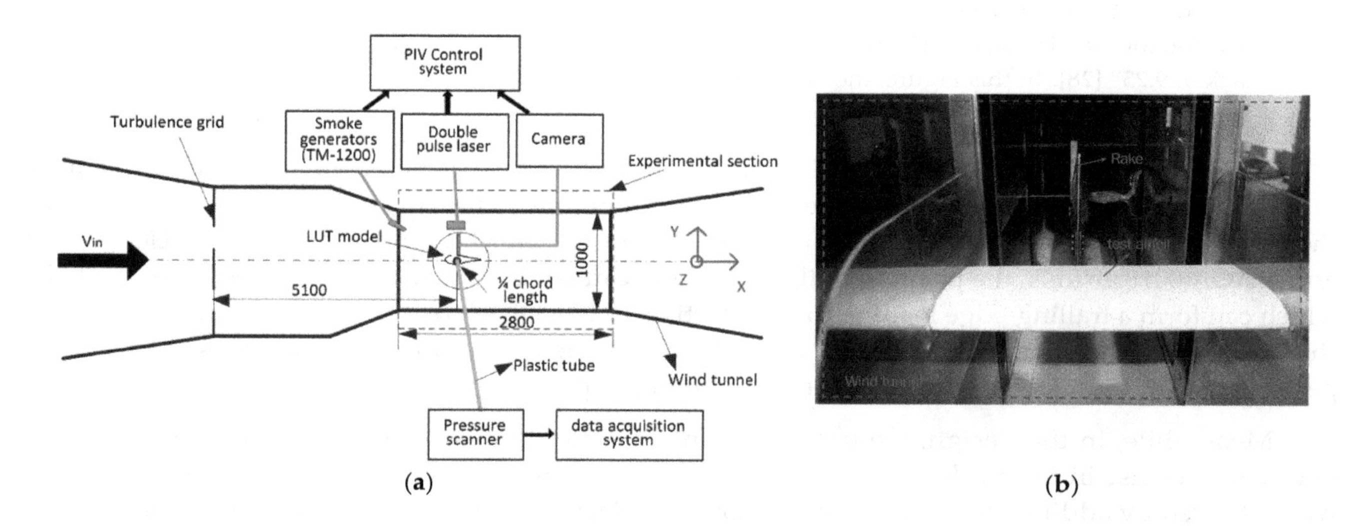

Figure 5. Experimental scheme of the wind tunnel and the testing of the LUT airfoil: (**a**) schematic of the wind tunnel and the experimental devices; and (**b**) experiment section. The cross-section of the wind tunnel is 400 mm × 1000 mm and the length is 2800 mm.

The coordinate system of the wind tunnel is defined by the x-, y-, and z-axes, which represent the freestream, perpendicular direction, and span of the airfoil, respectively.

3.2. Test LUT Airfoil Model

In this study, the test LUT airfoil was designed by the authors. The cross-section width of the LUT model is the same as that specified above, the chord of the LUT is 200 mm, and the span is 400 mm. As shown in Figure 6, to investigate the aerodynamic performance of the LUT airfoil under a static condition, 50 pressure taps with a diameter of 0.7 mm are distributed on the LUT airfoil model, where the distribution average density of the pressure taps is 0.7:0.19:0.4 from the leading edge to the trailing edge of the LUT model.

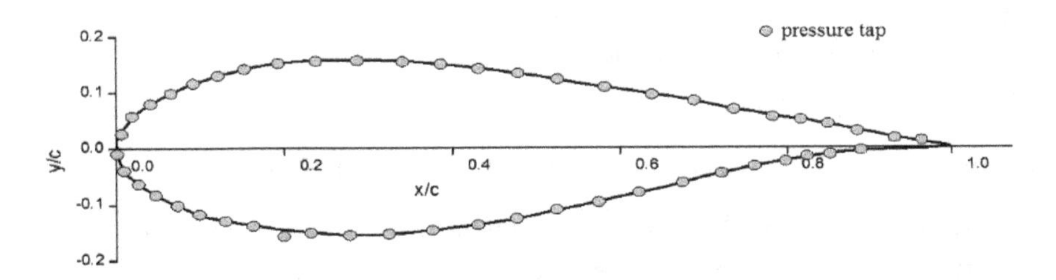

Figure 6. The profile and the distribution of the pressure taps for the test LUT airfoil. The total number of pressure taps is 50 on the surface of the airfoil, and the density of the pressure taps is closely distributed at the leading edge and the trailing edge of the test LUT airfoil.

3.3. Pressure Measurement and PIV (Particle Image Velocimetry) Devices

The pressure measurement devices include an electronic scanning measurement system (DSY104), pressure scanner (PSI9816), and angle of attack control mechanism. The accuracy of the pressure measurement and the PSI9816 pressure scanning are less than ±0.2% FS (Full Scale) and ±0.05% FS, respectively. The accuracy of the angle of attack control is $\pm 2'$. The Particle Image Velocimetry (PIV) systems include a double pulse laser with a maximum operating frequency of 15 Hz, a camera resolution of 2048 × 2048, a field range of 312 mm × 312 mm, and an image resolution of 152.66 μm/pixel. The tracer particles were produced by large smoke generators (TM-1200).

The pressure on the LUT airfoil surface is given by the pressure coefficient $C_p = (P_i - P_\infty)/(0.5\rho V_\infty^2)$, where P_i is the pressure of the pressure tap i, P_∞ is the freestream static pressure, V_∞ is the freestream speed, and ρ is the air density. In this work, the drag coefficient was calculated by the wake investigation method. The wake rake had a height of 300 mm and was installed 0.9 c downstream from the test airfoil trailing edge. The calculation formula of the drag coefficient is:

$$\mathrm{c_d} = \int_{wake\ rake\ height} c_x' \mathrm{d}(y/c) \tag{15}$$

where y is the y-coordinate value of the total pressure tap for the wake rake, whose integral function is:

$$c_x' = 2\left(\frac{p_1}{p_\infty}\right)^{1/\gamma}\left(\frac{p_{01}}{p_0}\right)^{(\gamma-1)/\gamma}\left[\frac{1-(p_1/p_{01})^{(\gamma-1)/\gamma}}{1-(p_\infty/p_0)^{(\gamma-1)/\gamma}}\right]^{0.5}\cdot\left[\left(\frac{1-(p_\infty/p_{01})^{(\gamma-1)/\gamma}}{1-(p_\infty/p_0)^{(\gamma-1)/\gamma}}\right)^{0.5}\right] - c_{x0}' \tag{16}$$

where C_x' is the local drag coefficient of the total pressure tap for the wake rake, p_1 is the static pressure at the wake of the model, p_∞ is the freestream static pressure, p_{01} is the total pressure of the wake rake, p_0 is the freestream total pressure, C_{x0}' is the C_x' arithmetic mean at the wake of the airfoil model, and γ is the specific heat ratio, which typically takes the value $\gamma = 1.4$.

The lift coefficient $\mathrm{C_L}$ was calculated by the relationship of the coordinate system, the drag coefficient $\mathrm{C_d}$, and the normal force coefficient $\mathrm{C_n}$:

$$\mathrm{C_L} = \mathrm{C_n} \cdot cos\alpha - (\mathrm{C_d} - \mathrm{C_n} \cdot sin\alpha) \cdot \tan\alpha \tag{17}$$

In the wind tunnel experiment, it was necessary to eliminate errors of measurement. Therefore, the pressure values of the pressure taps and the wake rake were referenced when the freestream velocity was 0 m/s and the angle of attack was 0°. Uncertainty analysis of the coefficients of both the lift and the drag was evaluated with the accuracy of the PSI9816 scanner.

4. Results and Discussion

4.1. *Static LUT Airfoil Performance at Low Reynolds Number*

In this section, the performance of the clean, static LUT airfoil is described for $R_e = 3 \times 10^5$ and 5×10^5 (these two low Reynolds numbers are suitable for the typical working environment of VAWTs), respectively.

4.1.1. Pressure Coefficient Distribution on the Surface for the LUT Airfoil

In Figure 7, the distribution of the pressure coefficients on the LUT airfoil surface is shown for some specific angles of attack at $R_e = 3 \times 10^5$ and at $R_e = 5 \times 10^5$. In this figure, the lines in black, red, and blue denoting the pressure coefficient (C_p) correspond to angles of attack of C_L/C_d-max, C_L-max, and post-stall, respectively.

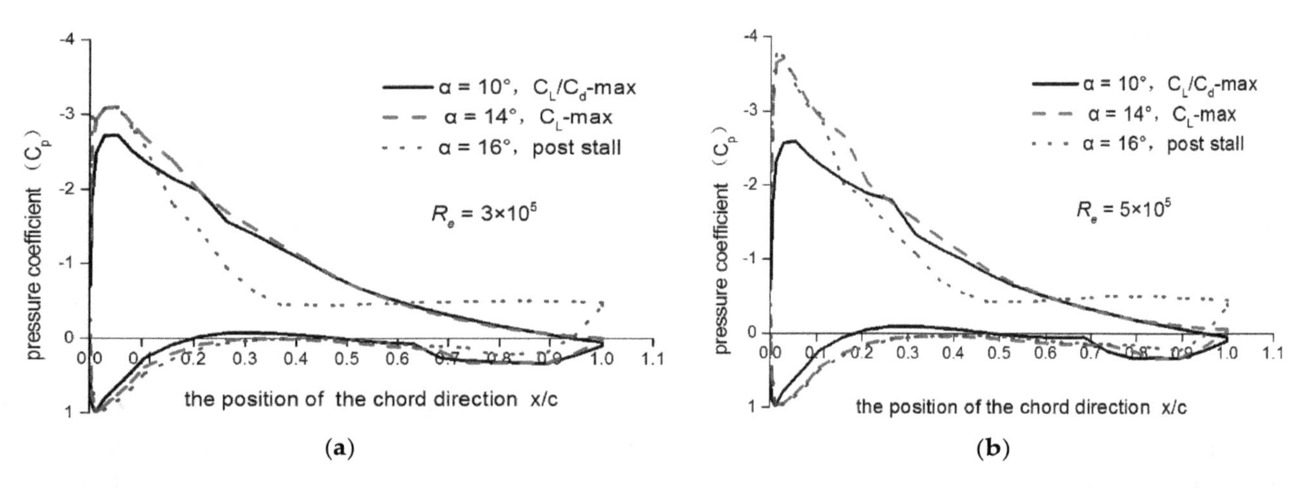

Figure 7. Pressure coefficient distribution of the LUT airfoil: (**a**) the case of $R_e = 3 \times 10^5$; and (**b**) the case of $R_e = 5 \times 10^5$. The data were measured when the surface of the LUT airfoil was clean. The lines in black, red, and blue show the pressure coefficients corresponding to angles of attack of C_L/C_d-max, C_L-max, and post-stall, respectively.

In the case of $R_e = 3 \times 10^5$, the LUT airfoil has a maximum lift coefficient of $C_{L\text{-max}} = 1.47$ at an angle of attack of $\alpha = 14°$ and a maximum lift–drag ratio of $C_L/C_{d\text{-max}} = 64.4$ at an angle of attack of $\alpha = 10°$. After an angle of attack of $\alpha = 14°$, the LUT airfoil starts to stall, and separation occurs on the suction surface at about 39% of the chord with an angle of attack of $\alpha = 16°$. Therefore, the pressure coefficient on the suction surface of the LUT airfoil keeps a constant value from about 39% of the chord to the trailing edge. In the case of $R_e = 5 \times 10^5$, the LUT airfoil has a maximum lift coefficient of $C_{L\text{-max}} = 1.5$ at an angle of attack of $\alpha = 14°$ and a maximum lift–drag ratio of $C_L/C_{d\text{-max}} = 79.8$ with an angle of attack of $\alpha = 10°$. After an angle of attack $\alpha = 14°$, the LUT airfoil starts to stall, and separation occurs on the suction surface at about 48% of the chord when the angle of attack is 16°. Moreover, after 48% of the chord, the pressure coefficient on the suction surface of the LUT airfoil keeps a constant value; that is, the LUT airfoil is fully stalled.

Comparing the cases of $R_e = 3 \times 10^5$ and $R_e = 5 \times 10^5$, the suction peak value increases and moves to the leading edge of the LUT airfoil with the increase in R_e. This indicates that the region of adverse pressure gradient on the suction surface of the LUT airfoil is extended when $R_e = 5 \times 10^5$. However, due to the effect of the turbulence, the flow separation is delayed [32]. Additionally, the pressure was not measured at large angles of attack due to the measuring range of the wake rake. In the future, the performance of the LUT needs to be studied at large angles of attack.

Figure 8 shows the changes in the pressure coefficient distribution with increasing angles of attack at different values of R_e for the LUT airfoil. In the case of $R_e = 3 \times 10^5$, the free transition position is located at 53% and 35% of the chord when the angle of attack is 0° and 8°, respectively. Until an angle of attack of $\alpha = 21°$, the pressure of the upper surface from the leading edge to the trailing edge remains constant due to the complete separation of the flows on the upper surface; that is, leading-edge separation is occurring. Combined with Figure 7, the transition position on the upper surface slightly develops in the range $8° < \alpha < 14°$ and is located at about 30% of the chord, indicating that the LUT is characterized by low drag, because the desirable pressure gradient exists at about 30% of the chord on the upper surface of the LUT airfoil [33], while the pressure coefficient on the lower surface is not obviously developed in the range $8° < \alpha < 14°$. Comparing the case of $R_e = 3 \times 10^5$ to the case of $R_e = 5 \times 10^5$, it is observed that the free transition position is more toward the leading edge with the increase in R_e. An angle of attack of $\alpha = 8°$, for example, results in a transition position located at about 29% of the chord when R_e is 5×10^5. In particular, in the case of $R_e = 5 \times 10^5$, the pressure on the lower surface significantly drops at an angle of attack of $\alpha = 21°$. This shows that obvious separation on the lower surface exists in the LUT airfoil at an angle of attack of $\alpha = 21°$.

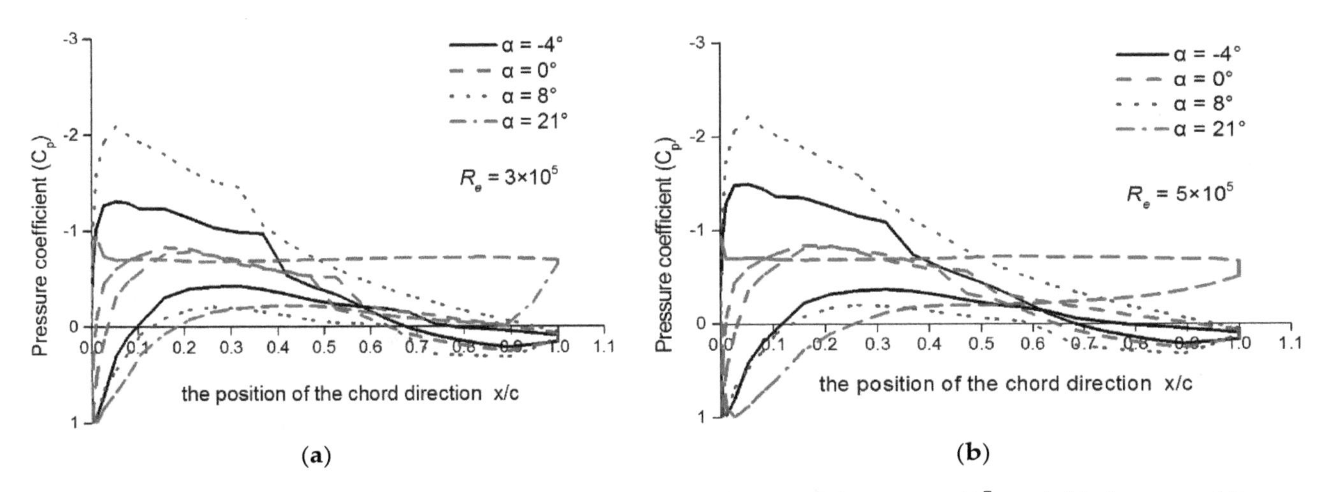

Figure 8. Pressure distribution for the LUT airfoil: (**a**) the case of $R_e = 3 \times 10^5$; and (**b**) the case of $R_e = 5 \times 10^5$. The data were measured when the surface of the LUT airfoil was clean. As shown in this figure, the lines in black, red, blue, and pink show the pressure coefficients corresponding to the angles $\alpha = -4°$, $\alpha = 0°$, $\alpha = 8°$, and $\alpha = 21°$, respectively.

The above discussions indicate that the distribution of the pressure coefficient of the LUT is better before a stall angle of $\alpha = 14°$ for both the case of $R_e = 3 \times 10^5$ and $R_e = 5 \times 10^5$; moreover, the free transition position at about 30% chord almost keeps constant before the stall, it shows that the pressure distribution on the upper surface is steady between $R_e = 3 \times 10^5$ and $R_e = 5 \times 10^5$ for the LUT airfoil. However, in the case of $R_e = 5 \times 10^5$, the pressure distribution on the lower surface is poor where the separation of the flows is obvious after a stall angle of $\alpha = 14°$.

4.1.2. Lift Coefficient of the LUT Airfoil

As shown in Figure 9, we investigated the relationship between the lift coefficient and the angles of attack at two Reynolds numbers, $R_e = 3 \times 10^5$ and $R_e = 5 \times 10^5$, obtained from Equation (17). In this Figure, the lines in black and red of the lift coefficient correspond to the results of the experiment and the numerical simulation (design results) for the LUT airfoil, respectively. In the case of $R_e = 3 \times 10^5$, the lift coefficient of the LUT airfoil linearly increases in the extending direction of the angles of attack from $\alpha = -9°$ to $\alpha = 12°$ (this region is called the linear region). At an angle of attack of $\alpha = 14°$, the lift coefficient reaches its maximum value at $C_{L\text{-max}} = 1.47$, and then the curve of the lift coefficient displays a gentle downward trend until an angle of attack of $\alpha = 21°$, a phenomenon called a stall. This indicates that the LUT airfoil has good stall performance.

In the case of $R_e = 5 \times 10^5$, the trend for the lift coefficient curve is the same as that of $R_e = 3 \times 10^5$ at the angles of attack from $\alpha = -9°$ to $\alpha = 12°$. However, the maximum lift coefficient $C_{L\text{-max}}$ is 1.5 at an angle of attack of $\alpha = 14°$. Comparing the results of the experiment to the results of the numerical simulation (design results), in the linearly increasing areas, the values of the lift coefficient are almost the same for the two values of R_e. However, after the stall angle of attack, the values of the lift coefficient for $R_e = 5 \times 10^5$ have a slight difference and are larger than the value of the lift coefficient for $R_e = 3 \times 10^5$. While the numerical simulation results are larger than those of the experiment, these differences are very small. Therefore, the results of the numerical simulation (design results) are in agreement with the results of the experiment reported in this paper.

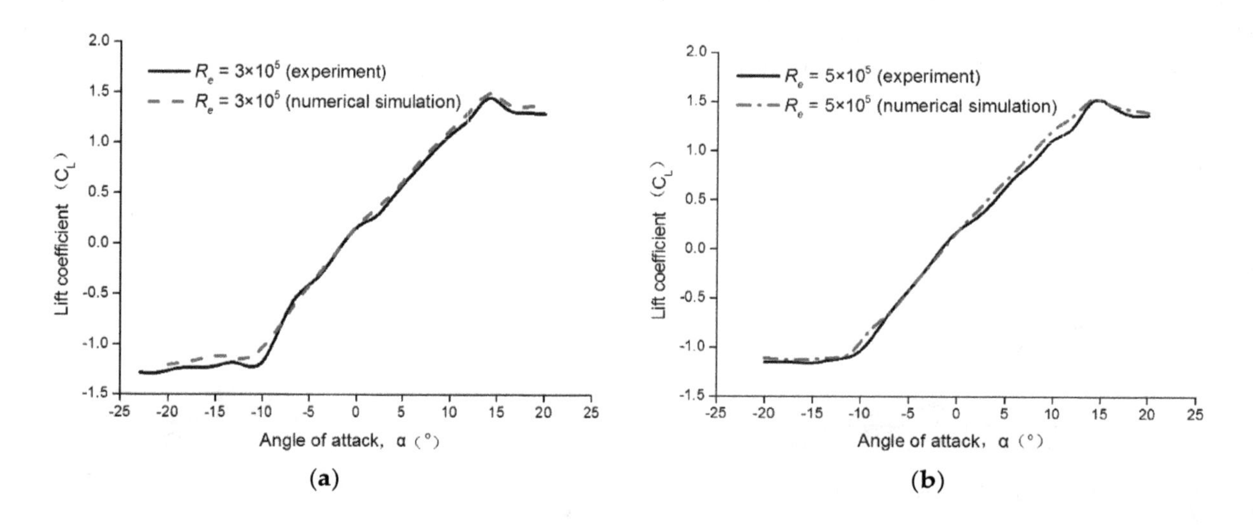

Figure 9. Lift coefficient for the LUT airfoil at different values of R_e: (**a**) the case of $R_e = 3 \times 10^5$; and (**b**) the case of $R_e = 5 \times 10^5$. The lines in black and red of the lift coefficient correspond to the experimental results and the numerical simulation results, respectively.

4.1.3. Drag Coefficient of the LUT Airfoil

The relationship between the drag coefficient and the angles of attack are displayed for the two Reynolds numbers, $R_e = 3 \times 10^5$ and $R_e = 5 \times 10^5$, in Figure 10, where the drag coefficient was obtained via Equations (15) and (16).

In Figure 10, the lines in black and red of the drag coefficient correspond to the experimental results and the numerical simulation results (design results), respectively. In the cases of $R_e = 3 \times 10^5$ and $R_e = 5 \times 10^5$, the values of the drag coefficient are smaller and remain almost constant in the range $-10° < \alpha < 10°$, because the most attached flow is on the surface of the LUT airfoil. The numerical simulation results agree with the experimental results in this region. However, the values of the drag coefficient increase rapidly when the angle of attack is greater than $\alpha = 13°$ or less than $\alpha = -10°$. The results of the experiment and the numerical simulation show a slight difference, but these errors are acceptable. Therefore, the results of the numerical simulation (design results) are in agreement with the results of the experiment reported in this paper.

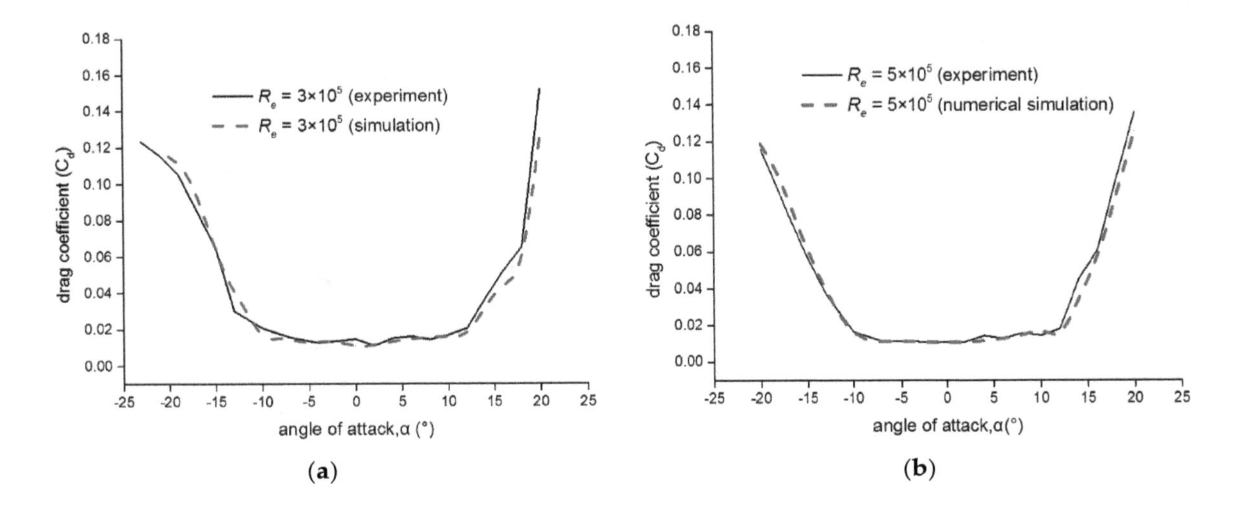

Figure 10. Drag coefficient for the LUT airfoil at different values of R_e: (**a**) the case of $R_e = 3 \times 10^5$; and (**b**) the case of $R_e = 5 \times 10^5$. The lines in black and red of the drag coefficient correspond to the experimental results and the numerical simulation results, respectively.

Figures 9 and 10 demonstrate that the results of the experiment agree with the results of the numerical simulation at low angles of attack. However, at high angles of attack where the flow separation occurs, there are some slight differences between the results of the experiment and the numerical simulation. On the one hand, the characteristics of the lift and drag of the LUT airfoil are obtained through integral surface pressure in this experiment, while the results of the numerical simulation with the *SST k-ω* turbulence model are acquired with solving the Reynolds-averaged Navier–Stokes equation; on the other hand, even though the current numerical methods have been greatly improved, for example, the predicted result of the *SST k-ω* turbulence model is accurate, but the *SST k-ω* turbulence model cannot completely solve the turbulence problem [34,35]. In addition, the wall of the LUT model is infinitely smooth in the numerical simulation, but in the current experiment, although the surface of the LUT model has very low roughness, the roughness of the surface is limited. Therefore, the results of the experiment and the numerical simulation have some differences in the region of the flow separation, while the results of the experiment and the numerical simulation are almost the same at low angles of attack.

Additionally, the maximum lift coefficient of the LUT airfoil has slow change with the increase in R_e from $R_e = 3 \times 10^5$ to $R_e = 5 \times 10^5$.

4.1.4. Lift–Drag Ratios of the LUT Airfoil

Figure 11a presents the lift coefficient versus the drag coefficient at different values of R_e for the LUT airfoil, where the lift coefficient and drag coefficient are described in and were obtained according to Sections 4.1.1 and 4.1.2. The lines in black and red of the lift–drag ratios correspond to $R_e = 3 \times 10^5$ and $R_e = 5 \times 10^5$, respectively. The results show that the LUT airfoil has the maximum lift–drag ratio $C_L/C_{d\text{-max}} = 64.4$ at an angle of attack of $\alpha = 10°$ when $R_e = 3 \times 10^5$, and it has the maximum lift–drag ratio $C_L/C_{d\text{-max}} = 79.8$ at an angle of attack of $\alpha = 10°$ when $R_e = 5 \times 10^5$. Moreover, combined with Figure 10, the lift–drag ratios linearly increase in the range $-10° < \alpha < 10°$. Therefore, the LUT airfoil presents a wider drag bucket, which is a desired characteristic for the airfoil of VAWTs due to the law of motion of the VAWTs [19]. Figure 11b shows the lift–drag ratios versus the lift coefficients, where the colors of the lines have the same meaning as in Figure 11a. The lift–drag ratios of the LUT increase almost linearly with the increase in the lift coefficient in the range $-1 \leq C_L \leq 1.25$.

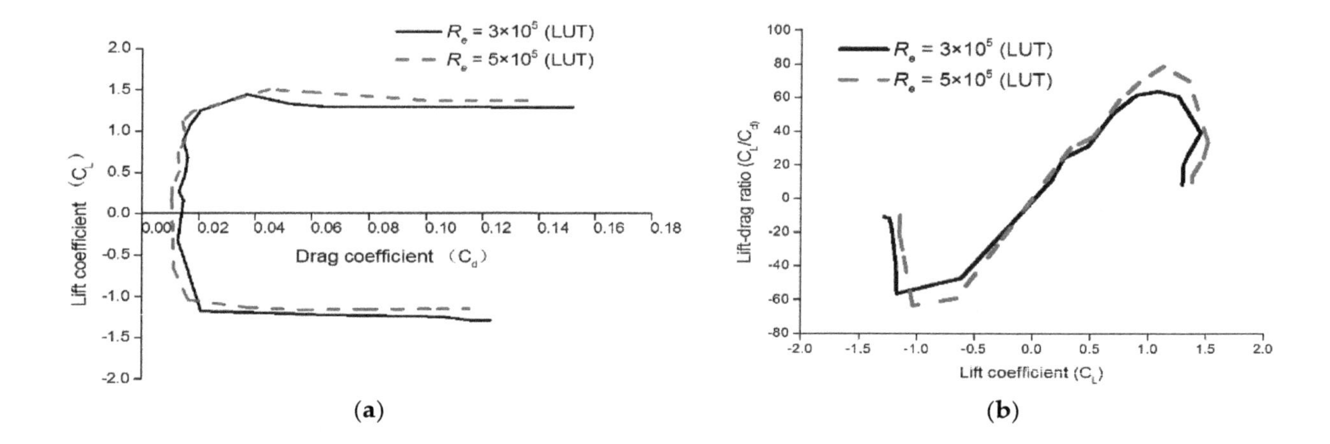

Figure 11. The characteristics of lift–drag ratios for the LUT airfoil at different values of R_e: (**a**) the relationship between the lift coefficient and the drag coefficient; and (**b**) the relationship between lift–drag ratios and the lift coefficient. The lines in black and red of the lift–drag ratios correspond to $R_e = 3 \times 10^5$ and $R_e = 5 \times 10^5$, respectively.

In Figure 12, the flow separation on the surface of the LUT is displayed at different angles of attack when the R_e is 3×10^5, measured using PIV technology. Figure 12 shows that the separation point gradually moves from the trailing edge to the leading edge with the increase in the angle of attack, which is consistent with the results in Figures 7 and 8. Moreover, the separation vortex becomes high due to the gradual flow separation. Especially at the angle of attack $\alpha = 22°$ in Figure 12g, the leading-edge separation results in the formation of the wake vortex and the detached vortex. At this moment, the pressure distribution on the upper surface of the LUT airfoil keeps a constant value, as shown in Figure 8a. Figure 12d–f displays the development of the flow separation on the upper surface of the LUT airfoil, which results in decreasing pressure on the upper surface of the airfoil, as shown in Figure 7a. However, the development of the vortices on the upper surface of the LUT airfoil are slow; combined with Figure 8a, it indicates that the LUT airfoil has good stall performance. Meanwhile, the changes in flow separation on the lower surface of the LUT airfoil are presented in Figure 12a–c when the angle of attack is a negative value.

The pressure distribution on the LUT airfoil surface at a negative angle of attack is shown in Figure 8. However, due to the limited field range of the PIV here, it is difficult to observe the development of the wake vortices. Meanwhile, at a large negative angle of attack, the flow separation on the lower surface is obvious, and the separation point on the lower surface moves toward the leading edge with increasing angles of attack.

However, because the placement of the PIV equipment was only convenient for photographing the upper surface of the LUT model, the phenomenon of flow separation was not obviously surveyed. Consequently, we could only survey the vortices near the trailing edge of the LUT.

Table 1 shows the main geometrical and aerodynamic parameters of a part of the commonly used and special-purpose airfoils for straight-bladed VAWTs when R_e is 3×10^5 and 5×10^5. In addition, the MI-VAWT1 airfoil [9] is also a dedicated airfoil for VAWTs; it has a 21% thickness and 0.44% camber. Compared with the results for Table 1, the geometrical parameters of the LUT airfoil balance the strength of the blade and the aerodynamic performance for a VAWTs with the LUT airfoil. Especially, the LUT airfoil performs better in the maximum lift coefficient and the maximum lift–drag ratios at a low R_e, it is very important, because VAWTs typically operate at a low R_e [19]. However, the stall angle of attack of the LUT airfoil is smaller than that of the airfoil in Table 1.

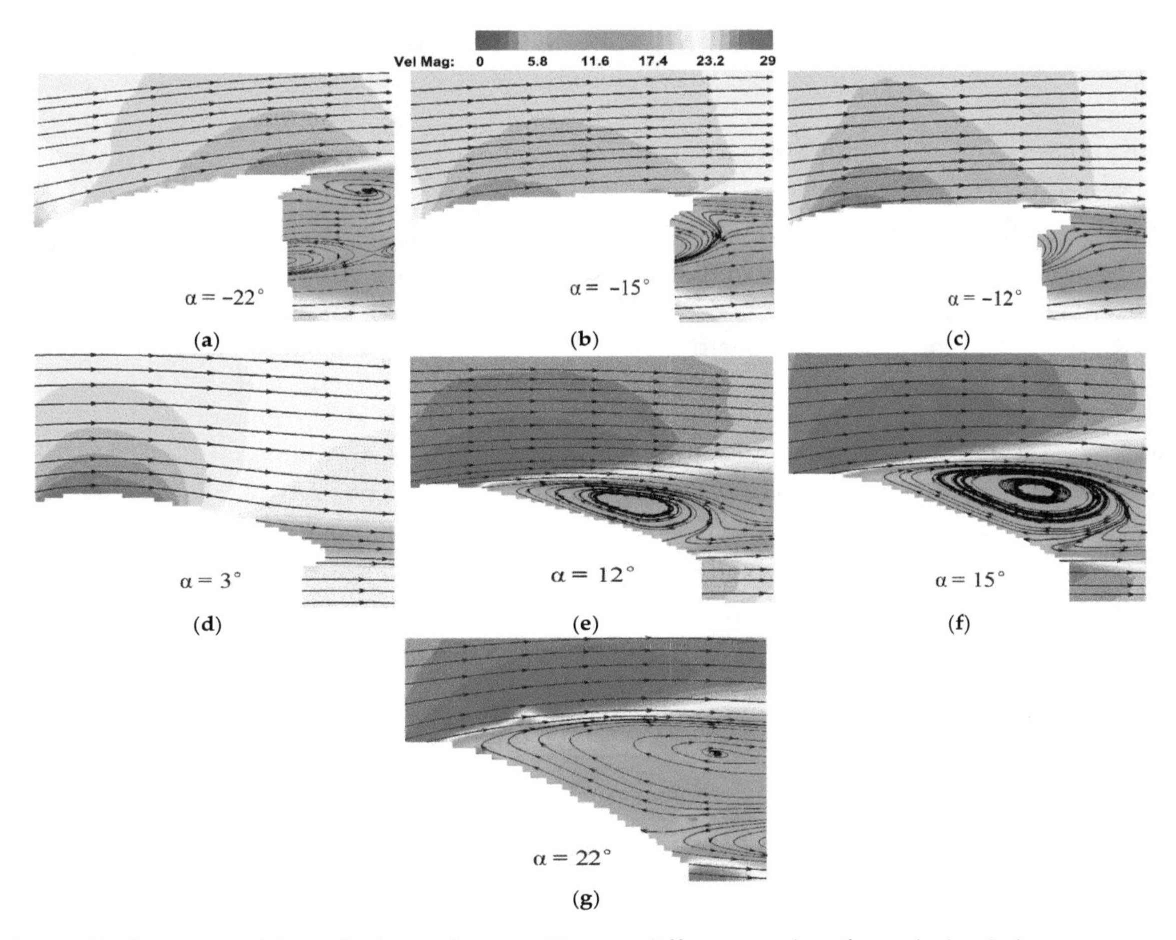

Figure 12. Contours of the velocity and stream lines at different angles of attack: (**a**–**g**) the contours of the velocity at different angles of attack. This figure shows the transient contours from the angle of attack $\alpha = -20°$ to the angle of attack $\alpha = 22°$ at $R_e = 3 \times 10^5$. However, only the varying velocity on the upper surface of the LUT and the larger vortex are captured in this figure due to the picture being limited by the ability of the Particle Image Velocimetry (PIV) equipment.

Table 1. Main geometrical parameters and aerodynamic parameters of a part of commonly used airfoils for straight-bladed VAWTs.

Airfoil	R_e (1×10^5)	Max. Thickness	(x/c) Max. Thickness	Max. Camber	(x/c) Max. Camber	Radius Leading Edge (r_0/c)	Max. C_L/C_d (α)	Max. C_L (α)
Du06-W-200 [28]	3	19.8%	31.1%	0.5%	84.6%	1.7442%	58.05 (9.75°)	1.4153 (16.75°)
	5						71.5 (9.5°)	1.4384 (18°)
NACA0018 [28]	3	18%	30%	0	0	3.1017%	57.09 (8.25°)	1.24 (18.75)
	5						65.8 (9.25°)	1.2624 (16.5°)
NACA0015 [28]	5	15%	30%	0	0	2.3742%	66.43 (7.5°)	1.2731 (16.75°)
DU12W262 [21]	-	26.2%	37%	-	-	-	-	-
AIR013 [20]	-	34.9%	-	2.29%	-	7.8%	-	-

Note: "-" indicates that the data of the corresponding locations are not presented in the literature.

5. Conclusions

In this study, we designed an asymmetrical airfoil for straight-bladed VAWTs based on the law of motion of the VAWTs by employing the SQP optimization method; we called this design the LUT airfoil. Meanwhile, the aerodynamic performance of the LUT airfoil was investigated at two Reynolds numbers, 3×10^5 and 5×10^5, under the conditions of a low-speed, low-turbulence wind tunnel at the Northwestern Polytechnical University, China; the flow field around the LUT airfoil was observed by PIV technology.

The LUT airfoil has a 20.77% moderate thickness and 1.11% moderate camber. Moreover, the position of the maximum camber is close to the trailing edge, which forms a trailing-edge loading to improve the efficient lift of the LUT airfoil. The maximum thickness of the LUT airfoil is achieved by restricting the thickness of the upper surface while adding the thickness of the lower surface. It is conducive to delay flow separation on the upper surface. Moreover, the moderate thickness helps to ensure the structural strength of the blade and to improve the performance of VAWTs with the LUT airfoil at high tip speed ratio.

Ultimately, the main objectives of achieving a higher maximum lift coefficient, higher maximum lift–drag coefficient, and gentle stall characteristics were achieved. Compared with the airfoil commonly used for VAWTs, the LUT airfoil displays a better aerodynamic performance at a low R_e. Additionally, the maximum lift coefficient of the LUT airfoil is steady with the increase in R_e between 3×10^5 and 5×10^5. The LUT airfoil retains the desirable characteristic of low drag over a wide range of angles of attack in the range $-10° < \alpha < 10°$, that is, the LUT airfoil has a wide drag bucket. The distribution of the pressure coefficient agrees with the results obtained using PIV. Theoretically, due to the larger 3.84% leading-edge radius and other geometric features, the LUT airfoil can reduce the sensitivity to roughness.

In future work, the LUT airfoil's sensitivity to roughness and noise level will be tested by wind tunnel experiments.

Author Contributions: Methodology, S.L., Y.L. and Q.W.; Writing—Original Draft Preparation, S.L. and X.Z.; and Writing—Review and Editing, Y.L., C.Y., D.L., W.Z. and T.W.

Acknowledgments: The authors would like to acknowledge and thank the Natural Science Foundation of GANSU (grant No. 1508RJYA098); the National Natural Science Foundation of China (No. 51506088 and No. 51766009); the National Basic Research Program of China (No. 2014CB046201); the people who provided many good suggestions for this paper; and the Northwestern Polytechnical University and China Aerodynamics Research and Development Center for providing the experiment instruments and wind tunnel.

Nomenclature

H-VAWT	H-type Vertical Axis Wind Turbines
VAWT	Vertical Axis Wind Turbines
HAWT	Horizontal Axis Wind Turbines
LUT	The name of the newly designed airfoil (Lanzhou University of Technology)
DTU	Technical University of Denmark
SST	Shear Stress Transport
R_e	Reynolds number
α	angle of attack, °
θ	azimuthal angle, °
ω	rotational speed of the wind turbine, rad/s
V_{in}	freestream velocity, m/s
R	rotor radius, m
C	airfoil chord length, mm
SQP	Sequential Quadratic Programming
σ	penalty factor
C_L/C_d	lift–drag ratios
C_L	lift coefficient
C_d	drag coefficient
r_0	leading-edge radius
C_p	Pressure coefficient
PIV	Particle Image Velocimetry
FS	Full Scale

Appendix A

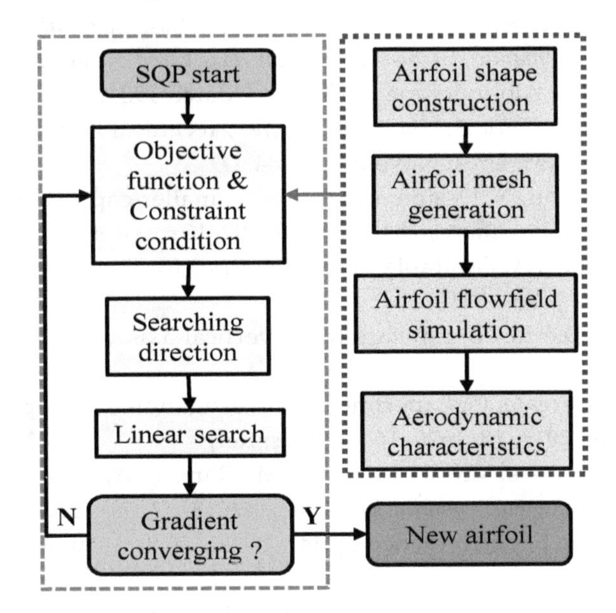

Figure A1. The optimization flow of the airfoil profile based on the SQP method.

References

1. Ho, A.; Mbistrova, A.; Corbetta, G. *The European Offshore Wind Industry-Key Trends and Statistics 2015*; Technical Report; European Wind Energy Association: Brussels, Belgium, February 2016.
2. Battisti, L.; Benini, E.; Brighenti, A.; Dell'Anna, S.; Raciti Castelli, M. Small wind turbine effectiveness in the urban environment. *Renew. Energy* **2018**, *129*, 102–113. [CrossRef]
3. Guo, Y.; Liu, L.; Gao, X.; Xu, W. Aerodynamics and Motion Performance of the H-Type Floating Vertical Axis Wind Turbine. *Appl. Sci.* **2018**, *8*, 262. [CrossRef]
4. Kumar, R.; Raahemifar, K.; Fung, A.S. A critical review of vertical axis wind turbines for urban applications. *Renew. Sustain. Energy Rev.* **2018**, *89*, 281–291. [CrossRef]
5. Li, Q.; Maeda, T.; Kamada, Y.; Ogasawara, T.; Nakai, A.; Kasuya, T. Investigation of power performance and wake on a straight-bladed vertical axis wind turbine with field experiments. *Energy* **2017**, *141*, 1113–1123. [CrossRef]
6. Aslam Bhutta, M.M.; Hayat, N.; Farooq, A.U.; Ali, Z.; Jamil, S.R.; Hussain, Z. Vertical axis wind turbine—A review of various configurations and design techniques. *Renew. Sustain. Energy Rev.* **2012**, *16*, 1926–1939. [CrossRef]
7. Sutherland, H.J.; Berg, D.E.; Ashwill, T.D. *A Retrospective of VAWT Technology*; Technical Report SAND2012-0304; Sandia National Laboratories: Albuquerque, NM, USA, January 2012.
8. Kanyako, F.; Janajreh, I. Numerical Investigation of Four Commonly Used Airfoils for Vertical Axis Wind Turbine. In *ICREGA'14-Renewable Energy: Generation and Application*; Hamdan, M., Hejase, H., Noura, H., Fardoun, A., Eds.; Springer: Cham, Switzerland, 2014; pp. 443–454. [CrossRef]
9. Islam, M. Analysis of Fixed-Pitch Straight-Bladed VAWT with Asymmetric Airfoils. Ph.D. Thesis, University of Windsor, Windsor, ON, Canada, 2008.
10. Kirke, B.K. Evaluation of Self-Starting Vertical Axis Wind Turbines for Stand-Alone Appllications. Ph.D. Thesis, Griffith University, Brisbane, Australia, 1998.
11. Rainbird, J.M.; Bianchini, A.; Balduzzi, F.; Peiró, J.; Graham, J.M.R.; Ferrara, G.; Ferrari, L. On the influence of virtual camber effect on airfoil polars for use in simulations of Darrieus wind turbines. *Energy Convers. Manag.* **2015**, *106*, 373–384. [CrossRef]
12. Mohamed, M.H.; Ali, A.M.; Hafiz, A.A. CFD analysis for H-rotor Darrieus turbine as a low speed wind energy converter. *Eng. Sci. Technol. Int. J.* **2015**, *18*, 1–13. [CrossRef]
13. Carrigan, T.J.; Dennis, B.H.; Han, Z.X.; Wang, B.P. Aerodynamic Shape Optimization of a Vertical-Axis Wind Turbine Using Differential Evolution. *ISRN Renew. Energy* **2012**, *2012*, 1–16. [CrossRef]
14. Asr, M.T.; Nezhad, E.Z.; Mustapha, F.; Wiriadidjaja, S. Study on start-up characteristics of H-Darrieus vertical axis wind turbines comprising NACA 4-digit series blade airfoils. *Energy* **2016**, *112*, 528–537. [CrossRef]

15. Chen, J.; Chen, L.; Xu, H.; Yang, H.; Ye, C.; Liu, D. Performance improvement of a vertical axis wind turbine by comprehensive assessment of an airfoil family. *Energy* **2016**, *114*, 318–331. [CrossRef]
16. Sheldahl, R.E.; Klimas, P.C. *Aerodynamic Characteristics of Seven Sysmmetrical Airfoil Sections Through 180-Degree Angle of Attack for Use in Aerodynamic Analysis of Vertical Axis Wind Turbines*; Technical Report SAND80-2114; Sandia National Laboratories: Albuquerque, NM, USA, March 1981.
17. Klimas, P.C. *Tailored Airfoils for Vertical Axis Wind Turbines*; Technical Report SAND84-1062; Sandia National Laboratories: Albuquerque, NM, USA, November 1984.
18. Islam, M.; Fartaj, A.; Carriveau, R. Design analysis of a smaller-capacity straight-bladed VAWT with an asymmetric airfoil. *Int. J. Sustain. Energy* **2011**, *30*, 179–192. [CrossRef]
19. Islam, M.; Ting, D.S.-K.; Fartaj, A. Desirable Airfoil Features for Smaller-Capacity Straight-Bladed VAWT. *Wind Eng.* **2007**, *31*, 165–196. [CrossRef]
20. Ferreira, C.S.; Geurts, B. Aerofoil optimization for vertical-axis wind turbines. *Wind Energy* **2015**, *18*, 1371–1385. [CrossRef]
21. Ragni, D.; Ferreira, C.S.; Correale, G. Experimental investigation of an optimized airfoil for vertical-axis wind turbines. *Wind Energy* **2015**, *18*, 1629–1643. [CrossRef]
22. Batista, N.C.; Melício, R.; Mendes, V.M.F.; Calderón, M.; Ramiro, A. On a self-start Darrieus wind turbine: Blade design and field tests. *Renew. Sustain. Energy Rev.* **2015**, *52*, 508–522. [CrossRef]
23. Bedon, G.; Betta, S.D.; Benini, E. Performance-optimized airfoil for Darrieus wind turbines. *Renew. Energy* **2016**, *94*, 328–340. [CrossRef]
24. Hand, B.; Kelly, G.; Cashman, A. Numerical simulation of a vertical axis wind turbine airfoil experiencing dynamic stall at high Reynolds numbers. *Comput. Fluids* **2017**, *149*, 12–30. [CrossRef]
25. Sargent, R.W.H.; Ding, M. A New SQP Algorithm for Large-Scale Nonlinear Programming. *SIAM J. Optim.* **2001**, *11*, 716–747. [CrossRef]
26. Andrei, N. Sequential Quadratic Programming Continuous (SQP). In *Nonlinear Optimization for Engineering Applications in GAMS Technology*, 1st ed.; Springer: Cham, Switzerland, 2017; Volume 121, pp. 269–288, ISBN 978-3-319-58355-6.
27. Howell, R.; Qin, N.; Edwards, J.; Durrani, N. Wind tunnel and numerical study of a small vertical axis wind turbine. *Renew. Energy* **2010**, *35*, 412–422. [CrossRef]
28. Airfoil Tools. Available online: http://airfoiltools.com/ (accessed on 1 September 2018).
29. Timmer, W.A.; Van Rooij, R.P.J.O.M. Summary of the Delft University Wind Turbine Dedicated Airfoils. *J. Sol. Energy Eng.* **2003**, *125*, 488–496. [CrossRef]
30. Fuglsang, P.; Bak, C. Development of the Risø wind turbine airfoils. *Wind Energy* **2004**, *7*, 145–162. [CrossRef]
31. Meng, X.; Hu, H.; Yan, X.; Liu, F.; Luo, S. Lift improvements using duty-cycled plasma actuation at low Reynolds numbers. *Aerosp. Sci. Technol.* **2018**, *72*, 123–133. [CrossRef]
32. Anderson, J.D. *Fundamentals of Aerodynamics*, 5th ed.; McGraw-Hill Education: New York, NY, USA, 2011; ISBN 9780073398105.
33. Somers, D.M. *Design and Experimental Results for the S825 Airfoil Period of Performance: 1998–1999 Design and Experimental Results for the S825 Airfoil*; Technical Report NREL/SR-500-36346; National Renewable Energy Laboratory: Golden, CO, USA, 2005.
34. Balduzzi, F.; Bianchini, A.; Maleci, R.; Ferrara, G.; Ferrari, L. Critical issues in the CFD simulation of Darrieus wind turbines. *Renew. Energy* **2016**, *85*, 419–435. [CrossRef]
35. Zhang, X.; Wang, G.; Zhang, M.; Liu, H.; Li, W. Numerical study of the aerodynamic performance of blunt trailing-edge airfoil considering the sensitive roughness height. *Int. J. Hydrogen Energy* **2017**, *42*, 18252–18262. [CrossRef]

4

Performance Characteristics of an Orthopter-Type Vertical Axis Wind Turbine in Shear Flows

Rudi Purwo Wijayanto [1,2,*], Takaaki Kono [3,*] and Takahiro Kiwata [3]

1 Graduate School of Natural Science and Technology, Kanazawa University, Kanazawa 920-1192, Japan
2 The Agency of the Assessment and Application of Technology (BPPT), Jakarta 10340, Indonesia
3 Institute of Science and Engineering, Kanazawa University, Kanazawa 920-1192, Japan; kiwata@se.kanazawa-u.ac.jp
* Correspondence: rudi.purwo@stu.kanazawa-u.ac.jp (R.P.W.); t-kono@se.kanazawa-u.ac.jp (T.K.)

Abstract: To properly conduct a micro-siting of an orthopter-type vertical axis wind turbine (O-VAWT) in the built environment, this study investigated the effects of horizontal shear flow on the power performance characteristics of an O-VAWT by performing wind tunnel experiments and computational fluid dynamics (CFD) simulations. A uniform flow and two types of shear flow (advancing side faster shear flow (ASF-SF) and retreating side faster shear flow (RSF-SF)) were employed as the approaching flow to the O-VAWT. The ASF-SF had a higher velocity on the advancing side of the rotor. The RSF-SF had a higher velocity on the retreating side of the rotor. For each type of shear flow, three shear strengths ($\Gamma = 0.28$, 0.40 and 0.51) were set. In the ASF-SF cases, the power coefficients (C_p) were significantly higher than the uniform flow case at all tip speed ratios (λ) and increased with Γ. In the RSF-SF cases, C_P increased with Γ. However, when $\Gamma = 0.28$, the C_P was lower than the uniform flow case at all λ. When $\Gamma = 0.51$, the C_P was higher than the uniform flow case except at low λ; however, it was lower than the ASF-SF case with $\Gamma = 0.28$. The causes of the features of C_P were discussed through the analysis of the variation of blade torque coefficient, its rotor-revolution component and its blade-rotation component with azimuthal angle by using the CFD results for flow fields (i.e., horizontal velocity vectors, pressure and vorticity). These results indicate that a location where ASF-SFs with high Γ values dominantly occur is ideal for installing the O-VAWT.

Keywords: orthopter; vertical axis wind turbine; power coefficient; torque coefficient; shear flow; wind tunnel; CFD; delayed detached-eddy simulation

---

1. Introduction

Since the 2000s, interest in installing small wind turbines (SWTs) in the built environment has been growing [1–9]. Wind conditions in the built environment are complex in nature and are characterized by lower wind speeds and higher turbulence because of the presence of obstructions [8,9]. For SWTs to be able to make up their costs within their lifetimes, they should have high efficiency and be placed at sites with high wind speeds, such as coastal sites or high-elevation inland sites. However, in the built environment, keeping the rotational speed of an SWT's rotor as low as possible is preferable from the viewpoint of aerodynamic noise [10,11]. Therefore, the optimal tip-speed ratio of an SWT in the built environment should be as low as possible, while the maximum power coefficient of the SWT should be as high as possible.

Wind turbines are classified into horizontal-axis wind turbines (HAWTs) and vertical-axis wind turbines (VAWTs), based on the orientation of their rotation axes. Generally, in the built environment, VAWTs are preferable to HAWTs because VAWTs do not suffer, as much as HAWTs, from reduced energy outputs from frequent wind direction changes [12]. Wind turbines are further classified into lift-type wind turbines and drag-type wind turbines, based on the aerodynamic force component that acts on a blade and dominantly contributes to the rotor rotation. With regard to lift-type VAWTs, a lot of research on Darrieus-type VAWTs, including straight-bladed and helical-bladed ones, has been conducted [13–16]. With regard to drag-type VAWTs, a lot of research on Savonius-type VAWTs has been conducted [17–21]. In general, the optimal tip-speed ratio of a drag-type VAWT is less than 1.0 [11], which is much smaller than that of a lift-type VAWT. Moreover, although the maximum power coefficient of a drag-type VAWT is generally much smaller than that of a lift-type VAWT, the power coefficient of a drag-type VAWT is generally greater than that of a lift-type VAWT at a low tip-speed ratio, of less than 1.0. Therefore, a drag-type VAWT is favorable in the built environment and was researched by our research group.

Our group [22,23] researched a drag-type VAWT called the orthopter-type VAWT (O-VAWT). The O-VAWT is a variable-pitch VAWT; each of the flat-plate blades not only revolves around the main shaft but also rotates around its own blade axis, which is rotationally supported by a pair of connecting arms. We investigated the effects of the number and aspect ratio of the flat-plate blades on the power performance of the O-VAWT in a uniform flow by conducting wind tunnel experiments with an open test section and three-dimensional computational fluid dynamics (CFD) simulations. When the number of the blades was three and the aspect ratio of the blades was 1:1, the maximum power coefficient was 0.25 and the optimal tip-speed ratio was 0.4 [22]. Here, the optimal tip-speed ratios of Savonius-type VAWTs are in the range of 0.45 to 1.0 [21]. Except for several studies that were conducted using a wind tunnel with a very high blocking ratio of the closed test section, the maximum power coefficient of the Savonius-type VAWT was, at most, 0.25 [21]. That is, the O-VAWT has a lower optimal tip-speed ratio than Savonius-type VAWTs, although the maximum power coefficient is relatively high. Therefore, from the viewpoint of aerodynamic noise, the O-VAWT can be more favorable in the built environment as compared to Savonius-type VAWTs. Except for our studies, studies on the power performance of O-VAWTs are very limited. Shimizu et al. [24] investigated the effects of the aspect ratio of the blade on the power performance of an O-VAWT with two blades whose cross-sectional shape was an ellipse by conducting wind tunnel experiments. Bayeul-Line et al. [25] examined the effects of the blade's cross-sectional shape (elliptical and straight) and the initial blade stagger angle on the performance of an O-VAWT by conducting 2-dimensional CFD simulations. Cooper and Kennedy [26] examined the power performance of an O-VAWT with three blades whose cross-sectional shape was the upstream half of a NACA0010-65 section reflected about the mid-chord by conducting theoretical analysis with a multiple-stream tube model and field measurements. Our group [23] conducted wind tunnel experiments to compare the performance of an O-VAWT with elliptic blades and one with flat-plate blades. By considering the mechanical loss torque, we obtained the maximum power coefficient of 0.246 at a tip-speed ratio of 0.4 for the O-VAWT with elliptic blades and of 0.288 at a tip-speed ratio of 0.4 for the O-VAWT with flat-plate blades. It should be noted that these studies on O-VAWTs were conducted in conditions where the approaching flows had uniform distribution.

To properly conduct a micro-sitting of an O-VAWT in the built environment, it is important to understand the effects of the strong shear approach flow with on the performance of the O-VAWT. Figure 1 illustrates the approaching wind flow to a building. As the wind flow approaches the building, the wind speed decreases, and the pressure increases. Then, the wind flow proceeds along the upwind face of the building and separates at the corners on the roof and the side walls. As the separated wind flow is not obstructed by the building, the pressure decreases, and the wind speed increases. Near the upwind corners, the wind speed increases more than that of the approaching wind. In addition, reverse flow regions are formed between the separated shear layer and the building's walls. As a result, strong shear flows are formed vertically over the roof surface and horizontally over the side walls. Due to the mixing of momentum, the shear becomes weaker as the flow proceeds downstream. To utilize the increased wind speed over the roof of a building, the effects of building shapes and wind directions on the wind conditions have been investigated (e.g., [27]). Furthermore, the effects of wind conditions, such as wind speed, turbulent intensity, and skew angles, on the potential energy yield and the power performance of a wind turbine have been studied (e.g., [28,29]). In this study, we investigated the effects of horizontal shear flow on the performance of the O-VAWT by conducting wind tunnel experiments and three-dimensional CFD simulations. A uniform flow and two types of shear flow were employed as the approaching flow to the O-VAWT. One type had a higher velocity on the advancing side of the rotor. The other type had a higher velocity on the retreating side of the rotor. For each type of shear flow, we set three different shear strengths.

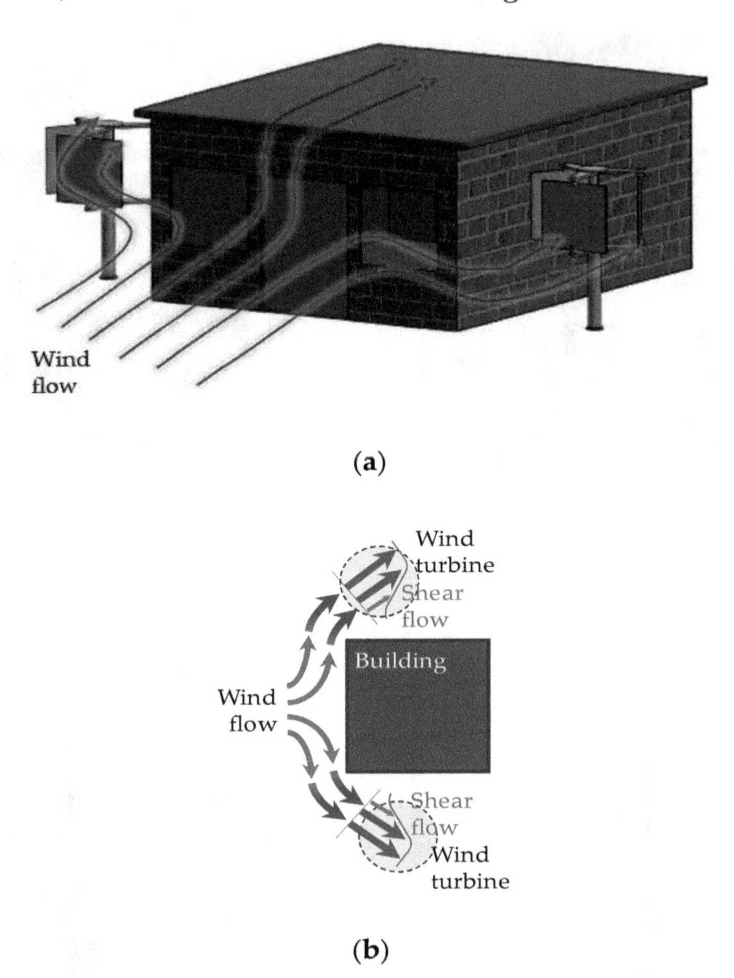

Figure 1. Illustration of the approaching wind flow to the wind turbine near upwind corners of a building. (**a**) Bird view; (**b**) enlarged top view.

2. Experimental Approach

2.1. Wind Turbine Model

In this paper, we employed a right-handed Cartesian coordinate system $(x_1, x_2, x_3) = (x, y, z)$, in which the z-direction was aligned in the vertical direction. The wind turbine used in this study was an O-VAWT with three flat-plate blades as shown in Figure 2. The blade had a height of $h = 4.00 \times 10^{-1}$ m, chord length of $c = 4.00 \times 10^{-1}$ m and thickness of 4.0×10^{-3} m. Each of the blades not only revolved around the main shaft but also rotated around its own blade axis, which was rotationally supported by a pair of connecting arms. The distance between the main shaft and one of the blade axes was $R = 2.55 \times 10^{-1}$ m. In addition, each of the blade axes was connected with the main shaft by a chain via sprockets. Since the ratio of the number of teeth on the sprocket of the main shaft to that of the blade axis was 1:2, each of the blades rotated around the own blade axis a half time while the rotor revolved around the main shaft one time. When seen from the top, the rotor revolved around the main shaft counterclockwise and each of the blades rotated around the own blade axis clockwise as shown in Figure 2c. Therefore, by using the angular velocity of the rotor revolution (ω), the angular velocity of the blade rotation is expressed as $-\omega/2$. Furthermore, as shown in Figure 2c, according to the azimuthal angle of a blade (φ), we call the range of $90° < \varphi < 270°$ the "upwind region" of the rotor, $0° < \varphi < 90°$ and $270° < \varphi < 360°$ the "downwind region" of the rotor, $180° < \varphi < 360°$ the "advancing side of the rotor" and $0° < \varphi < 180°$ the "retreating side" of the rotor. The O-VAWT is designed so that the drag force on a blade is large on the advancing side of the rotor while being small on the retreating side of the rotor.

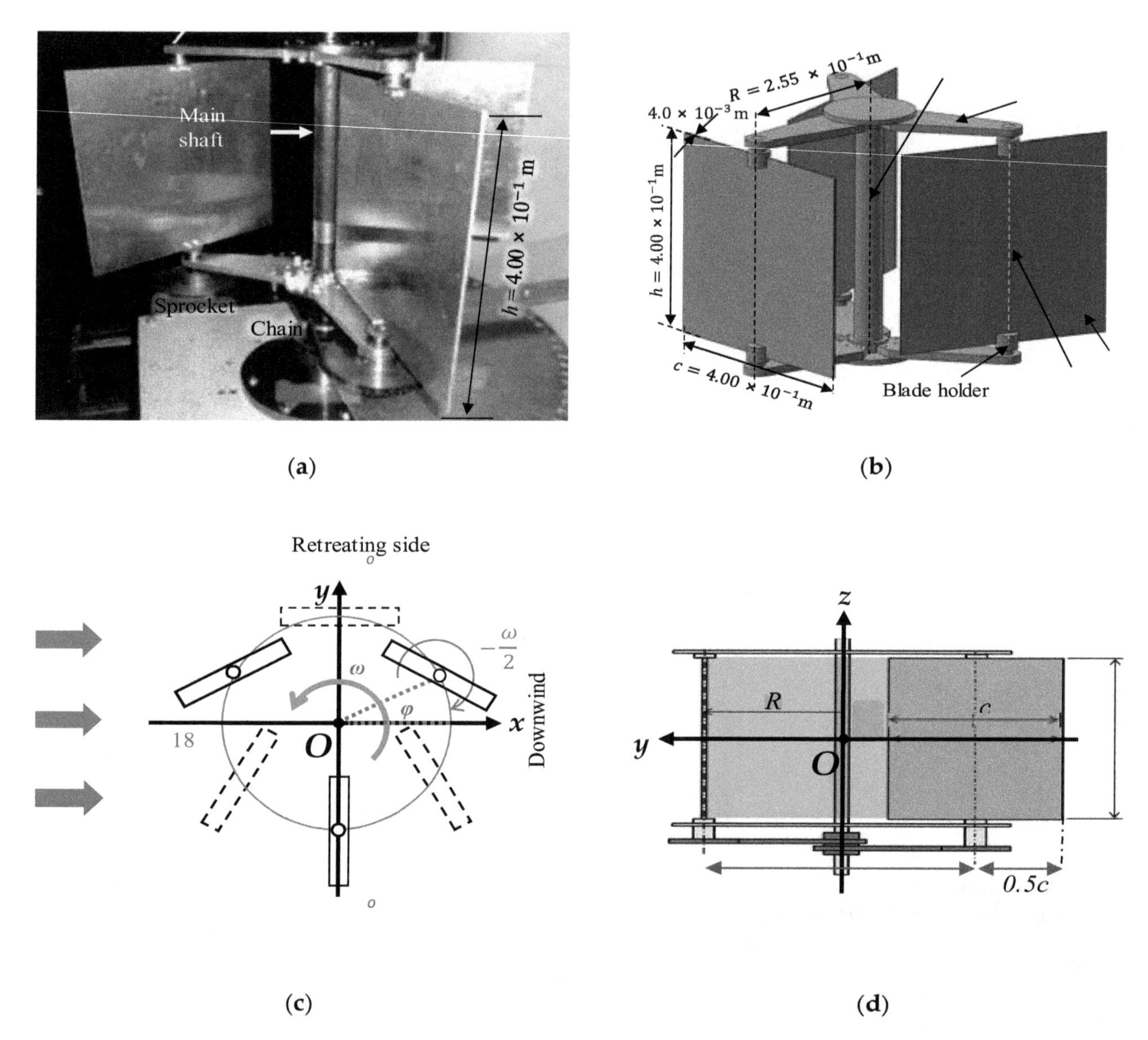

Figure 2. Orthopter-type vertical-axis wind turbine (O-VAWT). (**a**) A photograph of O-VAWT with three flat blades, main shaft, arm, the chains and sprockets; (**b**) an isometric view of O-VAWT with three flat blades; (**c**) motion of rotor and blades viewed from the top; (**d**) a projected swept area of the rotor viewed from the upwind side.

Figure 2d shows a projected swept area of the rotor (A), which is defined as:

$$A = (2R + 0.5\,c)\,h. \quad (1)$$

We define the diameter of the rotor as:

$$D = 2R. \quad (2)$$

The O-VAWT had a rotor's diameter of $D = 5.1 \times 10^{-1}$ m and a projected rotor's swept area of $A = 2.84 \times 10^{-1}$ m^2 can be considered as a micro wind turbine. A small scale wind turbine that has a diameter up to 1.25 m and the swept area up to 1.2 m^2 is categorized as a micro wind turbine [4].

2.2. *Experimental Setup for Uniform Flow Case*

Figure 3a shows the experimental setup for the uniform flow case. The experiments were conducted using a closed circuit wind tunnel with an open test section. The size of the cross section of the wind tunnel outlet was 1.25 m × 1.25 m. The blockage ratio which is defined as the ratio of the projected rotor's swept area to the wind tunnel outlet area was approximately 18%. The O-VAWT was set in the test section so that the rotor center was at the center of the cross-section of the wind tunnel outlet and 0.850 m downwind of the wind tunnel outlet. Here, we defined the rotor center as the point on the rotational axis of the rotor and at the mid-height of the blades. In addition, we set the origin of the coordinate system at the rotor center, as shown in Figure 2c,d. The rotor was driven by a motor (Mitsubishi Electric, GM-S) and its rotational speed (ω) was monitored by using a digital tachometer (Ono Sokki, HT-5500) and controlled by using an inverter (Hitachi, SJ200). The rotor torque was measured by using a torque meter (TEAC, TQ-AR), which was connected to the motor and shaft via couplings. The output signal of the torque meter was converted by a 16-bit analog-to-digital converter with a sampling interval of 0.5°, and 36,000 items (50 revolutions) of data were stored. To measure the reference wind speed U_∞, an ultrasonic anemometer (Kaijo Sonic, DA-650-3TH and TR-90 AH) was set approximately 2 m upwind of the wind tunnel outlet. The value of U_∞ was kept at 8 m/s. The value of tip speed ratio λ, which is defined as:

$$\lambda = R\omega/U_\infty, \tag{3}$$

was varied from 0.1 to 0.8 with an increment of 0.1.

2.3. *Experimental Setup for Shear Flow Cases*

To generate a horizontal shear flow, a perforated panel and a splitter plate were installed at the outlet of the wind tunnel as shown in Figure 3b,c and Figure 4. The perforated panel had a width of 1.0 m, a height of 1.5 m and a thickness of 2×10^{-3} m, and was set so that it covered half of the wind tunnel outlet in the horizontal wind direction. Due to the existence of the perforated panel, the pressure upwind of the panel increased and the wind flow rate through the wind-tunnel-outlet area covered by the panel decreased while that through the uncovered wind-tunnel-outlet area increased. When the perforated panel covered the wind-tunnel-outlet area upwind of the retreating side of the rotor, the wind speed of the generated shear flow was higher on the advancing side of the rotor. Hereafter, this type of shear flow is referred to as "advancing side faster shear flow" (ASF-SF). On the other hand, when the perforated panel covered the wind-tunnel-outlet area upwind of the advancing side of the rotor, the wind speed of the generated shear flow was higher on the retreating side of the rotor. Hereafter, this type of shear flow is referred to as "retreating side faster shear flow" (RSF-SF). The splitter plate was set vertically, parallel to the wind tunnel wall and at the center of the wind tunnel outlet to avoid the horizontal component of the wind velocity in the generated shear flow becoming significant. The splitter plate had a width of 0.88 m, a height of 1.25 m and a thickness of 4×10^{-3} m.

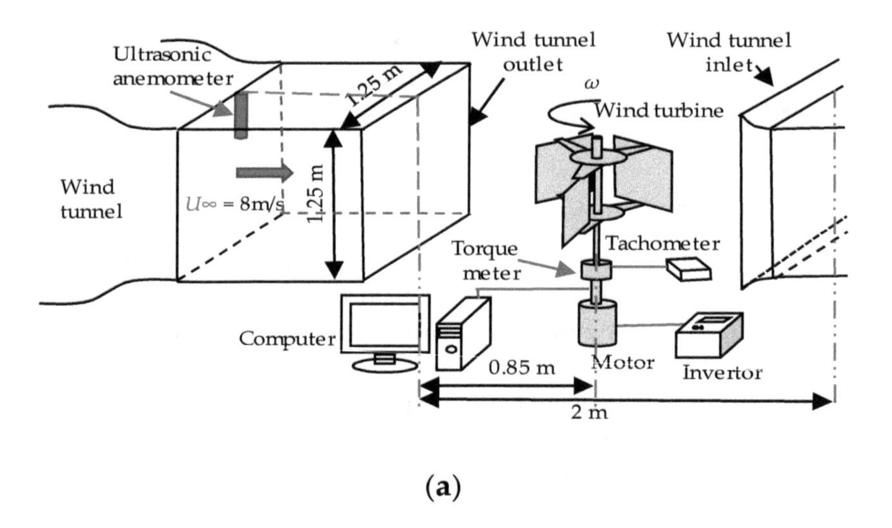

(a)

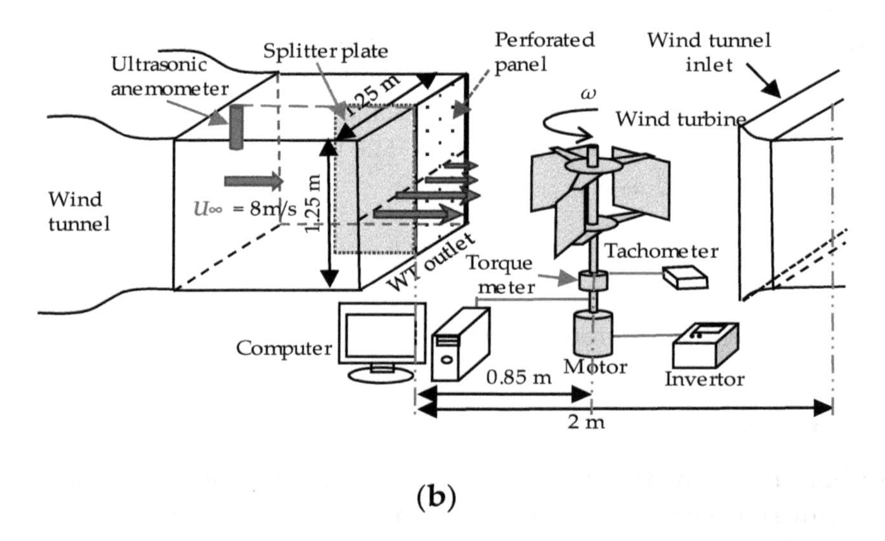

(b)

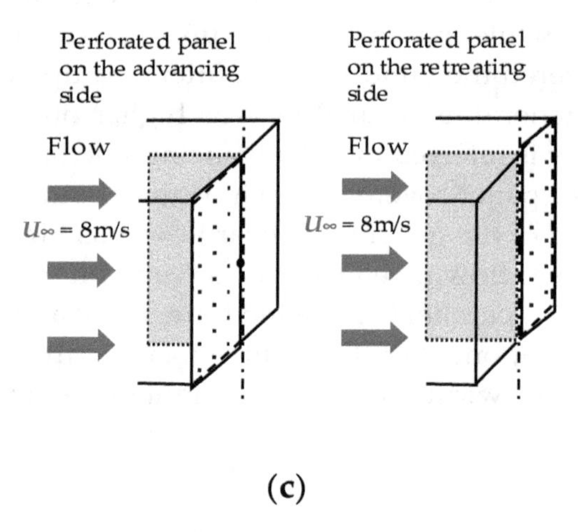

(c)

Figure 3. The experimental apparatus and measurement devices; (**a**) in uniform flow, (**b**) in shear flows and (**c**) the porous plate position at the nozzle exit of the wind tunnel in case of shear flows.

Figure 4. The perforated panel and splitter plate set at the outlet of the wind tunnel.

To investigate the effects of the strength of the shear flow on the performance of the O-VAWT, we generated three kinds of shear flows by using three perforated panels shown in Table 1. With regard to a staggered round-hole perforated panel, the shielding ratio Φ, which is the ratio of the area that shields the airflow to the whole area of the perforated panel, can be computed by:

$$\Phi = 1 - \frac{\pi d^2}{2\sqrt{3} L^2}, \tag{4}$$

where d is a diameter of a hole and L is the distance between the centers of adjacent holes.

Table 1. Perforated panels. Here, d is the diameter of a hole, L is the distance between the centers of adjacent holes and Φ is the shielding ratio.

Name	d [m]	L [m]	Φ [-]	Enlarged View
Perforated panel A	3×10^{-3}	4×10^{-3}	0.49	4mm
Perforated panel B	3×10^{-3}	4.5×10^{-3}	0.60	4.5mm
Perforated panel C	3×10^{-3}	5×10^{-3}	0.67	5mm

Except for the installation of the perforated panels and the splitter plate, the experimental setup for the measurement of the performance of the O-VAWT was the same as the uniform flow case. Prior to the measurement of the performance of the O-VAWT, we measured the horizontal profiles of the generated shear flows at 0.10 m downwind of and at the center height of the wind tunnel outlet by using an x-type hot-wire probe (Kanomax, 0252R-T5).

2.4. Torque and Power Coefficients

Due to the difficulty of evaluating the mechanical losses of the bearings, the sprockets and the chains, this study considers only the aerodynamic torque generated by the blades as the rotor torque of the O-VAWT. The aerodynamic torque generated by the blades was computed by:

$$T_B = T_{wB} - T_{woB}, \tag{5}$$

where T_{wB} was the measured aerodynamic torque generated by the rotor when the blades were not removed; and T_{woB} was the measured aerodynamic torque generated by the rotor when the blades were removed. It is worth noting that, generally, when the blades were not removed from the rotor, the rotor generated positive torque while the motor acted as a load to keep the value of ω constant. Conversely, at all tip speed ratios, when the blades were removed from the rotor, the rotor generated negative torque while the motor acted as the driving force of the rotor revolution. Therefore, at all tip speed ratios, T_B was higher than T_{wB}.

The power coefficient describes that fraction of the power in the wind that may be converted by the turbine into mechanical work [30] and is defined in this study as:

$$C_P = \frac{T_B\omega}{0.5\rho A U_0^3}, \tag{6}$$

and the torque coefficient is defined as:

$$C_T = \frac{T_B}{0.5\,\rho A U_0^2 R}. \tag{7}$$

Here, the U_0 is the time-mean stream-wise velocity, $\overline{u}(x, y, z)$, averaged over the projected rotor's swept area at $x = 0$ and is computed by:

$$U_0 = \frac{\int_{-(R+0.5c)}^{R} \overline{u}(0, y, 0)dy}{2R+0.5c}. \tag{8}$$

3. Numerical Approach

The CFD software utilized to simulate the wind flow field was ANSYS Fluent 17.2 [31,32]. The numerical approach was based on our previous paper [22,33], in which the CFD simulations with the delayed detached eddy simulation (DDES) turbulence model of flow around the O-VAWT were conducted and the validities of the grid resolution and the time-step size were confirmed.

3.1. Governing Equations and Discretization Method

The flow field around the wind turbine was assumed to be incompressible and isothermal. The DDES turbulence mode treats near-wall region in a manner like a Reynolds-averaged Navier–Stokes (RANS) turbulence model and treats the rest of the flow field in a manner like a large-eddy simulation (LES) turbulence model [34]. This model has the potential to achieve higher accuracy than RANS models and save a large number of computing resources compared with pure LES models. The governing equations for the CFD simulation with the DDES turbulence model based on the Spalart–Allmaras (SA) model are the continuity equation:

$$\frac{\partial u_i}{\partial x_i} = 0, \tag{9}$$

the Navier–Stokes equation:

$$\frac{\partial u_i}{\partial t} + \frac{\partial u_i u_j}{\partial x_j} = -\frac{1}{\rho}\frac{\partial p}{\partial x_i} + \nu\frac{\partial}{\partial x_j}\left(\frac{\partial u_i}{\partial x_j} + \frac{\partial u_j}{\partial x_i} - \frac{2}{3}\delta_{ij}\frac{\partial u_i}{\partial x_j}\right) - \frac{\partial}{\partial x_j}\left[\tilde{\nu}\left(\frac{\partial u_i}{\partial x_j} + \frac{\partial u_j}{\partial x_i}\right) - \frac{2}{3}k\delta_{ij}\right], \tag{10}$$

and the transport equation for the kinematic eddy viscosity $\tilde{\nu}$:

$$\frac{\partial \tilde{\nu}}{\partial t} + \frac{\partial \tilde{\nu} u_i}{\partial x_i} = C_{b1}\tilde{\nu}\left(S + \frac{\tilde{\nu}}{\kappa^2 d_{DDES}{}^2}\left(1 - \frac{\chi}{1 + \chi f_{v1}}\right)\right) + \frac{1}{\sigma_{\tilde{\nu}}}\left[\frac{\partial}{\partial x_j}\left\{(\nu + \tilde{\nu})\frac{\partial \tilde{\nu}}{\partial x_j}\right\} + C_{b2}\left(\frac{\partial \tilde{\nu}}{\partial x_j}\right)^2\right] - C_{w1} f_w\left(\frac{\tilde{\nu}}{d_{DDES}}\right)^2 \tag{11}$$

where u_i is the wind-velocity component in the x_i direction; p is the pressure; ν is the kinematic viscosity; t is the time; ρ is the air density; k is the turbulence kinetic energy; δ_{ij} is the Kronecker delta; d_{DDES} is the DDES length scale; χ is $(\tilde{\nu}/\nu)$; S is a scalar measure of the deformation tensor; f_{v1} and f_w are damping functions; and C_{b1}, C_{b2}, C_{w1}, $\sigma_{\tilde{\nu}}$ and κ are constants.

The DDES length scale is computed by:

$$d_{DDES} = d - f_d max\,(0, d - c_{des}\Delta_{max}) \tag{12}$$

where d is the distance to the closest wall; c_{des} is the empirical constant; and Δ_{max} is the maximum edge length of the local computational cell, i.e., $\Delta_{max} = max\,(\Delta_x, \Delta_y, \Delta_z)$. The switching between the RANS and the LES mode depends on the following shielding function;

$$f_d = 1 - tanh\left((8r_d)^3\right) \tag{13}$$

and

$$r_d = \frac{\tilde{\nu}}{\sqrt{u_{i,j}u_{i,j}\kappa^2 d^2}} \tag{14}$$

where $u_{i,j}$ is the velocity gradient. The damping functions and closure coefficients are as follows:

$$f_{v1} = \frac{\chi^3}{c_{v1}^3 + \chi^3},\; f_w = g\left[\frac{1 + c_{w3}^6}{g^6 + c_{w3}^6}\right]^{1/6},\; g = r + c_{w2}(r^6 - r),\; c_{w1} = \frac{c_{b1}}{\kappa^2} + \frac{(1 + c_{b2})}{\sigma_{\tilde{\nu}}}, \tag{15}$$

$$c_{b1} = 0.1355,\; c_{b2} = 0.622,\; c_{v1} = 7.1,\; c_{w2} = 0.3,\; c_{w3} = 2.0,\; \sigma_{\tilde{\nu}} = \frac{2}{3},\; c_{des} = 0.65,\; \kappa = 0.4187. \tag{16}$$

The governing equations are discretized by the finite-volume method. The advection terms of the Navier–Stokes equations are discretized by the bounded central-difference scheme. The advection term of the transportation equations for $\tilde{\nu}$ is discretized by a second-order upwind scheme. Other spatial derivatives are discretized by the central-difference scheme. The time integration is performed using the second-order implicit method.

3.2. Numerical Setup

The computational domain, the computational meshes and the boundary conditions are shown in Figure 5. As the same as the experimental setup, the origin of the coordinate system is defined at the center of the O-VAWT. The modeled O-VAWT was comprised of three blades, one main shaft and two sets of connecting arms. To reduce the computational cost, other components, such as the chains and sprockets are omitted. The sizes of these components are the same as those used in the experiment. The computational domain consists of three blade domains, one rotor domain and one far-field domain. The blade domain included one of the blades and rotates around each blade axis. The rotor domain included these three blade domains, the main shaft and the connecting arms and rotates around the main shaft. The far-field domain was a stationary domain and its size was $23.5D \times 17D \times 5D$. Except for uniform flow cases, a splitter plate with the same thickness as the experiment was set at $y = 0$ and $x = -11.37D$ to $-1.67D$. In all domains, only unstructured meshes were used. Based on our previous mesh-resolution dependency tests [33], the number of computational cells in each domain were set as shown in Table 2. The total number of computational cells was approximately 10 million. All surfaces of the solid components were covered with boundary-layer meshes. The first grid nodes over the surface of the blades were $y+ < 1$ in all run cases.

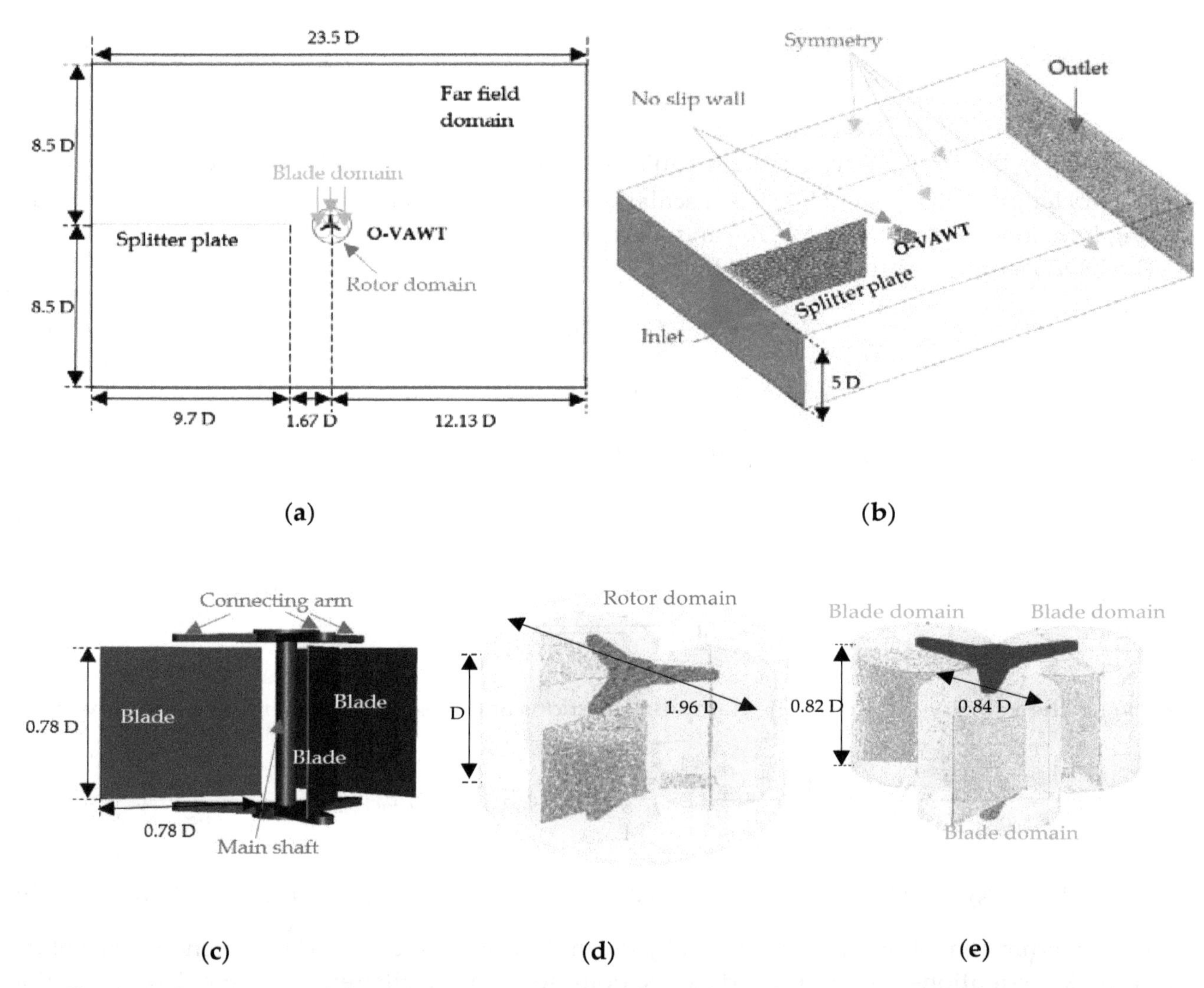

Figure 5. Computational domain, computational mesh and boundary conditions. (**a**) Top view of the computational domain; (**b**) bird's eye view of the computational domain with computational mesh and boundary conditions; (**c**) modeled rotor; (**d**) blade domains and (**e**) rotor domain.

Table 2. Numbers of grid sizes.

Domain	Mesh on This Research	Regular Mesh Used by ElCheikh [33]
Blade 1	2,300,530	1,610,660
Blade 2	2,303,725	1,614,903
Blade 3	2,304,607	1,588,363
Blade 4	-	1,614,661
Rotor	1,756,635	2,127,497
Far end	1,770,583	441,196
TOTAL	**10,436,080**	**8,997,280**

At the inlet boundary, the distributions of the stream-wise wind velocity shown in Table 3 were implemented. These distributions were set so that when the O-VAWT is absent, the distributions of the time-mean values of u at $x \approx -0.147\ D$, which corresponds to 0.1 m downwind of the wind tunnel outlet, matched well with those of the wind tunnel experiment, which were generated by using the three kinds of perforated panels with different shielding ratios Φ, as shown in Figure 6a,b. Here, U_H and U_L are the maximum and minimum streamwise velocities in the profile of a shear flow measured in the wind tunnel experiment; Γ is the velocity ratio defined as:

$$\Gamma = \frac{U_H - U_L}{U_H + U_L} \tag{17}$$

Table 3. Distribution of u at the inlet boundary. Here, Γ is the velocity ratio; Φ is the shielding ratio of a perforated panel; U_H and U_L are the maximum and minimum velocities in a shear flow.

Flow Type	Γ	Φ	U_H [m/s]	Range of U_H Region [m]	Velocity Distribution in Transition Region	Range of Transition Region [m]	U_L [m/s]	Range of U_L Region [m]
Uniform	0	-	8		8		8	
ASF-SF	0.51	0.67	13.2	$y \le -0.24$	$U_H \cdot(-y/0.24)^{0.3}$	$-0.24 \le y \le -0.01$	4.3	$y \ge -0.01$
ASF-SF	0.40	0.60	12.1	$y \le -0.22$	$U_H \cdot(-y/0.22)^{0.35}$	$-0.22 \le y \le 0.02$	5.2	$y \ge -0.02$
ASF-SF	0.28	0.49	10.8	$y \le -0.18$	$U_H \cdot(-y/0.18)^{0.2}$	$-0.18 \le y \le -0.01$	6.1	$y \ge -0.01$
RSF-SF	0.51	0.67	13.2	$y \ge 0.24$	$U_H \cdot(y/0.24)^{0.3}$	$0.01 \le y \le 0.24$	4.3	$y \le 0.01$
RSF-SF	0.40	0.60	12.1	$y \ge 0.22$	$U_H \cdot(y/0.24)^{0.35}$	$0.02 \le y \le 0.22$	5.2	$y \le 0.02$
RSF-SF	0.28	0.49	10.8	$y \ge 0.18$	$U_H \cdot(y/0.24)^{0.2}$	$0.01 \le y \le 0.18$	6.1	$y \le 0.01$

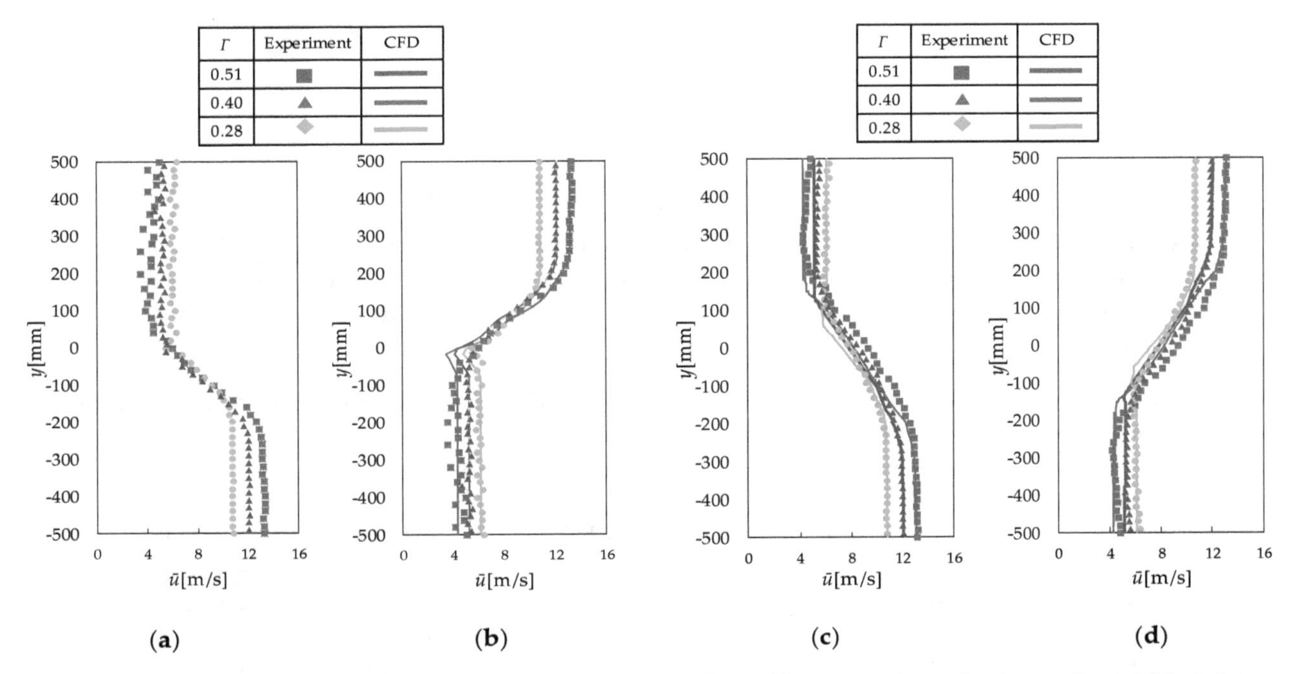

Figure 6. Horizontal distribution of the time-mean values of streamwise velocity at the mid-height of the wind tunnel outlet at $x \approx -0.147D$, which corresponds to 0.1 m downwind of the wind tunnel outlet, and $x = 0$, which corresponds to the position of the rotational axis of the O-VAWT, when the O-VAWT is absent. (**a**) ASF-SF cases at $x \approx -0.147D$, (**b**) RSF-SF cases at $x \approx -0.147D$, (**c**) ASF-SF cases at $x = 0$ and (**d**) RSF-SF cases $x = 0$.

It is worth noting that amplification factors of $\Gamma = 0.28$ and $\Gamma = 0.51$ are considered as 1.35 and 1.65, respectively, by computing U_H/U_∞. The amplification factor is defined as the ratio of wind speed in the case where there are buildings to wind speed in the case where these buildings are removed.

The separated shear flow from a building with an aspect ratio of 1:1:2 reaches an amplification factor of 1.2 [35]. In addition, the separated shear flow from a building with an aspect ratio of 1:1:6 reaches an amplification factor of 1.7 [36]. At $x = 0$, the values of Γ do no change, as shown in Figure 6c,d. However, due to the momentum diffusion, the horizontal gradients of the time-mean streamwise velocity become weaker. One of the reasons for the relatively large discrepancies between the profiles obtained by the experiment and the CFDs can be that the turbulence intensities of the CFDs are significantly small compared to those of the experiments, as shown in Figure 7. Even though very high values of turbulence intensity were set at the inlet boundary, the turbulence intensities dissipated rapidly and became very small at $x = 0$ as compared to those of the experiments. As the setting of high turbulence intensity at the inlet boundary leads to computational instability, we set no perturbation condition at the inlet boundary. Table 4 shows the values of U_0 computed by Equation (8) and the Reynolds number which is defined as:

$$Re = \frac{U_0 D}{\nu} \tag{18}$$

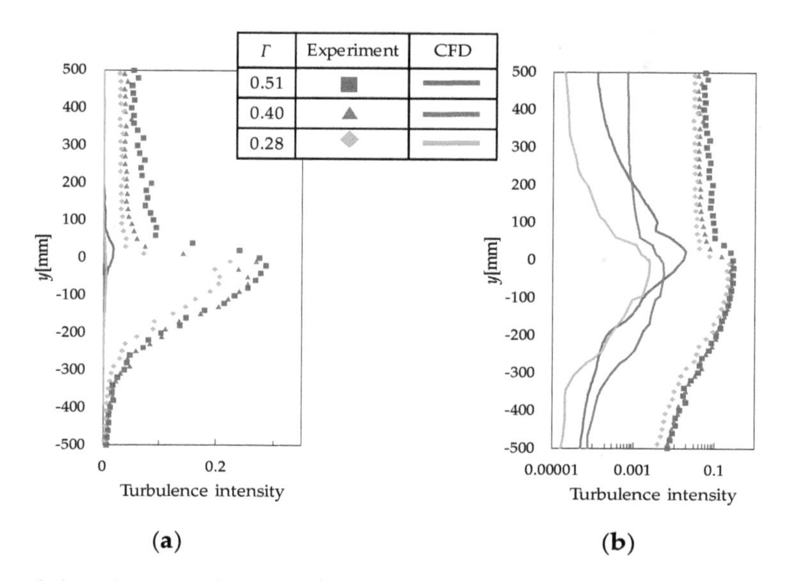

Figure 7. Horizontal distribution of the turbulence intensity of ASF-SF cases at the mid-height of the wind tunnel outlet at $x = 0$, which corresponds to the position of the rotational axis of the O-VAWT, when the O-VAWT was absent. (**a**) Linear scale on the horizontal axis and (**b**) logarithmic scale in the horizontal axis.

Table 4. Experimental and computational fluid dynamics (CFD) results for the values of U_0 and Re for the uniform flow, ASF-SF and RSF-SF cases.

Flow Type	Γ	U_0 [m/s]		Re	
		Experiment	CFD	Experiment	CFD
Uniform	0	8	8	2.79×10^5	2.79×10^5
ASF-SF	0.51	10.08	9.78	3.52×10^5	3.41×10^5
ASF-SF	0.40	9.38	9.22	3.28×10^5	3.22×10^5
ASF-SF	0.28	8.73	8.72	3.05×10^5	3.05×10^5
RSF-SF	0.51	7.76	7.48	2.71×10^5	2.61×10^5
RSF-SF	0.40	7.72	7.36	2.70×10^5	2.57×10^5
RSF-SF	0.28	7.52	7.33	2.62×10^5	2.56×10^5

At the outlet boundary, the pressure outlet condition with $p = 0$ was imposed. On the surface of the O-VAWT and the splitter plate, the no-slip boundary conditions were set. The sliding mesh technique was used to couple the rotational domains and the stationary domain. The direction of the rotor and the blade rotations are counterclockwise and clockwise, respectively, when viewed from the top (Figure 5a). By changing the rotational speed of the rotor ω, the tip speed ratio λ was set at 0.2, 0.4, 0.5, 0.6, or 0.8. The time step sizes were set as $dt = 0.5°/\omega$.

3.3. Torque and Power Coefficients

As mentioned in sub-Section 2.4, this study considers only the aerodynamic torque generated by the blades as the rotor torque of the O-VAWT. Since each of the blade axes was connected with the main shaft by a chain via sprockets, the aerodynamic torque on each of the blades about each of the blade axes was transmitted through the chain and contributed to the torque about the main shaft. Therefore, the rotor torque generated by a blade at an azimuthal angle φ is expressed as:

$$T_B(\varphi) = T_{B\_rev}(\varphi) + T_{B\_rot}(\varphi), \tag{19}$$

where $T_{B\_rev}(\varphi)$ is the conventional blade torque that is calculated by multiplying the rotor radius and the component of the aerodynamic force on the blade at φ in the rotor-revolution direction; and $T_{B\_rot}(\varphi)$ is the torque generated by the component of the aerodynamic force on the blade at φ in the blade-rotation direction about the blade axis. Hereafter, we call $T_B(\varphi)$ the "blade torque," $T_{B\_rev}(\varphi)$ the "rotor-revolution torque" and $T_{B\_rot}(\varphi)$ the "blade-rotation torque." The blade torque coefficient (C_{TB}), the rotor-revolution component (C_{TB_rev}) and the blade-rotation component (C_{TB_rot}) are defined as:

$$C_{TB}(\varphi) = \frac{T_B(\varphi)}{0.5\rho A U_0^2 R}, \tag{20}$$

$$C_{TB\_rev}(\varphi) = \frac{T_{B\_rev}(\varphi)}{0.5\rho A U_0^2 R}, \tag{21}$$

and

$$C_{TB\_rot}(\varphi) = \frac{T_{B\_rot}(\varphi)}{0.5\rho A U_0^2 R}. \tag{22}$$

It should be noted that C_{TB}, C_{TB_rev} and C_{TB_rot} are coefficients of one blade.

The CFD simulations were conducted for eight revolutions of the rotor. Using the data of the last two rotor revolutions, the torque coefficient C_T was computed by the following formula:

$$C_T = \frac{n}{N_e - N_s} \sum_{N=N_s+1}^{N_e} \int_0^{2\pi} C_{TB}\, d\varphi. \tag{23}$$

Here, n (= 3) is the number of the blades; N_s (= 6) is the number of the rotor revolutions before starting the computation of C_T; N_e (= 8) is the number of the rotor revolutions before finishing the computation of C_T. The power coefficient of C_P was computed by Equation (6).

4. Results and Discussion

In this section, the results of the wind tunnel experiments and the CFD simulations for the power performance of the O-VAWT, such as the dependency of the power and torque coefficients on the tip speed ratio, the variations of the torque coefficients with azimuthal angle, are presented for the uniform flow case and the shear flow cases. Subsequently, the causes of the features of the power performance of the O-VAWT are discussed based on the CFD results of the flow fields.

4.1. Performance in Uniform Flow

Figure 8a shows the power and torque coefficients (C_P and C_T) of the O-VAWT in the uniform flow. The CFD results are in good agreement with the experimental ones. The optimal tip speed ratio at which C_P becomes the maximum is less than unity; 0.4 in the experiments and 0.5 in the

CFD simulations. As mentioned in the introduction, this low optimal tip speed ratio is a favorable feature for the built environment from the viewpoint of aerodynamic noise. With increasing λ, C_T decreases monotonically from a small tip speed ratio ($\lambda = 0.2$). These tendencies are commonly found among drag-type wind turbines. The wind-tunnel experimental results by Shimizu et al. [24] for C_P and C_T of an O-VAWT with two elliptical cross-sectional blades show the same tendencies as our results. The value of the maximum C_P is 0.32 in our experiments and CFD simulations, while the value is 0.176 in Shimizu et al.'s experiments. The main factors for the better performance of our O-VAWT as compared to Shimizu et al.'s O-VAWT can be the number of blades and the cross-sectional shape of the blades. In our previous studies, the maximum Cp improved from 0.189 to 0.244 by changing the number of blades from two to three [33] and improved from 0.246 to 0.288 by changing the cross-sectional shape of blades from ellipse to rectangle [23].

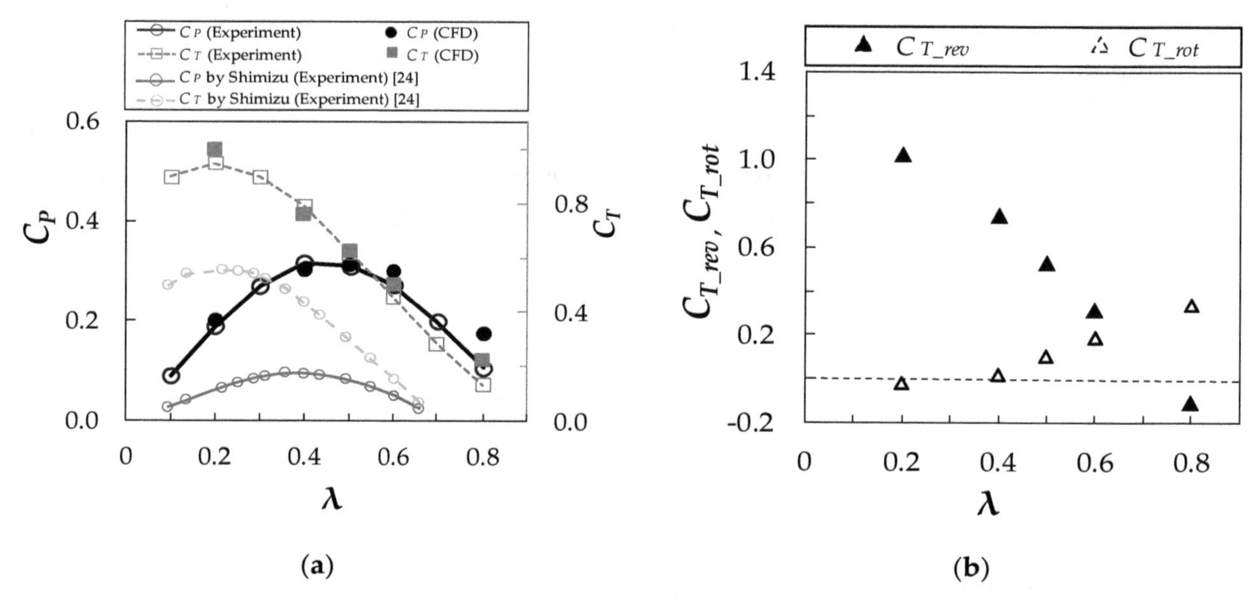

Figure 8. Performance of the O-VAWT in the uniform flow; (**a**) the variation of power coefficient (C_P) and torque coefficient (C_T) with tip speed ratios (λ) by the experiments and CFD simulations, (**b**) rotor-revolution (C_{T_rev}) and blade-rotation (C_{T_rot}) components of torque coefficients (C_T) computed by the CFD simulations. The wind-tunnel experimental results by Shimizu et al. [24] for C_P and C_T of an O-VAWT with two elliptical cross-sectional blades are added for reference.

Figure 8b shows C_{T_rev} and C_{T_rot} computed by the CFD results. The sum of C_{T_rev} and C_{T_rot} is C_T. As well as C_T, the value of C_{T_rev} decreases monotonically with an increase in λ. Conversely, the value of C_{T_rot} increases with an increase in λ. The values of C_{T_rev} are positive and larger than those of C_{T_rot} except for $\lambda = 0.8$. At $\lambda = 0.8$, C_{T_rev} is negative; however, C_{T_rot} is positive and its absolute value is larger than that of C_{T_rev}. As a result, C_T is positive at $\lambda = 0.8$.

Figure 9 shows the variations of blade torque coefficient (C_{TB}), its rotor-revolution component (C_{TB_rev}) and blade-rotation component (C_{TB_rot}) with respect to azimuthal angle (φ) at λ = 0.4 and 0.6. The value of C_{TB} is significantly large in the upwind region of the advancing side ($\varphi \approx 180°$ to $270°$) of the rotor, being the maximum at $\varphi \approx 210°$. In the range of φ where C_{TB} is significantly large, the contribution of C_{TB_rev} is dominant. Except for this range, C_{TB_rev} does not always positively contribute to C_{TB}. At φ where the value of C_{TB_rev} is negative, the value of C_{TB_rot} is generally positive and the rotation of the blade positively contributes to C_{TB}. Due to this positive contribution of C_{TB_rot} to C_{TB}, the value of C_{TB} is positive at almost all φ and the variation of C_{TB} of the O-VAWT with respect to φ is smaller as compared to that of a Savonius-type VAWT. The variation of C_{TB} of a Savonius-type VAWT with respect to φ in Figure 9 is a result of CFD simulation by Tian et al. [37]. The maximum C_P and the optimal λ of the Savonius-type VAWT were 0.258 and 1.0, respectively.

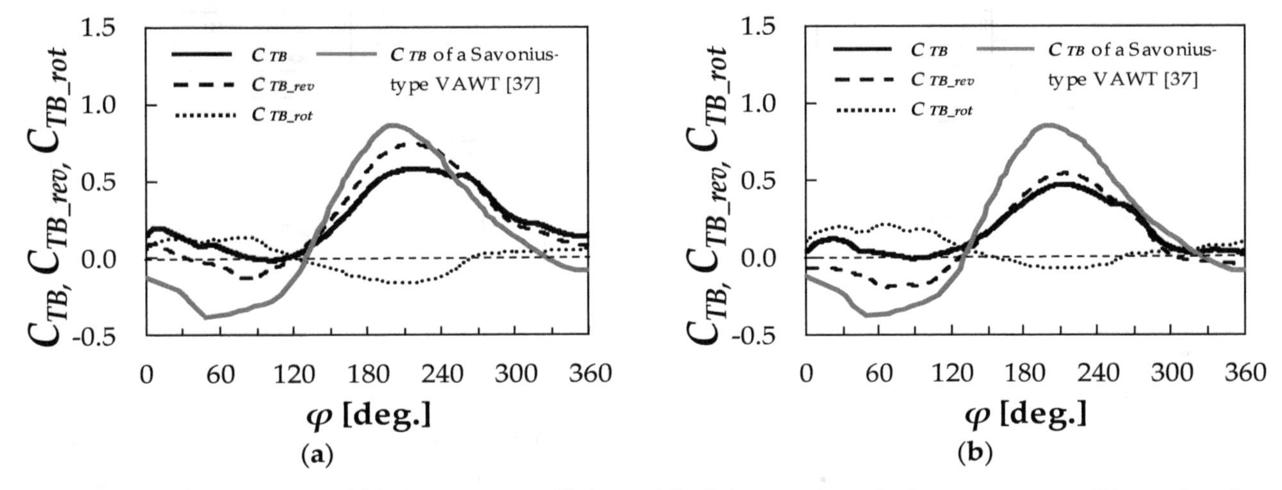

Figure 9. The variation of blade torque coefficients (C_{TB}), its rotor-revolution component (C_{TB_rev}) and its blade-rotation component (C_{TB_rot}) with azimuthal angle (φ) at; (**a**) $\lambda = 0.4$ and (**b**) $\lambda = 0.6$. Note that C_{TB}, C_{TB_rev} and C_{TB_rot} are coefficients of one blade. The CFD simulation result by Tian et al. [37] for the variation of C_{TB} of a Savonius-type VAWT with φ is added for comparison.

4.2. Performance in Shear Flows

Figure 10 compares the experimental results for C_P of the O-VAWT in the cases of the advancing side faster shear flow (ASF-SF), the retreating side faster shear flow (RSF-SF) and the uniform flow. In the cases of the ASF-SF, C_P is higher than in the case of the uniform flow at all λ and increases with an increase in Γ. In the cases of the RSF-SF, similar to the cases of the ASF-SF, C_P increases with an increase in Γ. However, when $\Gamma = 0.28$, the values of C_P are lower than those of the uniform flow case. Concerning the optimal λ, there is a trend that it shifts to higher λ in both cases of shear flows as compared to the uniform flow case, except for the RSF-SF with $\Gamma = 0.28$.

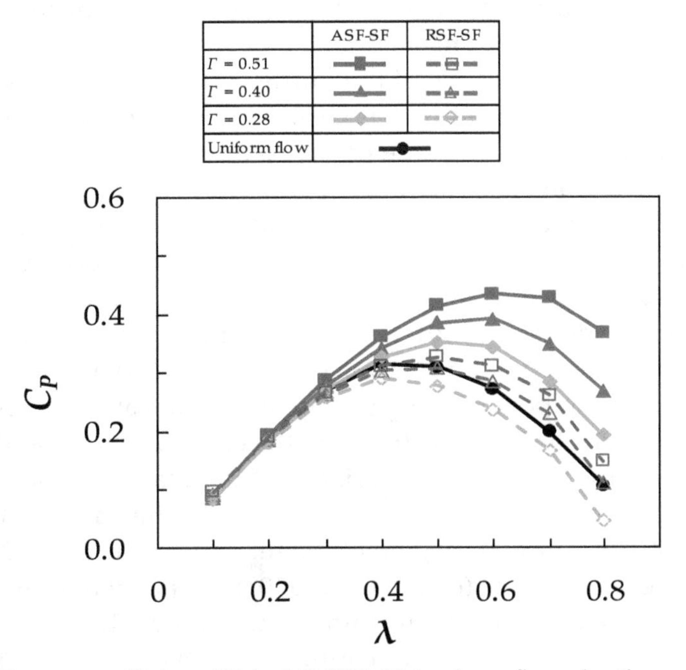

Figure 10. Power coefficient (C_P) of O-VAWT in shear flows by the experiments.

Both in the cases of the ASF-SF (Figure 11a) and the RSF-SF (Figure 11b), the CFD results for C_P–λ curves are in good agreement with the experimental ones. In the following discussion, we use the CFD results for the torque of the O-VAWT and the flow field to explain the effects of shear flows on the characteristics of the power performance.

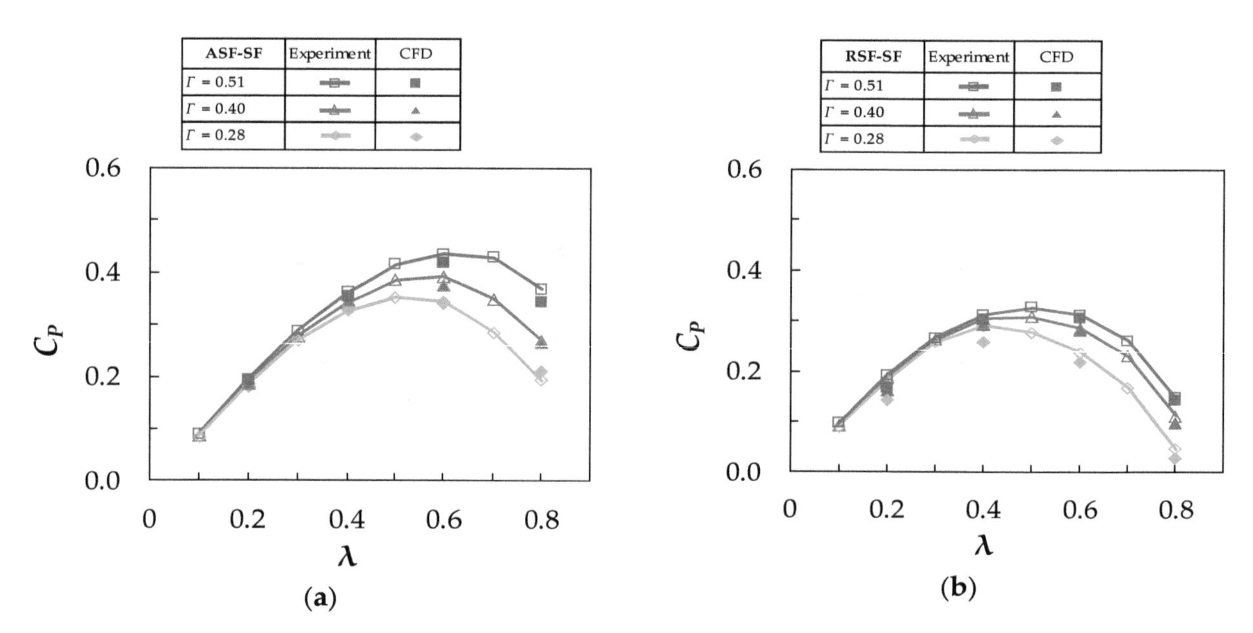

Figure 11. Power coefficient (C_P) of O-VAWT in shear flows by the experiments and the CFD simulations in the cases of; (**a**) the ASF-SF and (**b**) the RSF-SF.

Figure 12 shows the variations of blade torque coefficient (C_{TB}), its rotor-revolution component (C_{TB_rev}) and its blade-rotation component (C_{TB_rot}) with azimuthal angle (φ) in the cases of the ASF-SF. It is confirmed that these profiles are qualitatively the same between the cases of $\lambda = 0.4$ and $\lambda = 0.6$. Therefore, it is considered that the effects of the shear flow on the characteristics of the blade torque variations with φ do not significantly change around the optimal λ. As compared to the case of the uniform flow, C_{TB} is higher on most of the advancing side of the rotor ($\varphi \approx 210°$ to $330°$) and lower in the upwind region of the retreating side of the rotor, specifically at $\varphi \approx 120°$ to $150°$. In particular, with an increase in Γ, C_{TB} increases on most of the advancing side of the rotor ($\varphi \approx 210°$ to $330°$). The optimal φ at which C_{TB} is the maximum shifts to the downwind direction on the advancing side of the rotor ($\varphi \approx 240°$) as compared to the case of the uniform flow. The effects of the shear flow on the variations of C_{TB_rev} with φ is almost the same as C_{TB}. As compared to the case of the uniform flow, C_{TB_rev} is higher on most of the advancing side of the rotor ($\varphi \approx 210°$ to $330°$). In contrast, C_{TB_rot} is lower in most of the upwind region of the advancing side ($\varphi \approx 180°$ to $240°$) and slightly higher in the upwind region of the retreating side of the rotor, specifically at $120°$ to $150°$ as compared to the case of the uniform flow. With an increase in Γ, C_{TB_rot} decreases in the upwind region of the advancing side of the rotor ($\varphi \approx 180°$ to $240°$). Due to its negative values of C_{TB_rot}, the optimal φ at which C_{TB} is the maximum slightly shifts to the downwind direction as compared to C_{TB_rev}.

Figure 13 shows the variation of blade torque coefficient (C_{TB}), its rotor-revolution component (C_{TB_rev}) and its blade-rotation component (C_{TB_rot}) with azimuthal angle (φ) in the cases of the RSF-SF. Since these profiles are qualitatively the same between the cases of $\lambda = 0.4$ and $\lambda = 0.6$, it is considered that the effects of the shear flow on the characteristics of the blade torque variations with φ do not significantly change around the optimal λ. As compared to the case of the uniform flow, C_{TB} is significantly higher in most of the upwind region of the retreating side ($\varphi \approx 120°$ to $180°$) and lower on most of the advancing side of the rotor ($\varphi \approx 210°$ to $330°$). In particular, with an increase in Γ, C_{TB} increases in the upwind region of the retreating side of the rotor. The optimal φ at which C_{TB} is the maximum shifts to the upwind region of the retreating side of the rotor ($\varphi \approx 150°$) as compared to the case of the uniform flow. The effects of the shear flow on the variation of C_{TB_rev} with φ is almost the same as C_{TB}. As compared to the case of the uniform flow, C_{TB_rev} is higher in most of the upwind region of the retreating side ($\varphi \approx 100°$ to $160°$) and lower on most of the advancing side of the rotor ($\varphi \approx 210°$ to $330°$). In contrast, C_{TB_rot} is slightly lower in most of the upwind region of the retreating side ($\varphi \approx 90°$ to $150°$) and slightly higher in most of the upwind region of the advancing side of the rotor ($\varphi \approx 180°$ to $240°$) as compared to the case of the uniform flow. Furthermore, with an increase in Γ, C_{TB_rot} decreases in most of the upwind region of the retreating side ($\varphi \approx 100°$ to $160°$) and increases in most of the upwind region of the advancing side of the rotor ($\varphi \approx 180°$ to $240°$).

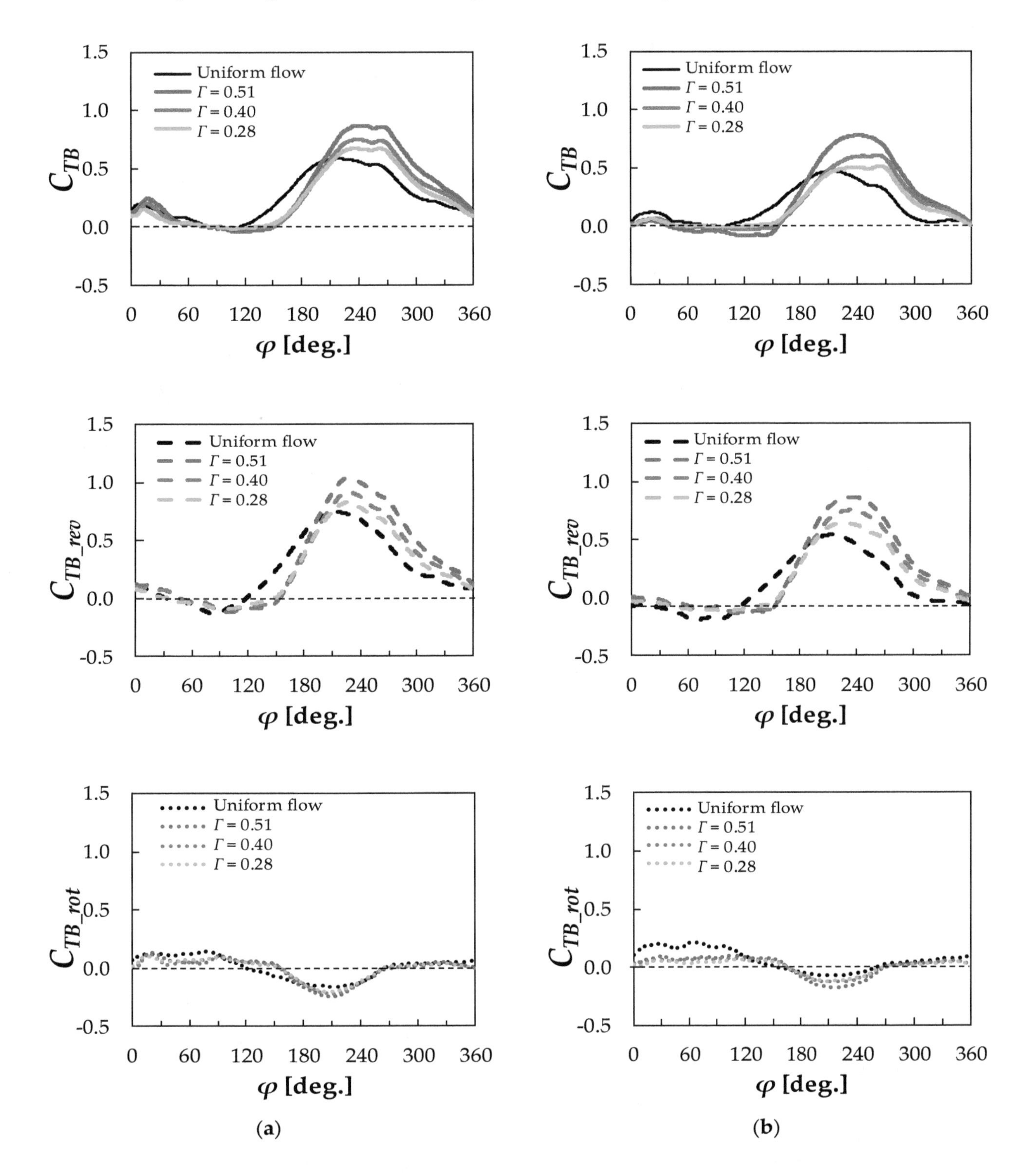

Figure 12. The variation of blade torque coefficient (C_{TB}), its rotor-revolution component (C_{TB_rev}) and its blade-rotation component (C_{TB_rot}) in the cases of the ASF-SF computed by the CFD simulations; (**a**) for $\lambda = 0.4$ and (**b**) for $\lambda = 0.6$. Note that C_{TB}, C_{TB_rev} and C_{TB_rot} are coefficients of one blade.

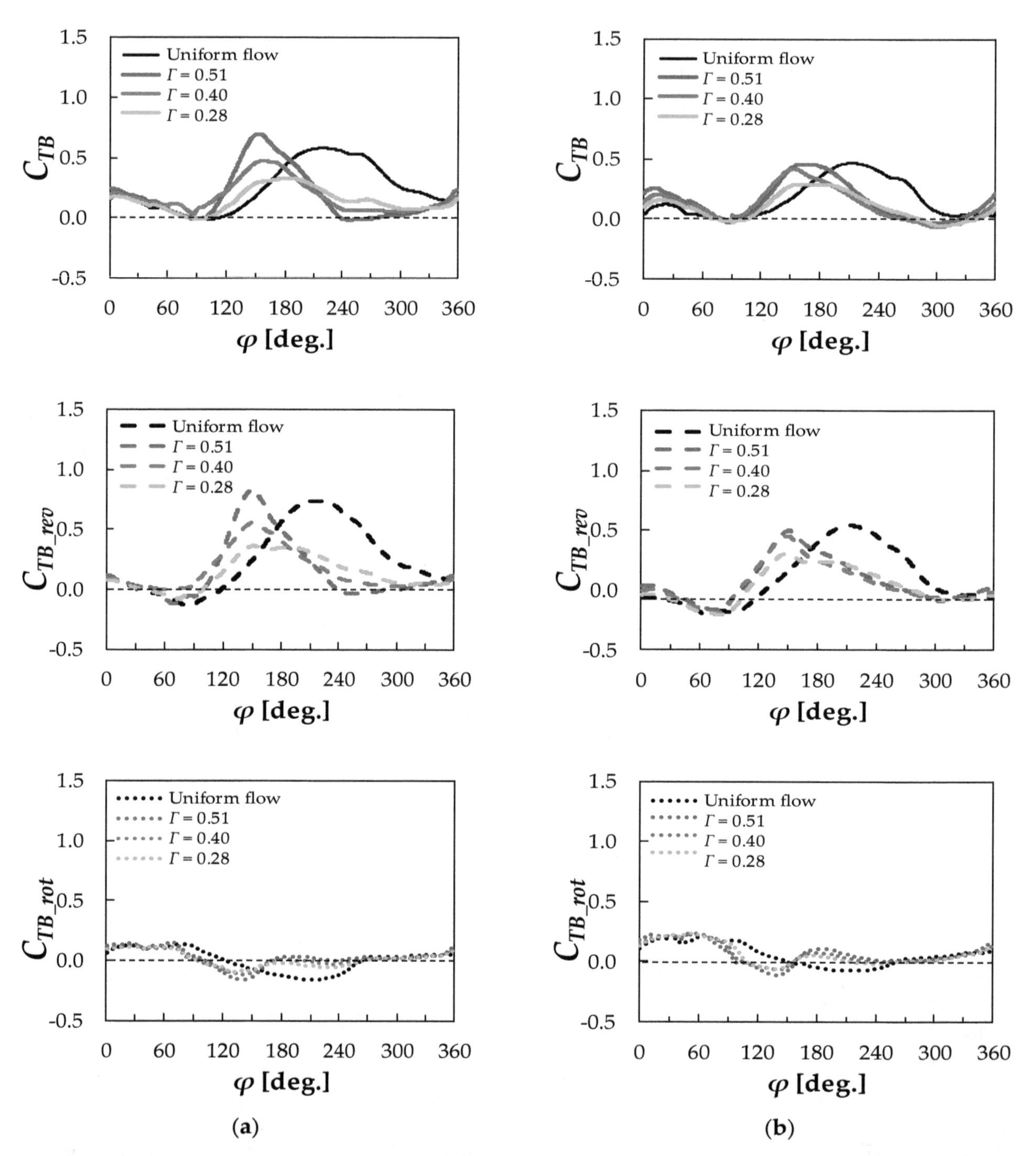

Figure 13. The variation of blade torque coefficient (C_{TB}), its rotor-revolution component (C_{TB_rev}) and its blade-rotation component (C_{TB_rot}) in the cases of the RSF-SF computed by the CFD simulations; (**a**) for $\lambda = 0.4$ and (**b**) for $\lambda = 0.6$. Noted that C_{TB}, C_{TB_rev} and C_{TB_rot} are coefficients of one blade.

4.3. Flow Characteristics

Figures 14a, 15a and 16a show the temporal sequence of the horizontal distributions of normalized horizontal velocity vectors, normalized pressure and normalized vorticity, respectively, at the mid-height of the O-VAWT at $\lambda = 0.4$ in the case of the uniform flow. At $\varphi = 180°$ to $270°$, the approaching flow to the blade has a large velocity component perpendicular to the blade, and the pressure on the upwind side of the blade is high. Due to this high pressure, C_{TB_rev} is significantly high in the range of $\varphi = 180°$ to $270°$ in Figure 9. In addition, at $\varphi = 210°$ to $270°$, due to the strong large vortex formed near the outer edge of the downwind side of the blade, the pressure is low near the vortex. This low pressure positively contributes to C_{TB_rev} (see Figure 9 at $\varphi = 210°$ to $270°$). By contrast, this low pressure negatively contributes to C_{TB_rot} (see Figure 9 at $\varphi = 210°$ to $270°$). At $\varphi = 300°$ and $330°$, the approaching flow to the blade has a small velocity component perpendicular to the blade and the pressure on the upwind side of the blade is not high. Therefore, C_{TB_rev} is lower as compared to the upwind region of the advancing side of the rotor (see Figure 9 at $\varphi = 180°$ to $270°$). At $\varphi = 0°$ to $120°$, the attack angle of the blade is positive (here, the counterclockwise direction is defined as positive) and the flow separates over the outer side of the blade. Therefore, on the upwind edge of the blade and on the inner side of the blade near its upwind edge, the pressure is relatively high. In contrast, on the outer side of the blade near its upwind edge, the pressure is relatively lower. This pressure distribution contributes negatively to C_{TB_rev} and positively to C_{TB_rot}.

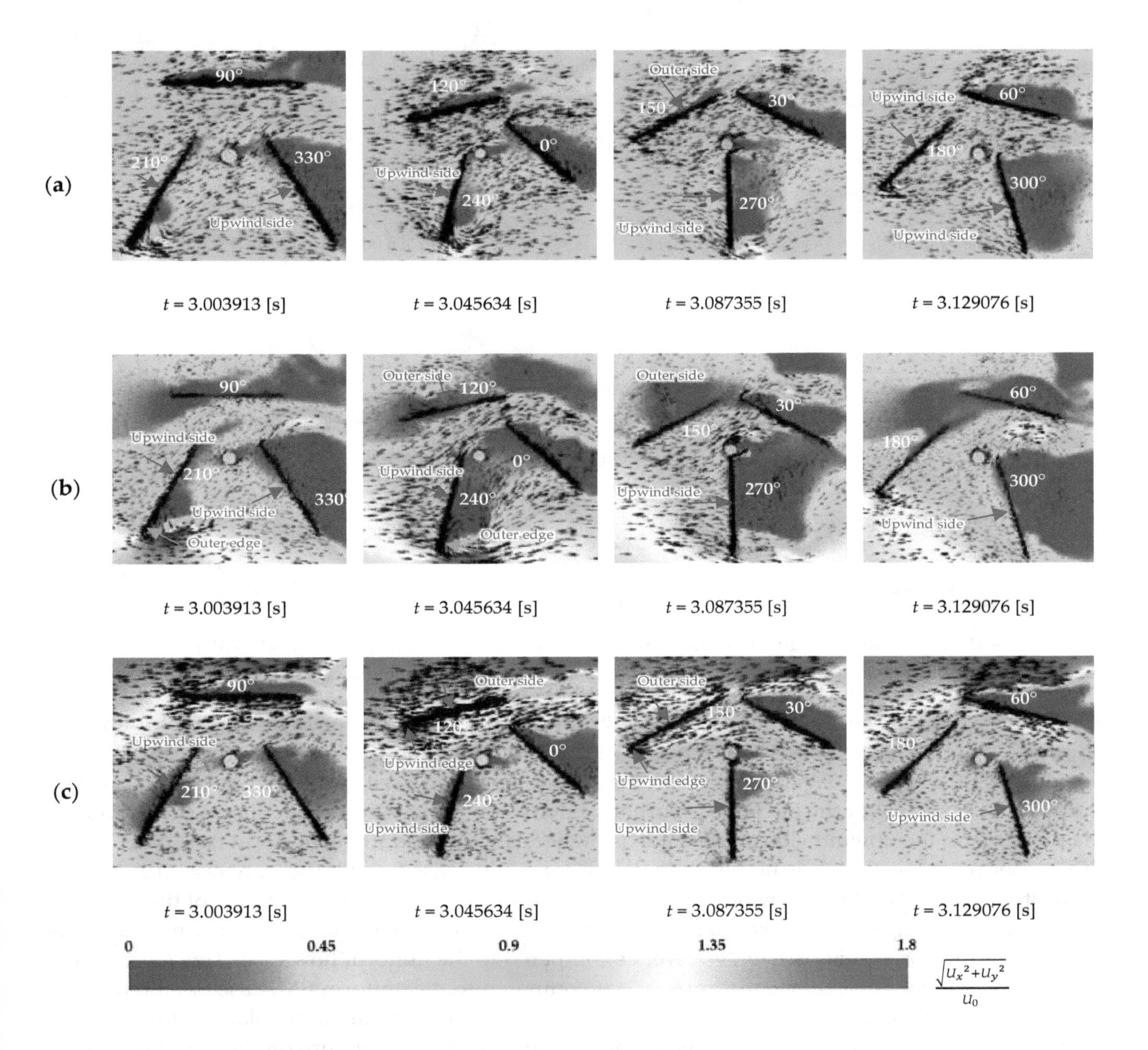

Figure 14. The temporal sequence of the horizontal distributions of normalized horizontal velocity vectors of the O-VAWT for $\lambda = 0.4$ in the case of; (**a**) the uniform flow, (**b**) the ASF-SF with $\Gamma = 0.51$ and (**c**) the RSF-SF with $\Gamma = 0.51$.

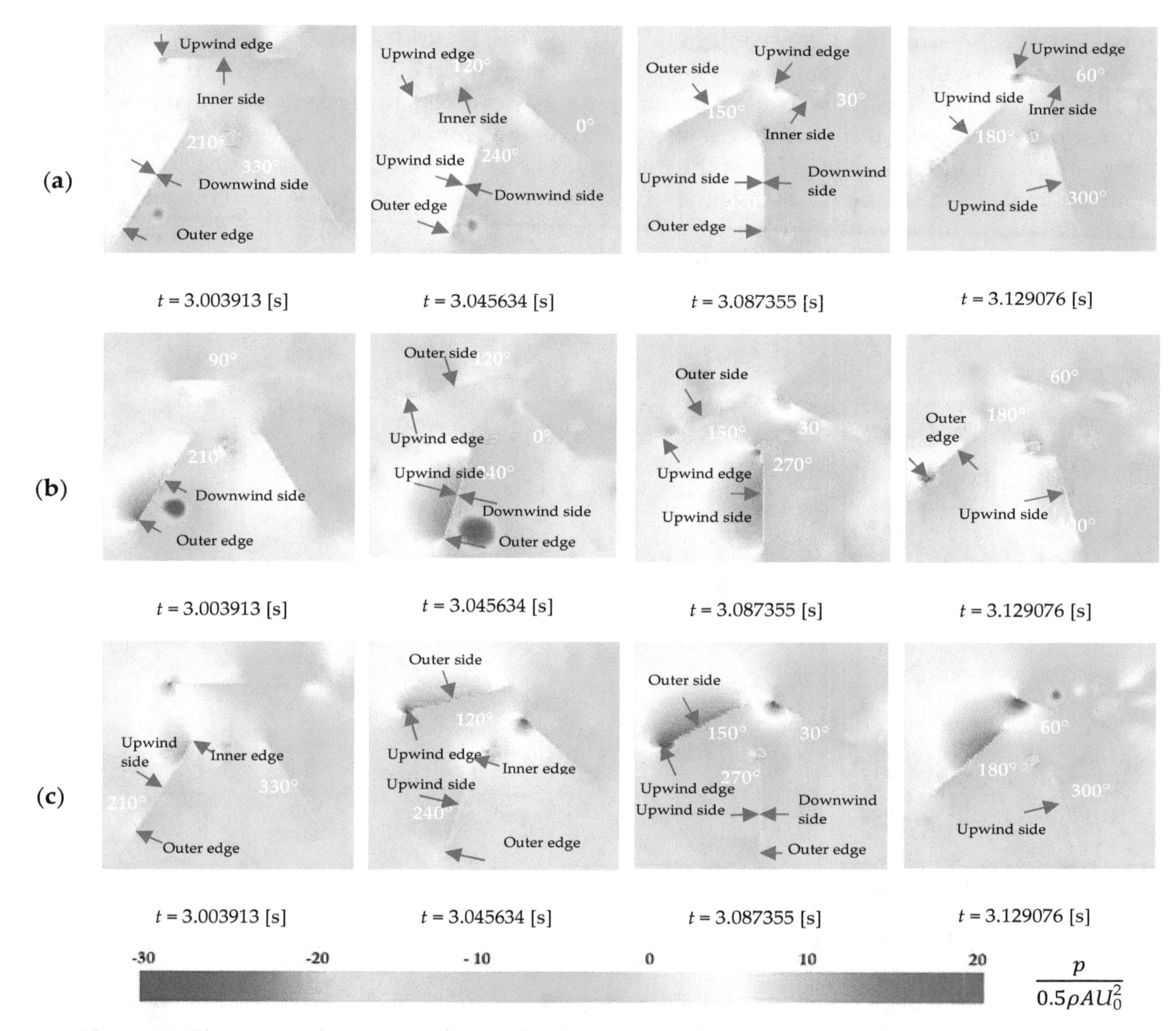

Figure 15. The temporal sequence of normalized pressure at the mid-height of the O-VAWT for λ = 0.4 in the case of; (**a**) the uniform flow, (**b**) the ASF-SF with Γ = 0.51 and (**c**) the RSF-SF with Γ = 0.51.

Figures 14b and 15b and Figure 16b show the temporal sequence of the horizontal distributions of normalized horizontal velocity vectors, normalized pressure and normalized vorticity, respectively, at the mid-height of the O-VAWT at $\lambda = 0.4$ in the case of the ASF-SF with $\Gamma = 0.51$. At $\varphi = 210°$ to $330°$, the approaching flow to the blade has a larger velocity component perpendicular to the blade, and the pressure on the upwind side of the blade is higher as compared to the uniform flow case. Due to this higher pressure, C_{TB_rev} is higher than the uniform flow case (see Figure 12 at $\varphi = 210°$ to $330°$). In addition, at $\varphi = 210°$ to $240°$, wind speed is significantly increased at the outer edge of the blade and a significantly stronger and larger vortex is formed near the outer edge on the downwind side of the blade. Near the vortex, the pressure is lower as compared to the uniform flow case. This low pressure contributes to higher C_{TB_rev} and lower C_{TB_rot} as compared to the uniform flow case (see Figure 12 at $\varphi = 210°$ to $240°$). At $\varphi = 120°$ and $150°$, the pressure on the outer side of the blade is lower due to the lower speed approaching flow to the blade. Furthermore, the pressure on the inner side of the blade near its upwind edge is higher due to the existence of the blade at $\varphi = 240°$ or $270°$, respectively. This pressure distribution contributes to lower C_{TB_rev} and higher C_{TB_rot} as compared to the uniform flow case (see Figure 12 at $\varphi = 120°$ to $150°$).

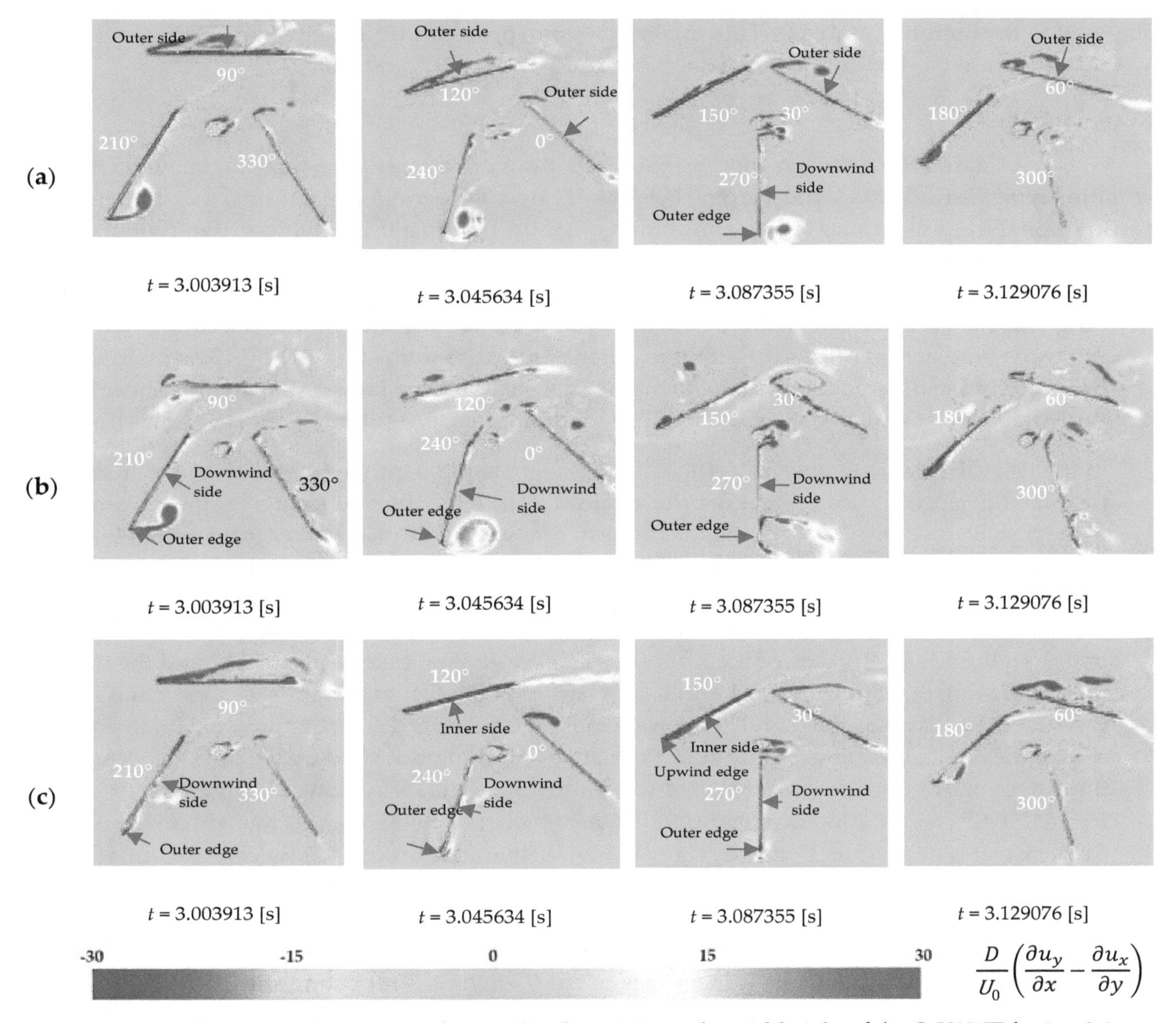

Figure 16. The temporal sequence of normalized vorticity at the mid-height of the O-VAWT for $\lambda = 0.4$ in the case of; (**a**) the uniform flow, (**b**) the ASF-SF with $\Gamma = 0.51$ and (**c**) the RSF-SF with $\Gamma = 0.51$.

Figures 14c, 15c and 16c show the temporal sequence of normalized horizontal distributions of horizontal velocity vectors, pressure and vorticity, respectively, at the mid-height of the O-VAWT at $\lambda = 0.4$ in the case of the RSF-SF with $\Gamma = 0.51$. At $\varphi = 120°$ and $150°$, the approaching flow to the blade has a larger velocity component perpendicular to the blade and the pressure on the outer side of the blade is higher as compared to the uniform flow case. Due to this higher pressure, C_{TB_rev} is higher than the uniform flow case (see Figure 13 at $\varphi = 120°$ and $150°$). Furthermore, wind speed is increased at the upwind edge of the blade and a stronger vortex is formed near the upwind edge of the blade on its inner side. Near the vortex, the pressure is lower as compared to the uniform flow case. This low pressure contributes to higher C_{TB_rev} and lower C_{TB_rot} as compared to the uniform flow case (see Figure 13 at $\varphi = 120°$ and $150°$). At $\varphi = 210°$ to $300°$, except in the vicinity of the upwind side of the inner edge of the blade at $\varphi = 210°$, the approaching flow to the blade has a smaller velocity component perpendicular to the blade and the pressure on the upwind side of the blade is lower as compared to the uniform flow case. Due to this lower pressure, C_{TB_rev} is lower than the uniform flow case (see Figure 13 at $\varphi = 210°$ and $300°$). In addition, at $\varphi = 210°$ and $240°$, the vortex formed near the outer edge of the downwind side of the blade is weaker and the pressure drop becomes smaller as compared to the uniform flow case. This smaller pressure drop contributes to lower C_{TB_rev} and higher C_{TB_rot} as compared to the uniform flow (see Figure 13 at $\varphi = 210°$ and $240°$).

5. Conclusions

We investigated the effects of horizontal shear flow on the performance characteristics of an orthopter-type vertical axis wind turbine (O-VAWT) by conducting wind tunnel experiments and computational fluid dynamics (CFD) simulations. In addition to a uniform flow, two types of shear flow were used as the approaching flow to the O-VAWT. One type was an advancing side faster shear flow (ASF-SF), which had a higher velocity on the advancing side of the rotor. The other type was a retreating side faster shear flow (RSF-SF), which had a higher velocity on the retreating side of the rotor. For each type of shear flow, we set three different velocity ratios ($\Gamma = 0.28$, 0.40 and 0.51), which were the ratios of the difference between the highest velocity and the lowest velocity in a shear flow to the sum of the highest and lowest velocities. The main findings are summarized as follows:

1. In the ASF-SF cases, the power coefficients (C_P) were significantly higher than the uniform flow case at all tip speed ratios (λ) and increased with Γ. The experimental results for the maximum C_P of the ASF-SF case with $\Gamma = 0.51$ and the uniform flow case were 0.43 when $\lambda = 0.6$ and 0.32 when $\lambda = 0.4$, respectively. Around the optimal λ, the blade torque coefficient (C_{TB}) on the advancing side of the rotor was, in general, significantly higher than the uniform flow case and increased with Γ, predominantly contributing to the increase in C_P. The CFD results for the maximum discrepancies of C_{TB} on the advancing side of the rotor between the ASF-SF case with $\Gamma = 0.51$ and the uniform flow case were 0.38 when $\lambda = 0.4$ and 0.44 when $\lambda = 0.6$. The high values of C_{TB} of the ASF-SF cases on the advancing side of the rotor were mainly caused by the higher pressure on the upwind side of the blade due to the higher speed of the approaching flow and by the lower pressure near the outer edge of the downwind side of the blade due to the formation of a larger vortex.
2. In the RSF-SF cases, C_P increased with Γ. However, when $\Gamma = 0.28$, C_P was lower than the uniform flow case at all λ. When $\Gamma = 0.51$, C_P was higher than the uniform flow case except at low λ; however, it was lower than the ASF-SF case with $\Gamma = 0.28$. The experimental results for the maximum C_P of the RSF-SF case with $\Gamma = 0.28$, the RSF-SF case with $\Gamma = 0.51$ and the ASF-SF case with $\Gamma = 0.28$ were 0.29 when $\lambda = 0.4$, 0.33 when $\lambda = 0.5$ and 0.35 when $\lambda = 0.5$, respectively. Around the optimal λ, the blade torque coefficient (C_{TB}) on the retreating side of the rotor was, in general, higher than the uniform flow case and increased with Γ, predominantly contributing to the increase in C_P. The CFD results for the maximum discrepancies of C_{TB} on the retreating side of the rotor between the RSF-SF case with $\Gamma = 0.51$ and the uniform flow case were 0.60 when $\lambda = 0.4$ and 0.36 when $\lambda = 0.6$. The high values of C_{TB} of the RSF-SF cases on the retreating side of the rotor were mainly caused by the higher pressure on the outer side of the blade on the upwind side of the rotor, due to the higher speed of the approaching flow. By contrast, C_{TB} on the advancing side of the rotor was, in general, lower than the uniform flow case, due to the lower pressure on the upwind side of the blade.
3. C_{TB} consists of the rotor-revolution component (C_{TB_rev}) and the blade-rotation component (C_{TB_rot}). In all the shear flow cases, as well as the uniform flow case, the contributions of C_{TB_rev} to C_{TB} were dominant. The dependencies of C_{TB_rev} and C_{TB_rot} on Γ had the opposite tendencies.

These findings are useful for micro-siting of an O-VAWT in the area where shear flows occur. A location where ASF-SFs with high Γ values dominantly occur is ideal for installing the O-VAWT. At a location where not only ASF-SFs but also RSF-SFs occur at high frequencies, higher Γ values are preferable. However, the shear flows utilized in this study are limited in their profiles and the relative positions to the rotor. To properly conduct the micro-siting of an O-VAWT in the area where various kinds of shear flows occur, such as the vicinity of a building, it is essential to understand the performance characteristics of the O-VAWT in the various kinds of shear flows. Therefore, in future research, we plan to investigate the effects of the broadness of the shear layer and the relative position of the shear flow to the rotor on the O-VAWT's performance characteristics. In addition, we plan to investigate the effects of the turbulence intensity of the approaching flow on the CFD results for the O-VAWT's performance characteristics by setting obstacles, which emit eddies, upwind of the O-VAWT to avoid the rapid dissipation of high turbulence intensity.

Author Contributions: R.P.W. performed the numerical simulations and prepared this manuscript being supervised by T.K. (Takaaki Kono). All authors contributed to the analyses of the data. T.K. (Takaaki Kono) and T.K. (Takahiro Kiwata) supervised the entire work. All authors have read and agreed to the published version of the manuscript.

Acknowledgments: This research was supported by the Program for Research and Innovation in Science and Technology (RISET-Pro) Kemenristekdikti. The authors are thankful to technician Kuratani and student Taiki Sugawara for their help with the experiment.

References

1. Bahaj, A.S.; Myers, L.; James, P.A.B. Urban energy generation: Influence of micro-wind turbine output on electricity consumption in buildings. *Energy Build.* **2007**, *39*, 154–165. [CrossRef]
2. Toja-Silva, F.; Colmenar-Santos, A.; Castro-Gil, M. Urban wind energy exploitation systems: Behaviour under multidirectional flow conditions—Opportunities and challenges. *Renew. Sustain. Energy Rev.* **2013**, *24*, 364–378. [CrossRef]
3. Ishugah, T.F.; Li, Y.; Wang, R.Z.; Kiplagat, J.K. Advances in wind energy resource exploitation in urban environment: A review. *Renew. Sustain. Energy Rev.* **2014**, *37*, 613–626. [CrossRef]
4. Tummala, A.; Velamati, R.K.; Sinha, D.K.; Indraja, V.; Krishna, V.H. A review on small scale wind turbines. *Renew. Sustain. Energy Rev.* **2016**, *56*, 1351–1371. [CrossRef]
5. Kumar, R.; Raahemifar, K.; Fung, A.S. A critical review of vertical axis wind turbines for urban applications. *Renew. Sustain. Energy Rev.* **2018**, *89*, 281–291. [CrossRef]
6. Stathopoulos, T.; Alrawashdeh, H.; Al-Quraan, A.; Blocken, B.; Dilimulati, A.; Paraschivoiu, M.; Pilay, P. Urban wind energy: Some views on potential and challenges. *J. Wind Eng. Ind. Aerodyn.* **2018**, *179*, 146–157. [CrossRef]
7. Toja-Silva, F.; Kono, T.; Peralta, C.; Lopez-Garcia, O.; Chen, J. A review of computational fluid dynamics (CFD) simulations of the wind flow around buildings for urban wind energy exploitation. *J. Wind Eng. Ind. Aerodyn.* **2018**, *180*, 66–87. [CrossRef]
8. Anup, K.C.; Whale, J.; Urmee, T. Urban wind conditions and small wind turbines in the built environment: A review. *Renew. Energy* **2019**, *131*, 268–283.
9. Carbó Molina, A.; De Troyer, T.; Massai, T.; Vergaerde, A.; Runacres, M.C.; Bartoli, G. Effect of turbulence on the performance of VAWTs: An experimental study in two different wind tunnels. *J. Wind Eng. Ind. Aerodyn.* **2019**, *193*, 103969. [CrossRef]
10. Wood, D. *Small Wind Turbines*; Green Energy and Technology; Springer: London, UK, 2011; ISBN 978-1-84996-174-5.
11. Manwell, J.F.; McGowan, J.G.; Rogers, A.L. *Wind Energy Explained: Theory, Design and Application*; John Wiley & Sons: Hoboken, NJ, USA, 2010; ISBN 9780470015001.
12. Allen, S.R.; Hammond, G.P.; McManus, M.C. Prospects for and barriers to domestic micro-generation: A United Kingdom perspective. *Appl. Energy* **2008**, *85*, 528–544. [CrossRef]
13. Islam, M.; Ting, D.S.K.; Fartaj, A. Aerodynamic models for Darrieus-type straight-bladed vertical axis wind turbines. *Renew. Sustain. Energy Rev.* **2008**, *12*, 1087–1109. [CrossRef]
14. Jin, X.; Zhao, G.; Gao, K.; Ju, W. Darrieus vertical axis wind turbine: Basic research methods. *Renew. Sustain. Energy Rev.* **2015**, *42*, 212–225. [CrossRef]
15. Ghasemian, M.; Ashrafi, Z.N.; Sedaghat, A. A review on computational fluid dynamic simulation techniques for Darrieus vertical axis wind turbines. *Energy Convers. Manag.* **2017**, *149*, 87–100. [CrossRef]
16. Kumar, P.M.; Sivalingam, K.; Narasimalu, S.; Lim, T.-C.; Ramakrishna, S.; Wei, H. A review on the evolution of darrieus vertical axis wind turbine: Small wind turbines. *J. Power Energy Eng.* **2019**, *7*, 27–44. [CrossRef]
17. Abraham, J.P.; Plourde, B.D.; Mowry, G.S.; Minkowycz, W.J.; Sparrow, E.M. Summary of Savonius wind turbine development and future applications for small-scale power generation. *J. Renew. Sustain. Energy* **2012**, *4*, 042703. [CrossRef]

18. Akwa, J.V.; Vielmo, H.A.; Petry, A.P. A review on the performance of Savonius wind turbines. *Renew. Sustain. Energy Rev.* **2012**, *16*, 3054–3064. [CrossRef]
19. Roy, S.; Saha, U.K. Review on the numerical investigations into the design and development of Savonius wind rotors. *Renew. Sustain. Energy Rev.* **2013**, *24*, 73–83. [CrossRef]
20. Kang, C.; Liu, H.; Yang, X. Review of fluid dynamics aspects of Savonius-rotor-based vertical-axis wind rotors. *Renew. Sustain. Energy Rev.* **2014**, *33*, 499–508. [CrossRef]
21. Alom, N.; Saha, U.K. Evolution and progress in the development of savonius wind turbine rotor blade profiles and shapes. *J. Sol. Energy Eng. Trans. ASME* **2019**, *141*, 1–15. [CrossRef]
22. Elkhoury, M.; Kiwata, T.; Nagao, K.; Kono, T.; ElHajj, F. Wind tunnel experiments and Delayed Detached Eddy Simulation of a three-bladed micro vertical axis wind turbine. *Renew. Energy* **2018**, *129*, 63–74. [CrossRef]
23. Kiwata, T. Vertical axis wind turbine with variable-pitch straight blades. In Proceedings of the International Conferemce on Jets, Wakes and Separated Flows, ICJWSF-2017, Cincinnati, OH, USA, 9–12 October 2017; pp. 1–6.
24. Shimizu, Y.; Maeda, T.; Kamada, Y.; Takada, M.; Katayama, T. Development of micro wind turbine (Orthoptere Wind Turbine). In Proceedings of the International Conference on Fluid Engineering JSME Centennial Grand Congress, Tokyo, Japan, 13–16 July 1997; pp. 1551–1556.
25. Bayeul-Laine, A.-C.; Simonet, S.; Dockter, A.; Bois, G. Numerical study of flow stream in a mini VAWT with relative rotating blades. In Proceedings of the 22nd International Symposium on Transport Phenomena Conference, Delft, The Netherlands, 8–11 November 2011; p. 13.
26. Cooper, P.; Kennedy, O.C.; Cooper, P.; Kennedy, O. Development and analysis of a novel vertical axis wind turbine development and analysis of a novel vertical axis wind turbine. In Proceedings of the Solar 2004: Life, The Universe and Renewables, Perth, Australia, 30 November–3 December 2004; pp. 1–9.
27. Kono, T.; Kogaki, T.; Kiwata, T. Numerical investigation ofwind conditions for roof-mounted wind turbines: Effects of wind direction and horizontal aspect ratio of a high-rise cuboid building. *Energies* **2016**, *9*, 907. [CrossRef]
28. Balduzzi, F.; Bianchini, A.; Ferrari, L. Microeolic turbines in the built environment: Influence of the installation site on the potential energy yield. *Renew. Energy* **2012**, *45*, 163–174. [CrossRef]
29. Ferreira, C.J.S.; Van Bussel, G.J.W.; Van Kuik, G.A.M. Wind tunnel hotwire measurements, flow visualization and thrust measurement of a VAWT in skew. *J. Sol. Energy Eng. Trans. ASME* **2006**, *128*, 487–497. [CrossRef]
30. Burton, T.; Jenkins, N.; Sharpe, D.; Bossanyi, E. *Wind Energy Handbook*, 2nd ed.; John Wiley & Sons: Hoboken, NJ, USA, 2011; ISBN 9780470699751.
31. ANSYS, Inc. *ANSYS Fluent Theory Guide, Release 17.2*; ANSYS, Inc.: Canonsburg, PA, USA, 2016.
32. ANSYS, Inc. *ANSYS Fluent User Guide, Release 17.2*; ANSYS, Inc.: Canonsburg, PA, USA, 2016.
33. ElCheikh, A.; Elkhoury, M.; Kiwata, T.; Kono, T. Performance analysis of a small-scale orthopter-type vertical axis wind turbine. *J. Wind Eng. Ind. Aerodyn.* **2018**, *180*, 19–33. [CrossRef]
34. Spalart, P.R.; Jou, W.H.; Strelets, M.; Allmaras, S.R. Comments on the feasibility of LES for wings, and on a hybrid RANS/LES approach. In Proceedings of the First AFOSR International Conference on DNS/LES, Ruston, LA, USA, 4–8 August 1997; pp. 137–147.
35. Tominaga, Y.; Mochida, A.; Yoshie, R.; Kataoka, H.; Nozu, T.; Yoshikawa, M.; Shirasawa, T. AIJ guidelines for practical applications of CFD to pedestrian wind environment around buildings. *J. Wind Eng. Ind. Aerodyn.* **2008**, *96*, 1749–1761. [CrossRef]
36. Wu, H.; Stathopoulos, T. Wind-tunnel techniques for assessment of pedestrian-level winds. *J. Eng. Mech.* **1993**, *119*, 1920–1936. [CrossRef]
37. Tian, W.; Mao, Z.; Zhang, B.; Li, Y. Shape optimization of a Savonius wind rotor with different convex and concave sides. *Renew. Energy* **2018**, *117*, 287–299. [CrossRef]

5

Impact of Economic Indicators on the Integrated Design of Wind Turbine Systems

Jianghai Wu, Tongguang Wang *, Long Wang and Ning Zhao

Jiangsu Key Laboratory of Hi-Tech Research for Wind Turbine Design, Nanjing University of Aeronautics and Astronautics, Nanjing 210016, China; jianghaiwu@nuaa.edu.cn (J.W.); longwang@nuaa.edu.cn (L.W.); zhaoam@nuaa.edu.cn (N.Z.)

* Correspondence: tgwang@nuaa.edu.cn

Abstract: This article presents a framework to integrate and optimize the design of large-scale wind turbines. Annual energy production, load analysis, the structural design of components and the wind farm operation model are coupled to perform a system-level nonlinear optimization. As well as the commonly used design objective levelized cost of energy (LCoE), key metrics of engineering economics such as net present value (NPV), internal rate of return (IRR) and the discounted payback time (DPT) are calculated and used as design objectives, respectively. The results show that IRR and DPT have the same effect as LCoE since they all lead to minimization of the ratio of the capital expenditure to the energy production. Meanwhile, the optimization for NPV tends to maximize the margin between incomes and costs. These two types of economic metrics provide the minimal blade length and maximal blade length of an optimal blade for a target wind turbine at a given wind farm. The turbine properties with respect to the blade length and tower height are also examined. The blade obtained with economic optimization objectives has a much larger relative thickness and smaller chord distributions than that obtained for high aerodynamic performance design. Furthermore, the use of cost control objectives in optimization is crucial in improving the economic efficiency of wind turbines and sacrificing some aerodynamic performance can bring significant reductions in design loads and turbine costs.

Keywords: wind turbine; wind turbine design; optimization; blade length; economic analysis

---

1. Introduction

Wind turbine design is a complex task comprising multiple disciplines, requiring a trade-off between many conflicting objectives. Many research articles have been published to achieve an optimal turbine design. Some use a single objective such as maximum annual energy production (AEP) or maximum AEP per turbine weight to carry out a single-objective optimization [1]. Others use multi-objective methods [2,3] or a multi-level system design [4] to accomplish a balance between different conflicting objectives, often drawn from different scientific and economic disciplines. The objective functions of wind turbine design can be divided in four main categories: Maximization of the energy production, minimization of the blade mass, minimization of the cost of energy, and multi-objective optimization [5].

Multi-objective optimization offers a set of Pareto Optimal design solutions and places the burden of choice on the shoulders of the decision maker. In contrast, a single objective lumps all different objectives into one and provides a unique design result to the decision maker, which seems to be more practical. The difficulty is that the objective must reflect the nature of the problem. Inappropriate design goal will lead to unfeasible results. For example, maximization of the power coefficient results in larger root chords and a very high blade twist [5], while minimization of mass/AEP may overemphasize the role of the tower [6]. Hence, the selection of objective is crucial in wind turbine design.

It is noticed that the development of a wind farm is essentially an investment activity and wind turbine design is an upstream stage of wind farm development. In this case, turbine optimization also needs to be performed from an engineering economics point of view. Many economic functions such as levelized cost of energy (LCoE), net present value (NPV), internal rate of return (IRR) and discounted payback time (DPT) have been used to evaluate the profitability of wind farms [7,8]. The LCoE represents the minimum energy price that meets the desired interest rate by the designers, NPV defines the total profit of the wind farm and takes into account the price of energy, IRR is the interest rate that sets the NPV function equal to zero and enables checking if a minimum rate of return set by the designers is met, and DPT determines the time required to cover the initial investment while taking into account the time value of the money [9,10]. In these economic indicators, LCoE is the most common one used as an objective in wind turbine design [11–13]. But sometimes LCoE alone is not a sufficient measure to determine a project's profitability or competitiveness. Investors need other parameters as inputs to make investment decisions [10]. However, there has been no research published using other economic indicators as objectives to perform a wind turbine design. Therefore, besides LCoE, economic functions like NPV, IRR and DPT will also be utilized as design objectives in this study. Insights into the differences between these metrics and the impact of these economic functions when used as optimization objectives will be assessed.

To carry out this research, a system-level optimization framework is used to perform an integrated design of a wind turbine. Blade length, the geometry of the blade and the tower height are selected as design variables. The purpose of this study is not to develop a design methodology, or present exact parameters for an optimized wind turbine, but rather to assess the effect of different objective functions on the properties of optimal wind turbines.

2. Methodology

2.1. Calculation Tools

Recent developments in the methodology of wind turbine design have mainly focused on simultaneous evaluation of aerodynamic and structural design [14–17]. In this paper a series of automated calculation models are coupled to perform an integrated optimization. The integrated methodology mainly builds upon that previously described in References [6,16], which includes the rotor aerodynamic analysis, blade structure design and cost model. These models are selected as a compromise between computational effort and calculation accuracy. The integrated design strategy has been successfully used in the design of a 1.5 MW stall regulated rotor [14], BONUS 1 MW and WM 600 wind turbine [18], and the investigation of 5 MW upwind and downwind turbines [16], proved to be effective in the wind turbine optimization.

In the next subsection we describe how the integrated modelling tool is implemented, giving details of how the models are coupled and how the optimization process is sequenced.

2.1.1. Power Production

Aerodynamic analysis is based on ordinary blade element/momentum (BEM) theory. A detailed description of the process with an extensive explanation of its fundamental equations can be found in Reference [19]. AEP is computed using a Rayleigh distribution in which mean speed varies with the hub height. An exponential law is used to calculate the average wind speed at different hub heights as follows:

$$V(h) = V(h_0)\left(\frac{h}{h_0}\right)^{\alpha} \quad (1)$$

where h_0 is the reference height, $V(h_0)$ is the average wind speed at h_0 and α is the wind shear exponent.

2.1.2. Loading

Load calculation is an important discipline within wind turbine technology. Design loads serve as inputs for the structural design of rotor blades and other turbine components, and essentially determine the cost of producing a wind turbine. Typically, load analysis is performed using dynamic simulation software such as "DNV Bladed" according to the IEC61400-1 standard [20] during the real-world wind turbine design process. However, this load estimation procedure is difficult to carry out in the optimization period for the following reasons:

1. For an accurate estimation of design loads, there are thousands of design load cases (DLCs) which must be simulated, causing a significant computational cost. Design optimization of a wind turbine requires estimations of hundreds of thousands of different turbine configurations. For this reason, a full IEC loads analysis is not computationally feasible.
2. The output loads are influenced by the controller algorithm. For example, coefficients in the PID controller and pitch rates under different operational conditions can have a significant effect. Typically, we must modify many parameters of the controller to find the best controller algorithm and configuration for a given wind turbine. This is another computationally expensive aspect which limits the usefulness of automated design load analysis.

Some approaches have been described to deal with this problem. For example, some essential load cases can be selected to reduce the computational cost [18]. It is also possible to use the static forces of rotor and tower at some critical conditions, with an amplification factor to correct for dynamic effects [16]. In accordance with the latter approach, we explored a series of load calculation results from commercial wind turbines, correlated the conditions a wind turbine experiences when ultimate loads occur (rotor speed, wind speed, pitch angle, azimuth angle, etc.), and then identified the appropriate static load cases (SLCs) to perform the loads estimation, which are summarized in Table 1.

Table 1. Static conditions for load estimation.

Static Load Cases	Wind Speed (m/s)	Rotor Speed (rpm)	Pitch Angle (deg)	Yaw Angle (deg)	Azimuth Angle (deg)
SLC 1.1	$V_r + 3\sigma$	$1.1\ \omega_r$	0~10	−8~+8	0~90
SLC 1.2	$V_{out} + 3\sigma$	ω_r	10~20	−8~+8	0~90
SLC 2.1	V_{e1}	0	90	90,270	0
SLC 2.2	V_{e50}	0	90	30,330	0

Note: V_r: rated wind speed; V_{out}: cut-out wind speed; σ: standard deviation of the wind speed, relates to short term turbulence; ω_r: rated rotor speed; V_{e1}: 10-min average extreme wind speed with a recurrence period of 1 year; Ve50: 10-min average extreme wind speed with a recurrence period of 50 years.

SLC 1.1 represents an extremely gusty wind condition. The ultimate loads on a wind turbine are more likely to happen in gusty conditions near the rated wind speed, typically due to a rapid increase of wind speed which the pitch angle can't catch up with. SLC 1.2 examines the operational condition IEC DLC 1.3, with an extreme turbulence model. In this case the maximum wind speed will increase to a very high value in a short period due to extreme turbulence at the cut-out wind speed. SLC 2.1 and 2.2 represent the parked survival cases of wind turbines under extreme conditions which occur every 1 year and every 50 years, respectively.

The predicted static thrust from the blade requires a dynamic amplification factor of 1.45 to fall within the range of dynamic loadings from "DNV Bladed" simulations. To check the feasibility of this assumption, we carried out several load estimations for different wind turbines with various blade lengths. Figure 1 shows the comparison of blade root flap-wise moments (Mf) between the SLCs and a more rigorous DLCs calculation procedure.

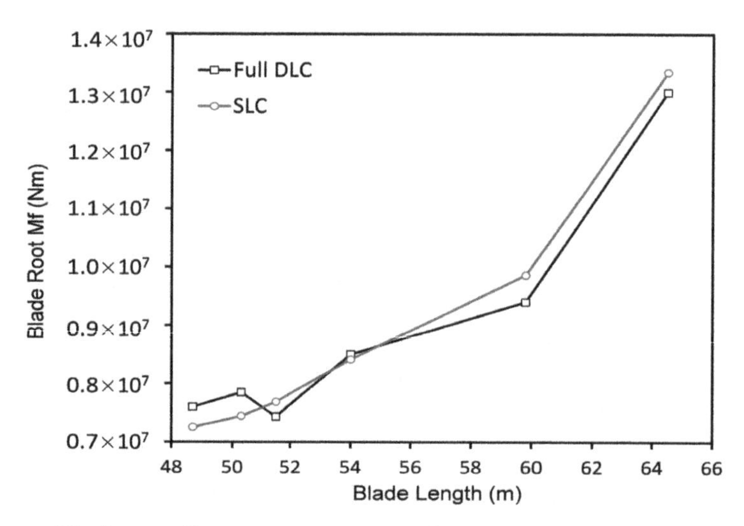

Figure 1. Blade root flap-wise moments of SLC and full DLC calculations.

Although there is some deviation between the calculated loads, the errors are within an acceptable range. Usually the output loads on a turbine blade will increase with the blade length, however, the load evaluation of real-world turbines shows that sometimes longer blades can have lower output loads. This is mostly due to variations in the turbine configuration and control algorithms. If this influence is eliminated, we can expect that the error between the load trends in the DLC and SLC calculations would be reduced. Moreover, since the load calculations as well as the cost model described in the next section inevitably need to be simplified, the primary goal of this study is not to present an exact geometry for an optimized wind turbine, but to assess the effect of different objective functions on the properties of optimal wind turbines. For this purpose, the SLC method is more suitable.

2.1.3. Cost Model

Besides AEP, capital expenditure (CAPEX), operational expenditures (OPEX) and financing are needed as inputs to perform overall economic analysis. These are calculated using the models described in this section.

The CAPEX herein represents the overall cost of a single wind turbine. It includes turbine manufacturing cost and the sharing portion of wind farm construction and maintenance costs. The manufacturing cost uses the individual component masses for the turbine to evaluate the costs of individual components. The blade mass is derived by a structural model using beam finite element theory and classical laminate theory to determine the effect of the design loads [2]. The blade structure is formed from a double web I-beam. The material used for the bulk of the blade is glass fiber with reinforced polyester and foam.

A set of physics-based models are used to estimate the sizes and costs of a subset of the major load-bearing components (the main shaft, gear box, bedplate, gearbox, tower) and parametric formulations representative of current wind turbine technology are used to evaluate sizes and costs for the remaining components (the hub and yaw system). For example, the hub is treated as a thin-walled, ductile, cast iron cylinder with holes for blade root openings and main shaft flanges. The bedplate module is separated into two distinct front and rear components. The rear frame is modelled as two parallel steel I-beams and the front frame is modelled as two parallel ductile cast iron I-beams. The physics-based models have internal iteration schemes based on system constraints and design criteria, which are described in Reference [21].

The balance-of-station costs vary with the rotor diameter and tower height, and the OPEX are accounted for by taking a small, fixed percentage of the capital cost as the annual maintenance cost [18]. Decommissioning costs are also considered in this model.

2.2. *Objective Function*

The appropriate choice of an objective function is critical to an optimization process. In order to perform the wind turbine optimization from an economic point of view and ensure the designs correctly assess fundamental trade-offs (primarily between energy capture and overall cost) we chose LCoE, NPV, IRR and DPT as design metrics. The models employed in this research are proposed in Reference [10] and shown in Table 2. Dynamic evaluation of these economic functions can also be found in References [22,23].

Table 2. Economic functions for the wind turbine optimization problem.

Function	Equation	Parameters
LCoE ($/kWh)	$\frac{1}{AEP}(\frac{CAPEX}{a}+OPEX)$	a—annuity factor ($a=\frac{1-(1+r)^{-n}}{r}$), r—interest rate, n—wind farm lifetime
NPV ($)	$(AEP\cdot p_{kWh}-OPEX)a-CAPEX$	p_{kWh}—market energy price
IRR (%)	$(AEP\cdot p_{kWh}-OPEX)\frac{1-(1+r_{IRR})^{-n}}{r_{IRR}}=CAPEX$	r_{IRR}—interest rate that zeroes the NPV equation
DPT (years)	$\frac{n\cdot CAPEX}{(AEP\cdot p_{kWh}-OPEX)\cdot a}$	

In the economic models, the total utilized energy output and the total cost over the lifetime of the wind turbine are both discounted to the start of operation.

2.3. *Design Variables and Constraints*

There are a total of 15 design variables used in the optimization process. B-Spline curves are used to create smooth distributions of chord, twist and relative thickness of the blade. Five control points are used to shape the chord distribution and a further four points are used to describe both twist and relative thickness distribution, as shown in Figure 2.

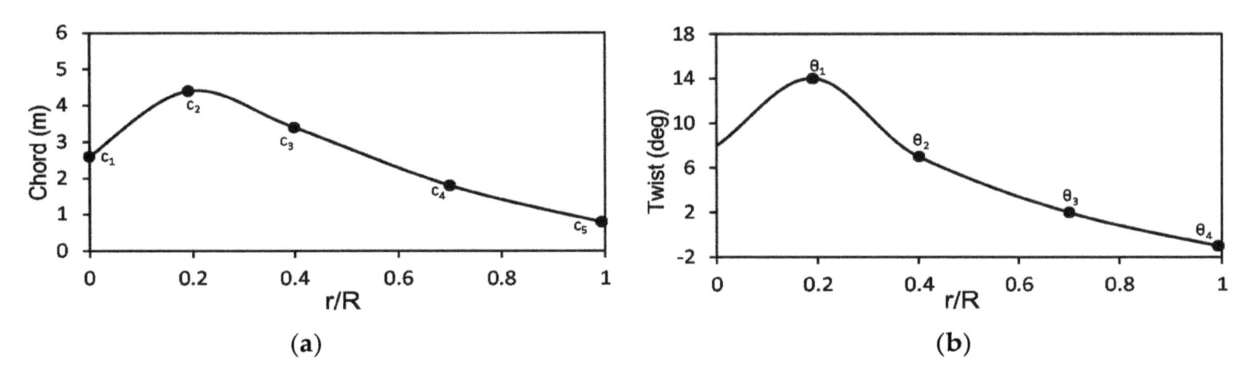

Figure 2. Design variables of blade geometry. (**a**) Chord distribution; (**b**) Twist distribution.

Blade length and tower height are also set as variables for the purpose of determining the most suitable rotor diameter and tower height for a specific wind speed site. Upper and lower limits for these variables are implemented to define the bounds of the design space.

The DU and NACA 64 series airfoils are employed in this study, as shown in Table 3. These airfoils are optimized for high speed wind condition with an advantage of high maximum lift coefficient and very low drag over a small range of operation condition. They have been used by various wind turbine manufacturers worldwide in over 10 different rotor blades for turbines with rotor diameters ranging from 29 m to over 100 m [24].

Blade structure design is set to meet the needs of strength and deflection constraints. The maximum velocity of the blade tip is 80 m/s. The rated rotational speed and the gearbox ratio change with the blade length in proportion to keep the generator operational speed the same. To have a safe blade-tower-clearance, a constraint for the blade out-of-plane deflection is used and the clearance between the ground and the blade tip is constrained by a minimum of 25 m [25].

Table 3. Airfoil family.

Airfoil Name	t/c Ratio
DU99-W-405	40%
DU99-W-350	35%
DU97-W-300	30%
DU91-W2-250	25%
DU93-W-210	21%
NACA 64-618	18%

To examine wind turbine designs at low speed sites, the average wind speed is set as 6 m/s at a reference height of 70 m. The wind shear exponent α is set to be 0.2 for normal wind profile model according to the IEC61400-1 standard [20]. The power transfer efficiency is 92% and losses such as those caused by wake interference, electrical grid unavailability and air density correction are estimated using an array loss factor and an availability factor. An overall correction factor of 0.64 is used in this research.

2.4. Genetic Algorithm

The genetic algorithm (GA) is a search procedure based on genetics and natural selection mechanisms, which has proved to be efficient and robust for wind turbine system [26], wind turbine layout in wind farms [27], and offshore wind turbine support structures [28].

In this study, multi island genetic algorithm (MIGA) is employed for the optimization. In MIGA, the population is divided into several subpopulations staying on isolated "islands," whereas traditional genetic algorithm operations are performed on each subpopulation separately. A certain number of individuals between the islands migrate after a certain number of generations. Thus, MIGA can prevent the problem of "premature" by maintaining the diversity of the population [29]. Table 4 presents the main parameters of MIGA.

Table 4. Main parameters of MIGA.

Item	Value
Sub-Population Size	10
Number of Islands	5
Number of Generations	150
Rate of Crossover	0.8
Rate of Mutation	0.01
Rate of Migration	0.3
Interval of Migration	5

3. Description of the Optimization Process

All sequential programming methods and the optimization processes are formulated in a framework. Figure 3 shows the full optimization methodology adopted in this study. Parameterization, geometry generation, AEP and load evaluation, structural design, mass and cost estimation, and economic analysis are presented in the flow chart.

Objectives such as LCoE and NPV are specified at the beginning of the optimization process. The AEP and design loads are evaluated by BEM theory, followed by calculation of the component masses and costs with the consideration of the design loads. The turbine capital cost and AEP together with the wind farm operational model are used to obtain the economic characteristic during the whole wind turbine life cycle as the final design target. The variables are changed in the optimization depending on the design objective. The maximum number of iterations is the stopping criterion. When stopping criterion is reached, the design result is picked out by the algorithm.

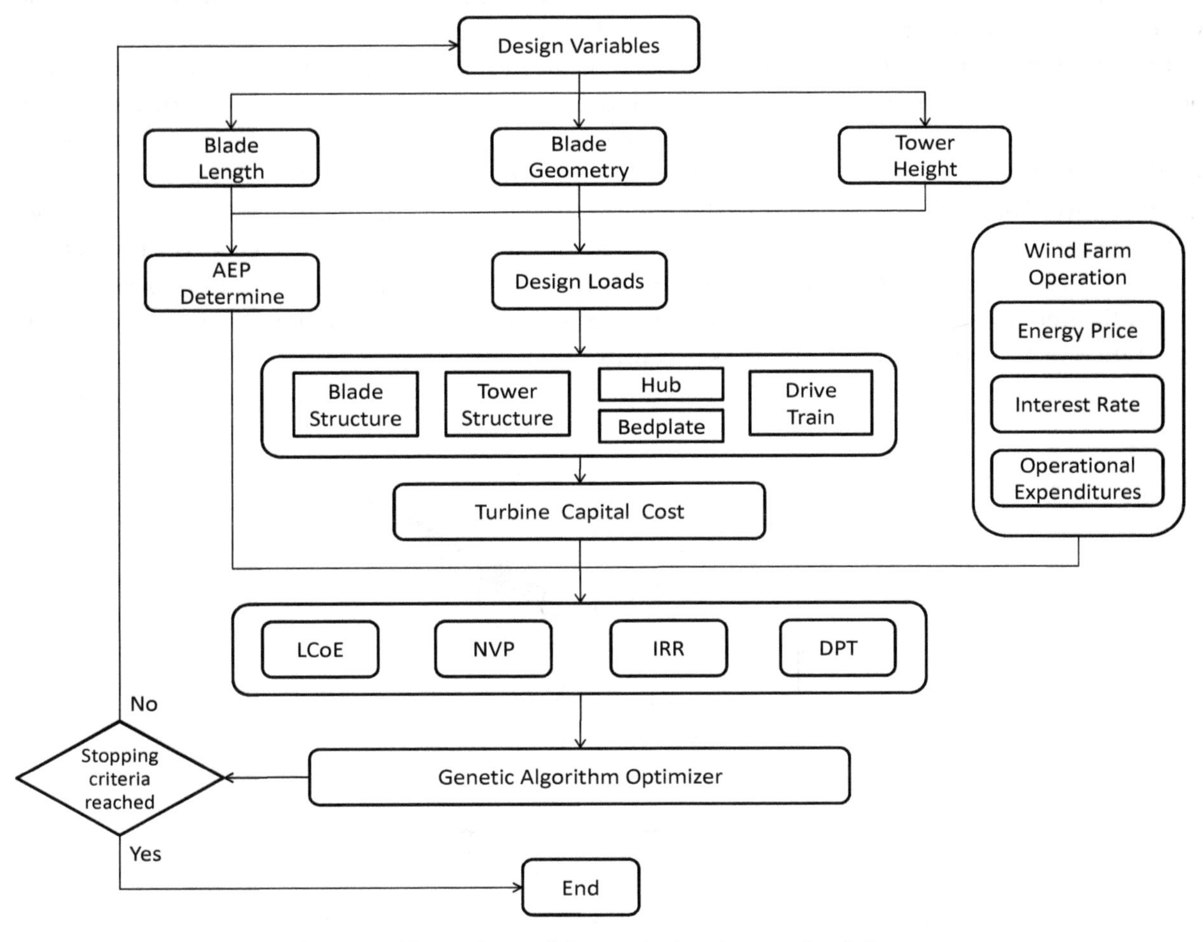

Figure 3. Flow chart of the optimization methodology.

4. Results and Discussion

4.1. 2 MW Case Study

Recently, increasing numbers of low wind speed sites have been developed, and the 2 MW turbine is dominating the market (In 2017, the average rated power of the new installed wind turbine in China is 2.1 MW [30] and China alone accounted for 37% of the world's new installed capacity [31]). Thus, this paper will focus on the optimization of 2 MW low wind speed turbines to capture the industrial trends of the state-of-the-art wind farm.

As the market energy price, labor costs, material costs, etc. vary between countries, the parameters considered for this case study are mainly derived from the industry's status in China. For example, the NPV is calculated with a market energy price of 8.8 cents/kWh, the weight unit price of the blade is set to be $5.88/kg and $1.54/kg for the tower. A discount rate of 8% and an economic lifetime of 20 years are employed in this study.

The results of the optimization are presented in the following sections. Optimization outputs using different design targets are examined. Fundamental differences and observations from each case are discussed below.

4.2. *Effects of Blade Length*

Increasing blade length will increase energy production, but it will also increase the costs of material, labor and delivery. Generally, the initial capital cost and maintenance cost will also increase.

To analyze how blade length impacts the economic effectiveness of a design, we recorded the trends in energy production, loading, mass, and costs from the properties of all the intermediate cases during the optimization. These data points are results for all metrics put together, show clearly the relationships between various turbine properties and the blade length.

The annual capacity factor is commonly used as an indicator of energy performance. It is defined as the energy generated during the year divided by wind turbine rated power multiplied by the number of hours in the year. Figure 4 shows the capacity factor against blade length. Each point in the scatter plot represents an individual turbine configuration evaluated during the genetic algorithm. Although wind energy is proportional to the square of blade length, due to the limitation of generator capacity, the output electrical power is not permitted to exceed the rated power beyond the rated wind speed. The total energy production increases with the blade length while the effect is reduced as the blade length increases. The capital expenditure also increases with blade length, however in this case the effect increases as blade length increases (Figure 5).

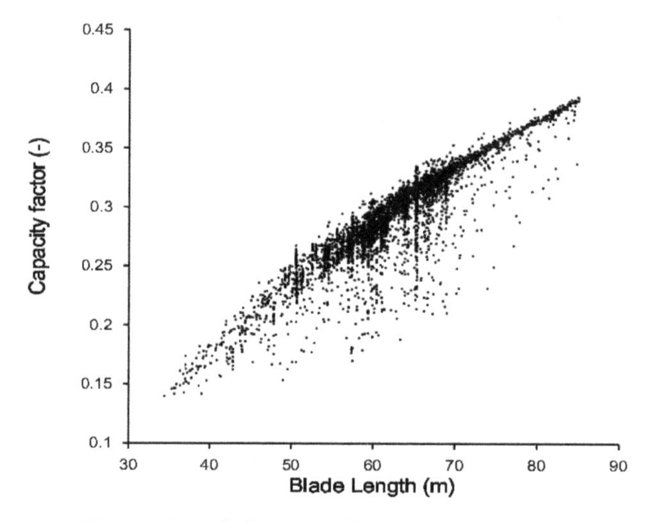

Figure 4. Equivalent full-loaded hours against blade length.

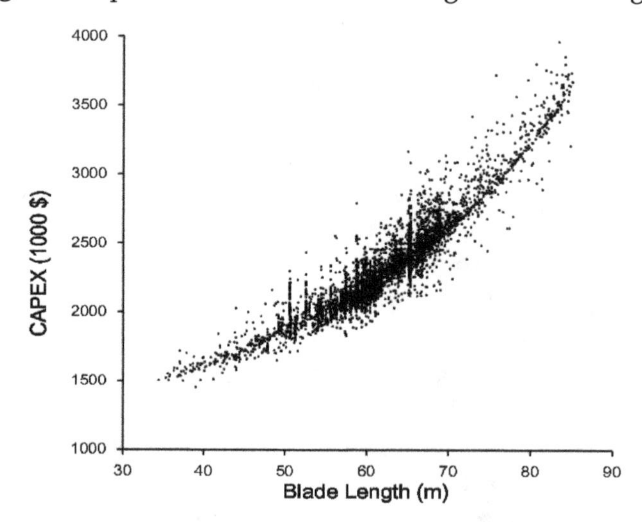

Figure 5. Initial capital cost against blade length.

The same comparison was made for blade length against blade root loading and against blade mass, as shown in Figures 6 and 7 respectively. Power law relationships are fitted to this data to quantify the correlations. The flap-wise bending moment (Mf) is the critical load in blade structure design. This research shows that the flap-wise bending moment of blade root increases with length, L, as $L^{2.84}$, while the mass of the blades scales with $L^{2.52}$. The enveloping functions on the outer border of the scattered points are also presented to indicate the boundary of the design results.

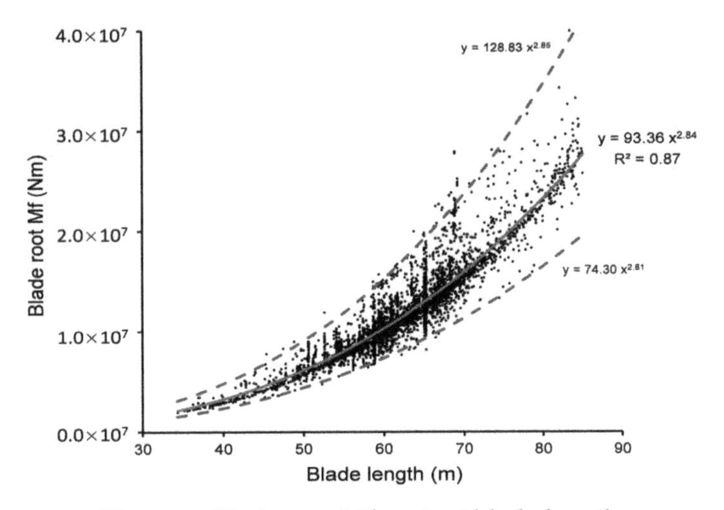

Figure 6. Blade root Mf against blade length.

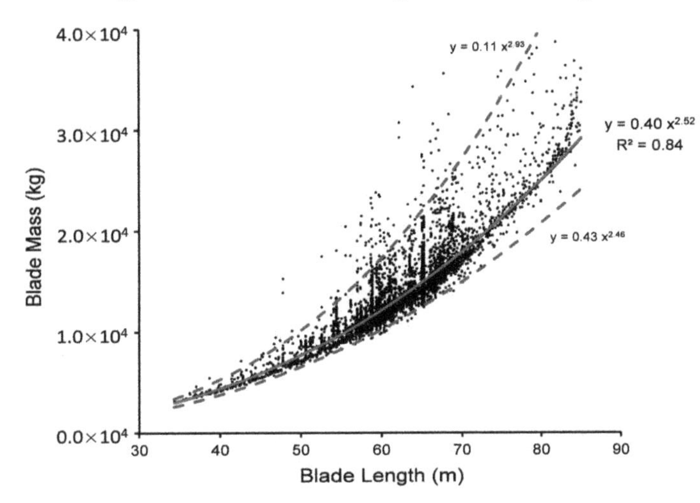

Figure 7. Blade mass against blade length.

Torsional moment (Mx) under fixed hub coordinates greatly influences the gearbox design so this is examined closely in this work. The optimization data trend predicts that Mx increases linearly with L as depicted in Figure 8. Meanwhile, Figure 9 shows the gearbox mass scales as $L^{1.78}$. The loading and mass of the remaining components such as the low speed shaft (Figure 10) and the bed-plate (Figure 11) each have different scale exponents as described in the figures.

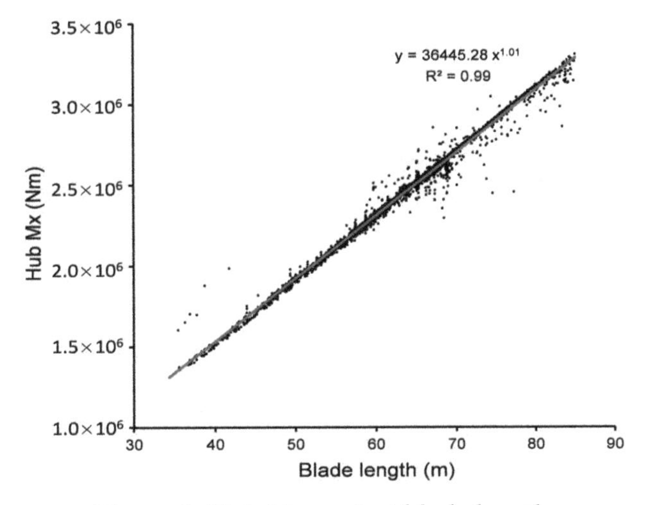

Figure 8. Hub Mx against blade length.

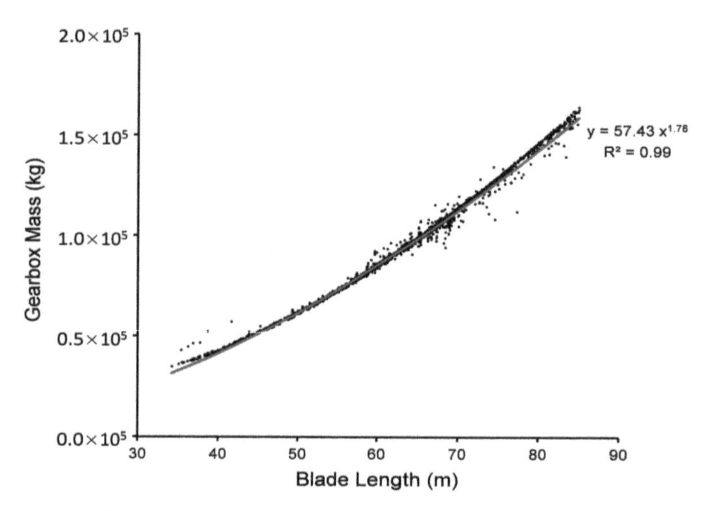

Figure 9. Gearbox mass against blade length.

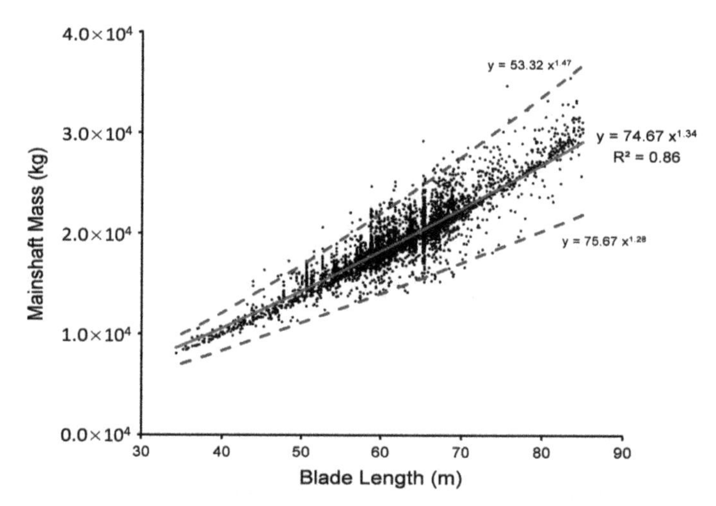

Figure 10. Mainshaft mass against blade length.

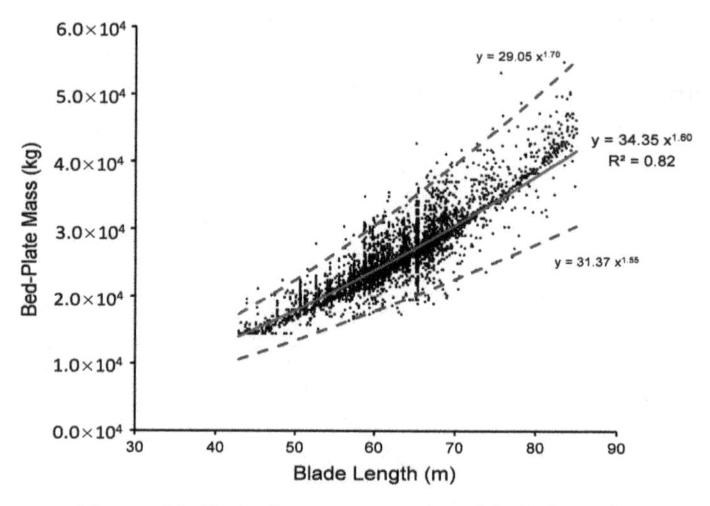

Figure 11. Bed-plate mass against blade length.

In the scatter plots of economic indicators as a function of blade length, the boundaries of these scatter cloud appear in the form of a parabola. The data points to an optimal blade length for each of these economic metrics are shown in Figures 12–15.

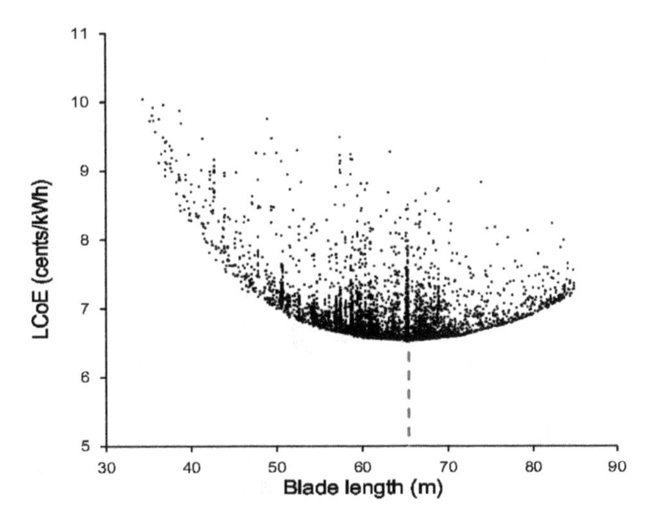

Figure 12. LCoE against blade length.

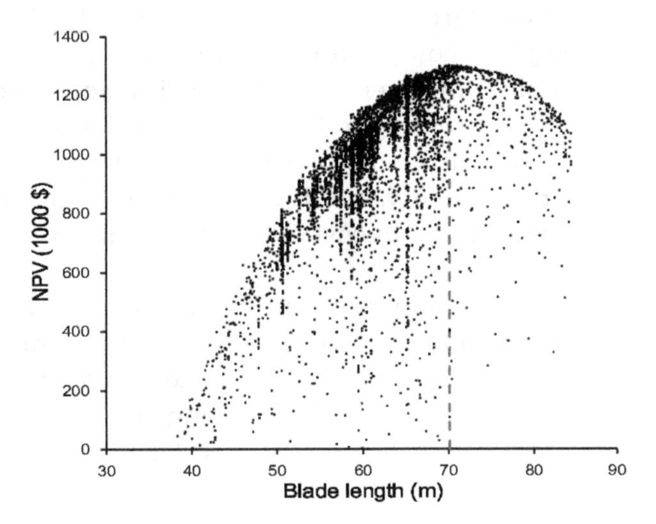

Figure 13. NPV against blade length.

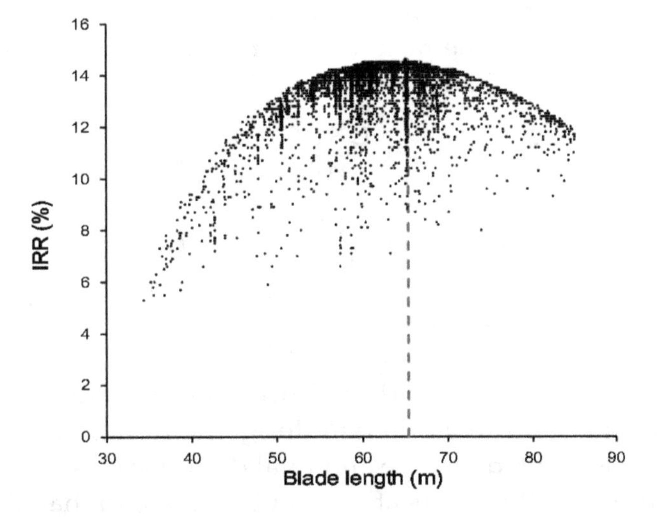

Figure 14. IRR against blade length.

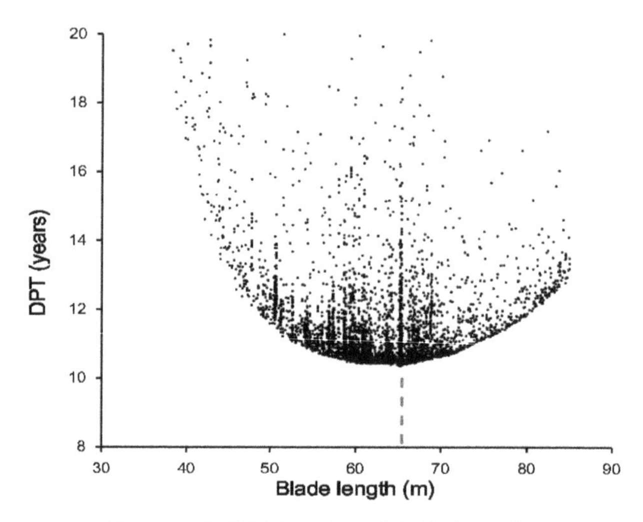

Figure 15. DPT against blade length.

A large proportion of the initial capital costs are unrelated to the rotor diameter or design loads, for example the generator, converter and control system costs. This leads to a higher LCoE and lower NPV when blade length is comparatively short. For example, a 30 m–38 m blade is uneconomical for a 2 MW turbine since the energy production is inadequate compared with the costs (Figure 13). As the blade length increases beyond 38 m, the benefits outweigh the economic costs. The economic efficiency improves until it reaches a maximum at a certain blade length. Beyond this length, the turbine may produce more energy, but due to higher manufacturing effort and increased loading on many components, the increase in cost exceeds the improvement in AEP, resulting in a lower economic efficiency.

As a result, LCoE first decreases then increases with blade length. The characteristic of NPV is just the opposite. Optimal blade length and corresponding values for LCoE and NPV can be observed. LCoE achieves a minimal value of 6.54 cents/kWh at a blade length of 65.4 m, while the NPV reaches a maximum value of $1,295,000 when the blade is 70.1 m long. The same configuration that had the lowest LCoE also presented the best values for the IRR and DPT. The highest IRR is 14.6% and the shortest DPT is 10.4 years.

4.3. Economic Functions

Based on the equations presented in Table 3, as previously assumed, OPEX are accounted for by taking a fixed percentage of the capital cost ($OPEX = f \cdot CAPEX$), after some mathematical transformation, it can be found that the task of minimization of LCoE, maximization of IRR and minimization of DPT can be considered as a task to minimize the ratio between CAPEX and the AEP:

$$LCoE|_{\min} \Rightarrow \frac{CAPEX}{AEP}(\frac{1}{a} + f)|_{\min} \tag{2}$$

$$IRR|_{\max} \Rightarrow \frac{AEP \cdot p_{kwh}}{CAPEX} - f|_{\max} \tag{3}$$

$$DPT|_{\min} \Rightarrow \frac{1}{\frac{AEP \cdot p_{kwh}}{CAPEX} - f}|_{\min} \tag{4}$$

Consequently, optimum LCoE, IRR and DPT occur at the same turbine configuration while optimum NPV is achieved at an entirely different design point. Maximum NPV is achieved when the margins between total discounted income and total discounted costs are maximized. As seen in Figures 12 and 13, the optimum NPV occurs at a larger blade length than the optimum LCoE.

Figure 16 illustrates the variation of total discounted incomes of different wind turbines as a function of their total discounted costs. The dashed line in the middle is the break-even line, while turbines above the line are economic and those below the line are not. Two points have been highlighted on the cost-income front. Point A, with total discounted costs of \$3.9 m, has the highest net profit margin between incomes and costs which implies the maximum NPV is achieved. The investment returns of designs with costs higher than point A decline. Point B, with total discounted costs of \$3.5 m has the maximum slope which indicates that the ratio between incomes and costs is maximized. This point corresponds to minimum LCoE. In this optimization problem, a wind turbine designed with the objective of minimum LCoE tends to choose a smaller-scale investment compared with the objective of maximum NPV.

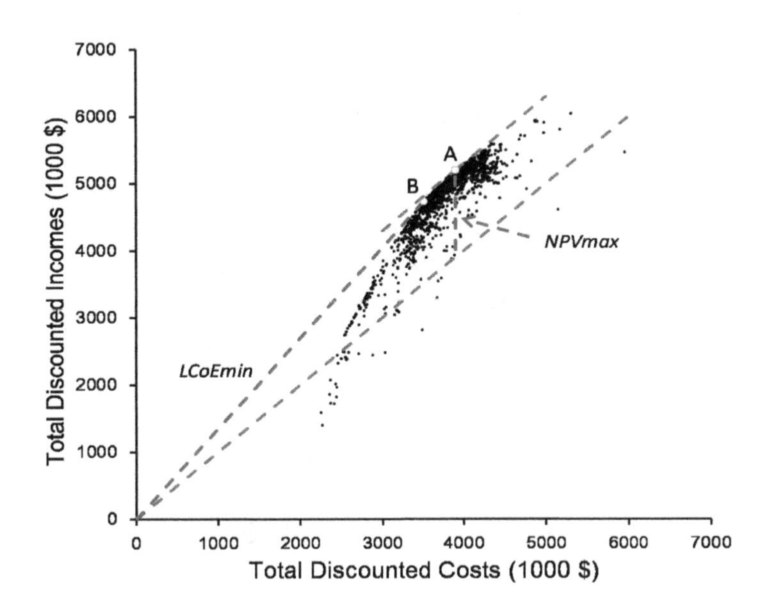

Figure 16. Total discounted incomes against total discounted costs.

The choice between LCoE and NPV is a complex economic problem. Investment size and incremental internal rate of return may also be taken into consideration to make the decision that most meets the desire of the designers. Further exploration is beyond the scope of this research. Nonetheless, the result obtained herein, that the optimal blade length of a certain turbine should be situated between the value of optimal LCoE and optimal NPV, may be valuable to wind turbine manufacturers.

4.4. Discussion of the Optimization Blades

In addition to the different blade lengths obtained from these metrics, the detailed turbine configurations show the effect of several other variables. In the following subsection, optimization results are presented to provide insight about the impact of blade geometry to the integrated turbine performance. Blade shapes of optimum NPV and optimum LCoE are plotted in Figure 17, and the turbine properties with respect to these blades are summarized in Table 5.

The design providing an optimal NPV has a larger chord distribution along the full blade span, but the difference remains almost constant up to the blade tips, revealing a certain degree of similarity between these two blades. Both blades have a relatively high thickness-to-chord ratio along the blade span. The minimum relative thickness is 21.6% for optimal NPV and 23% for optimal LCoE. Though large relative thickness will undoubtedly reduce blade aerodynamic performance, the beneficial aspect is that larger section thickness can increase the section moment of inertia, which is helpful in increasing both strength and stiffness, consequently reduces the blade mass. This configuration is mainly the result of a trade-off between aerodynamic performance and blade structure. Additionally, the reduced lift characteristic of high relative thickness airfoils may also be helpful in controlling the aerodynamic loads.

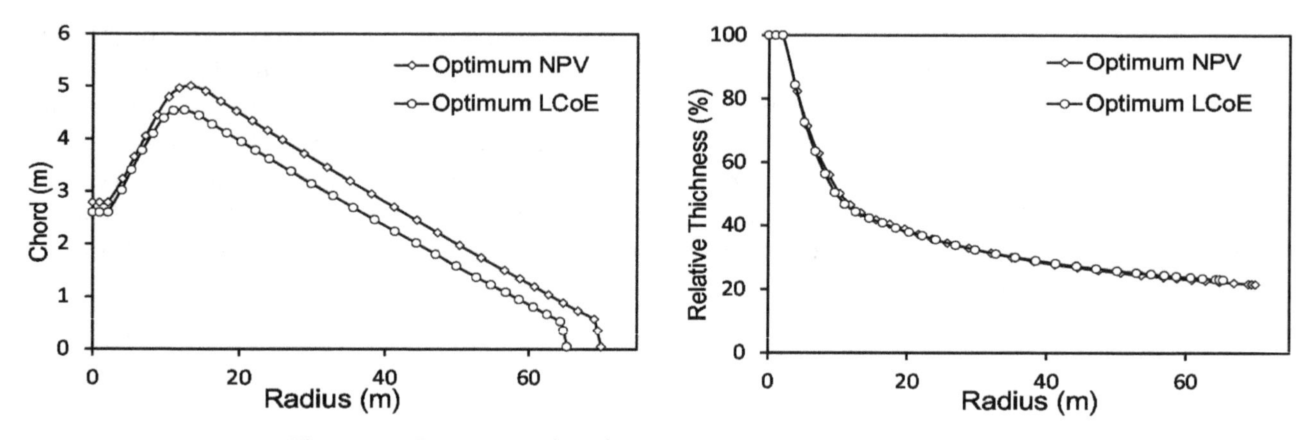

Figure 17. Design results of optimum NPV and optimum LCoE.

Generally, neither optimum LCoE design nor optimum NPV designs tend to pursue a maximum electricity production. The maximum power coefficients (Cp) of both blades are about 0.46, which is a relatively low value by current industrial standards. Note that the optimization procedure can achieve a Cp of 0.487, as shown by the High Cp design in Table 5. One of the main reasons for this is that the increase of Cp is always associated with an increase in design loads and consequently the initial capital cost.

Table 5. Turbine properties of designs produced for optimum NPV, LCoE and Cp.

Design Results	Blade Length (m)	Cp (-)	Capacity Factor (-)	Blade Root Mf (Nm)	Blade Mass (kg)	Hub Fx (Nm)
Optimum NPV	70.1	0.463	0.342	1.58×10^7	17,930	7.44×10^5
Optimum LCoE	65.4	0.457	0.312	1.27×10^7	14,552	6.80×10^5
High Cp	65.4	0.487	0.321	1.40×10^7	17,303	8.63×10^5
Design Results	**Tower Height (m)**	**Tower Mass (tonne)**	**CAPEX (1000 $)**	**LCoE (cents/kWh)**	**NVP (1000 $)**	
Optimum NPV	105	369	2725	6.62	1295	
Optimum LCoE	93	270	2377	6.54	1223	
High Cp	93	346	2625	6.84	1096	

In order to elucidate why optimization designs based on economic analysis lead to relatively lower AEP, we chose the High Cp design from Table 5, which shares the same blade length and tower height as optimum LCoE. A commercial 58 m blade with a maximum Cp of 0.49 is also presented for comparison, as is shown in Figure 18. This commercial blade holds a geometry layout more similar to High Cp design, shows that some industrial practices are using the design strategy of maximum aerodynamic performance.

The turbine performance of High Cp design differs greatly from the optimum LCoE design as presented in Table 5. Generally, the design loads and the mass of structural components rise considerably.

The minimum LCoE design has a much larger relative thickness distribution along most of the blade but a smaller chord distribution at the blade tip compared with the High Cp design. This is because the minimum LCoE sacrifices rotor performance to reduce thrust and decrease turbine costs. As a result, the design loads are significantly reduced. For example, the flap-wise moment of the blade root is reduced by 9%. Together with a larger blade thickness at the mid span, which is beneficial to the structural integrity of the blade, the blade mass is reduced from 17.3 tonnes to 14.5 tonnes, a decrease of 16%. The expense is that the maximum power coefficient is reduced from 0.487 to 0.457.

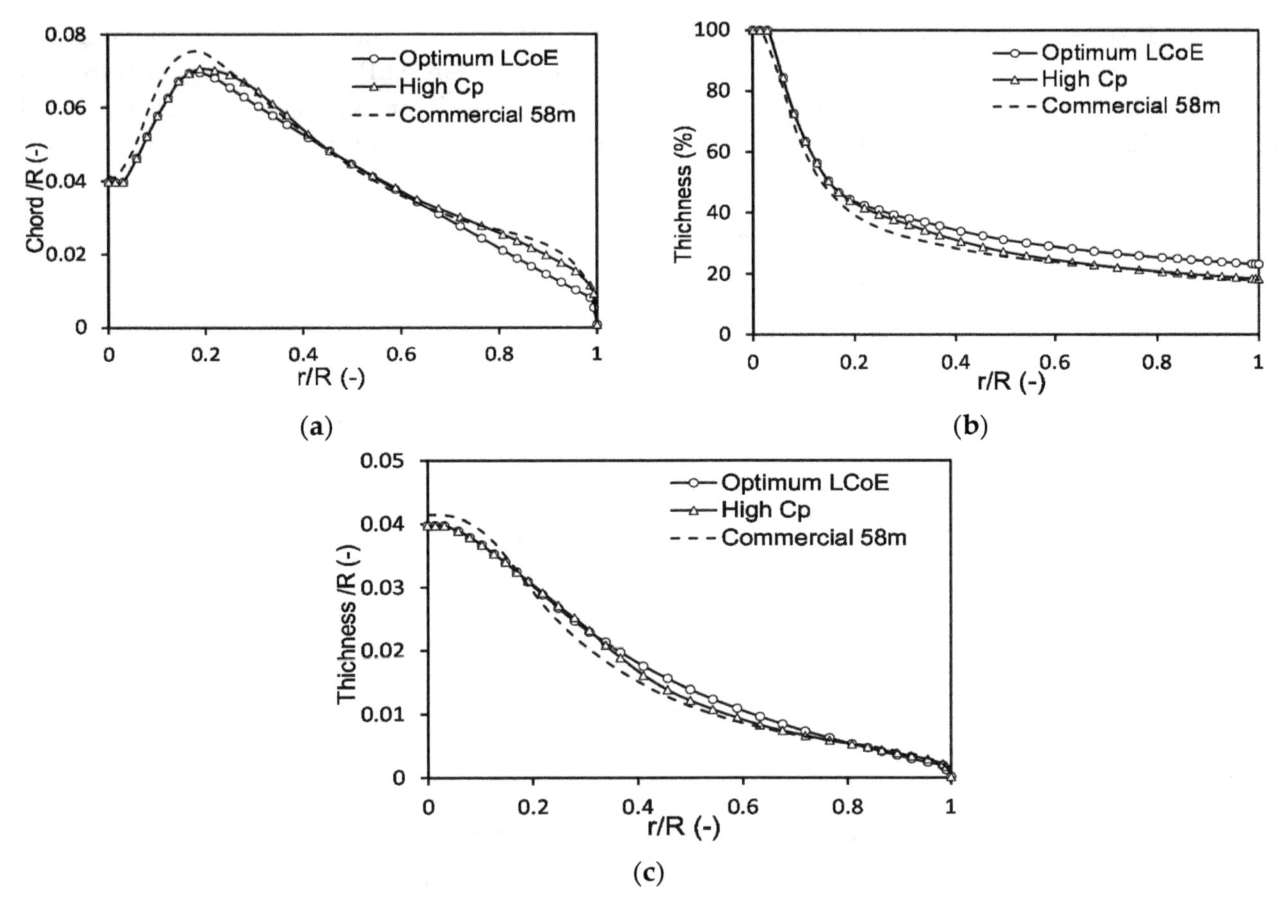

Figure 18. Comparison between Optimum LCoE, High Cp and Commercial 58 m blade. (**a**) Chord/R distribution; (**b**) Relative thickness distribution; (**c**) Thickness/R distribution.

0.457 may seem as unacceptable compared with 0.487, a decrease of 6.2%. However, the AEP is only degraded by 3%. Meanwhile, the design loads and structure masses were reduced significantly, accounting for a reduction in capital expenditure by 9.3%. The result produces 4.1% less LCoE and the NPV is increased. The LCoE design could be a more reasonable choice from the economic point of view.

4.5. Optimization of the Tower Height

At a certain wind shear exponent, the quality of the wind improves with increasing tower height. However, the higher tower results in a heavier structure which increases the cost. Therefore, the tower height of the turbine should match the site and rotor to achieve maximum economic efficiency [32,33]. An analysis was conducted to examine the impact of the tower height on the LCoE and NPV. Both are depicted graphically in Figures 19 and 20.

We took the rotors obtained from the optimization for examination. One is equipped with a 65.4 m blade and the other with 70.1 m blade. There are local optimum values of LCoE and NPV as a function of the tower height. The reason for this is similar to that which describes the influence of the blade length. The economic efficiency increases first and then decreases with the tower height, due to an increase in the costs of rotor, tower and the balance-of-station. Although the increased tower height can produce more energy, the increase in cost eventually outweighs the increase in profits. The optimum NPV design results in a higher tower than the optimum LCoE design, showing again that a minimum LCoE design is more geared towards controlling the cost.

Interestingly, we observed that a lower tower is a better match for the larger rotor in both cases. For example, NPV is optimized at a 110 m tower height for a 65.4 m blade rotor, but at a 105 m tower for a 70.1 m blade rotor, as shown in Figure 21. This trend is repeated in the optimal LCoE design.

Therefore, it seems from this result that, unlike the case for smaller rotor, the cost of the increase in tower height for larger wind turbine may overweight the increase in AEP due to its higher rotor thrust. The result may be influenced by the value selected for the wind shear exponent, but it is certain that more attention should be paid in upgrading the tower height of large rotors.

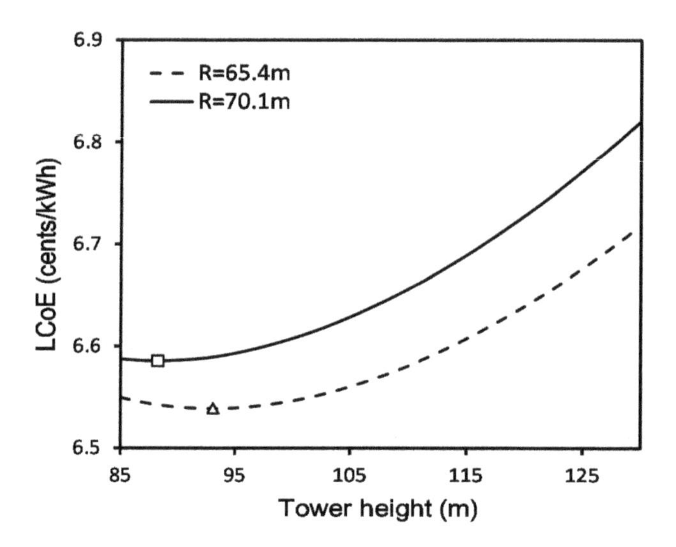

Figure 19. LCoE against tower height.

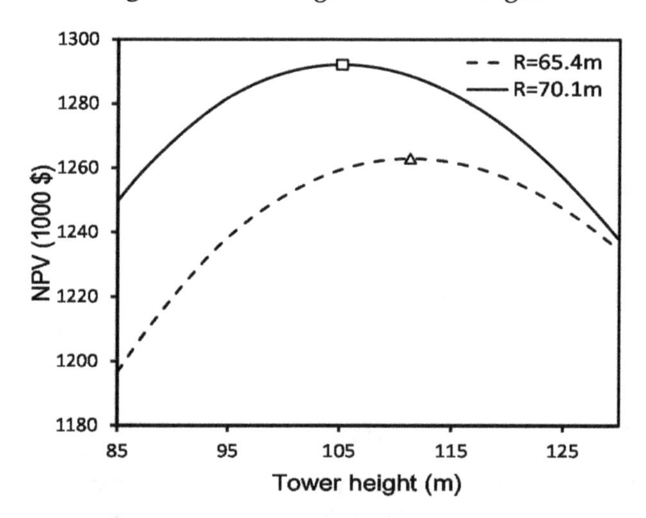

Figure 20. NPV against tower height.

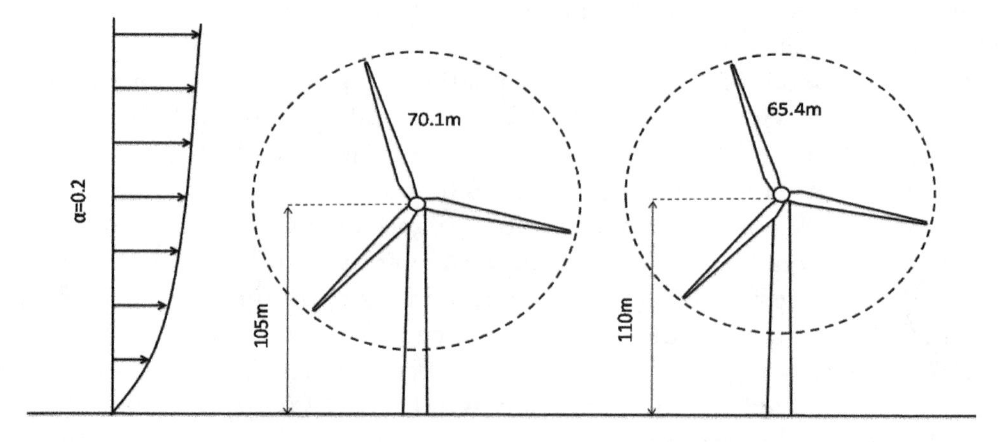

Figure 21. Optimal tower heights of different rotors.

5. Conclusions

In this paper, a system-level optimization analysis is used to perform an integrated design of a 2 MW low wind speed turbine. Besides LCoE, the economic functions NPV, IRR and DPT are also applied as objectives in wind turbine optimization. Theoretical analysis and design results indicate that in this study IRR and DPT effect the turbine design in the same way as LCoE, all leading to a minimization of the ratio of capital expenditures to AEP. However, optimum NPV implies the largest margin between incomes and costs, resulting in a longer blade than optimum LCoE. Though more economic factors should be included to make the final decision, the optimal blade length for a certain turbine should be situated between the value of optimal LCoE and optimal NPV according to the economic analysis in this study.

The blades obtained from these economic objectives seem to be aerodynamically uncompetitive due to their larger relative thickness and smaller chord distribution compare with high Cp design. However, the sacrifice of aerodynamic performance brings significant reduction in design loads and turbine costs. This optimization shows that sometimes control of construction and maintenance costs is more crucial than optimal aerodynamics in improving the economic efficiency of wind turbines.

Further developments in this work could consider other economic metrics, such as investment size and incremental internal rate of return, to achieve more practical results. Adding modal properties, local and global buckling in the structural design constraints for the blade and tower could also be considered for future work. Finally, through a more accurate estimation of the design loads and turbine capital costs, the economic performance of a wind turbine could be more accurately evaluated.

Author Contributions: Writing-Original Draft Preparation, J.W.; Writing-Review & Editing, T.W.; Supervision, L.W. and N.Z.

References

1. Johansen, J.; Madsen, H.A.; Gaunaa, M.; Bak, C.; Sørensen, N.N. Design of a wind turbine rotor for maximum aerodynamic efficiency. *Wind Energy* **2009**, *12*, 261–273. [CrossRef]
2. Wang, L.; Wang, T.G.; Wu, J.H.; Chen, G.P. Multi-objective differential evolution optimization based on uniform decomposition for wind turbine blade design. *Energy* **2017**, *120*, 346–361. [CrossRef]
3. Benini, E.; Toffolo, A. Optimal design of horizontal-axis wind turbines using blade-element theory and evolutionary computation. *J. Sol. Energy Eng.* **2002**, *124*, 357–363. [CrossRef]
4. Maki, K.; Sbragio, R.; Vlahopoulos, N. System design of a wind turbine using a multi-level optimization approach. *Renew. Energy* **2012**, *43*, 101–110. [CrossRef]
5. Chehouri, A.; Younes, R.; Ilinca, A. Review of performance optimization techniques applied to wind turbines. *Appl. Energy* **2015**, *142*, 361–388. [CrossRef]
6. Ning, A.; Damiani, R.; Moriarty, P.J. Objectives and constraints for wind turbine optimization. *J. Sol. Energy Eng.* **2014**, *136*, 041010. [CrossRef]
7. Savino, M.M.; Manzini, R.; Della Selva, V.; Accorsi, R. A new model for environmental and economic evaluation of renewable energy systems: The case of wind turbines. *Appl. Energy* **2017**, *189*, 739–752. [CrossRef]
8. Herbert-Acero, J.F.; Probst, O.; Réthoré, P.E.; Larsen, G.C.; Castillo-Villar, K.K. A review of methodological approaches for the design and optimization of wind farms. *Energies* **2014**, *7*, 6930–7016. [CrossRef]
9. Short, W.; Packey, D.J.; Holt, T. *A Manual for the Economic Evaluation of Energy Efficiency and Renewable Energy Technologies*; Technical Report NREL/TP-462-5173; National Renewable Energy Laboratory: Golden, CO, USA, 1995.
10. Rodrigues, S.; Restrepo, C.; Katsouris, G.; Pinto, R.T.; Soleimanzadeh, M.; Bosman, P.; Bauer, P. A multi-objective optimization framework for offshore wind farm layouts and electric infrastructures. *Energies* **2016**, *9*, 216. [CrossRef]

11. Mirghaed, M.R.; Roshandel, R. Site specific optimization of wind turbines energy cost: Iterative approach. *Energy Convers. Manag.* **2013**, *73*, 167–175. [CrossRef]
12. Ashuri, T.; Zaaijer, M.B.; Martins, J.R.; van Bussel, G.J.; van Kuik, G.A. Multidisciplinary design optimization of offshore wind turbines for minimum levelized cost of energy. *Renew. Energy* **2014**, *68*, 893–905. [CrossRef]
13. Sun, Z.Y.; Sessarego, M.; Chen, J.; Shen, W.Z. Design of the OffWindChina 5MW Wind Turbine Rotor. *Energies* **2017**, *10*, 777. [CrossRef]
14. Fuglsang, P.; Madsen, H.A. Optimization method for wind turbine rotors. *J. Wind Eng. Ind. Aerodyn.* **1999**, *80*, 191–206. [CrossRef]
15. Dykes, K.; Platt, A.; Guo, Y.; Ning, A.; King, R.; Parsons, T.; Petch, D.; Veers, P. *Effect of Tip-Speed Constraints on the Optimized Design of a Wind Turbine*; Technical Report NREL/TP-5000-61726; National Renewable Energy Laboratory: Golden, CO, USA, 2014.
16. Ning, A.; Petch, D. Integrated design of downwind land-based wind turbines using analytic gradients. *Wind Energy* **2016**, *19*, 2137–2152. [CrossRef]
17. Ashuri, T.; Zaaijer, M.B.; Martins, J.R.; Zhang, J. Multidisciplinary design optimization of large wind turbines—Technical, economic, and design challenges. *Energy Convers. Manag.* **2016**, *123*, 56–70. [CrossRef]
18. Fuglsang, P.; Bak, C.; Schepers, J.G.; Bulder, B.; Cockerill, T.T.; Claiden, P.; Olesen, A.; van Rossen, R. Site-specific Design Optimization of Wind Turbines. *Wind Energy* **2002**, *5*, 261–279. [CrossRef]
19. Hansen, M.O.L. *Aerodynamics of Wind Turbines*, 2nd ed.; Earthscan: London, UK, 2008.
20. *International Electrotechnical Committee IEC 61400-1: Wind turbines Part 1: Design Requirements*, 3rd ed.; IEC: Geneva, Switzerland, 2005.
21. Guo, Y.; Parsons, T.; King, R.; Dykes, K.; Veers, P. *An Analytical Formulation for Sizing and Estimating the Dimensions and Weight of Wind Turbine Hub and Drivetrain Components*; Technical Report NREL/TP-5000-63008; National Renewable Energy Laboratory: Golden, CO, USA, 2015.
22. Díaz, G.; Gómez-Aleixandre, J.; Coto, J. Dynamic evaluation of the levelized cost of wind power generation. *Energy Convers. Manag.* **2015**, *101*, 721–729. [CrossRef]
23. Tang, S.L.; Tang, H.G. The variable financial indicator IRR and the constant economic indicator NPV. *Eng. Econ.* **2003**, *48*, 69–78. [CrossRef]
24. Timmer, W.A.; van Rooij, R.P.J.O.M. Summary of the Delft University wind turbine dedicated airfoils. *J. Sol. Energy Eng. Trans. ASME* **2003**, *125*, 488–496. [CrossRef]
25. Ning, A.; Dykes, K. Understanding the benefits and limitations of increasing maximum rotor tip speed for utility-scale wind turbines. *J. Phys. Conf. Ser.* **2014**, *524*, 012087. [CrossRef]
26. Diveux, T.; Sebastian, P.; Bernard, D.; Puiggali, J.R.; Grandidier, J.Y. Horizontal axis wind turbine systems: Optimization using genetic algorithms. *Wind Energy* **2001**, *4*, 151–171. [CrossRef]
27. Gao, X.; Yang, H.; Lu, L. Optimization of wind turbine layout position in a wind farm using a newly-developed two-dimensional wake model. *Appl. Energy* **2016**, *174*, 192–200. [CrossRef]
28. Gentils, T.; Wang, L.; Kolios, A. Integrated structural optimisation of offshore wind turbine support structures based on finite element analysis and genetic algorithm. *Appl. Energy* **2017**, *199*, 187–204. [CrossRef]
29. Zhang, J.J.; Xu, L.W.; Gao, R.Z. Multi-island Genetic Algorithm Optimization of Suspension System. *Telkomnika* **2012**, *10*, 1685–1691. [CrossRef]
30. China Wind Energy Association. *China Wind Power Industry Map 2017*; CWEA: Beijing, China, 2018.
31. Global Wind Energy Council. *Global Wind Report*; GWEC: Brussels, Belgium, 2018.
32. Abdulrahman, M.; Wood, D. Investigating the Power-COE trade-off for wind farm layout optimization considering commercial turbine selection and hub height variation. *Renew. Energy* **2017**, *102*, 267–278. [CrossRef]
33. Alam, M.M.; Rehman, S.; Meyer, J.P.; Al-Hadhrami, L.M. Review of 600–2500 kW sized wind turbines and optimization of hub height for maximum wind energy yield realization. *Renew. Sustain. Energy Rev.* **2011**, *15*, 3839–3849. [CrossRef]

6

An Optimization Framework for Wind Farm Design in Complex Terrain

Ju Feng [1], Wen Zhong Shen [1,*] and Ye Li [2]

[1] Department of Wind Energy, Technical University of Denmark, 2800 Kgs. Lyngby, Denmark; jufen@dtu.dk
[2] School of Naval Architecture, Ocean and Civil Engineering, Shanghai Jiao Tong University, Shanghai 201100, China; ye.li@sjtu.edu.cn
* Correspondence: wzsh@dtu.dk

Featured Application: Analysis and optimization of wind farm design in complex terrain.

Abstract: Designing wind farms in complex terrain is an important task, especially for countries with a large portion of complex terrain territory. To tackle this task, an optimization framework is developed in this study, which combines the solution from a wind resource assessment tool, an engineering wake model adapted for complex terrain, and an advanced wind farm layout optimization algorithm. Various realistic constraints are modelled and considered, such as the inclusive and exclusive boundaries, minimal distances between turbines, and specific requirements on wind resource and terrain conditions. The default objective function in this framework is the total net annual energy production (AEP) of the wind farm, and the Random Search algorithm is employed to solve the optimization problem. A new algorithm called Heuristic Fill is also developed in this study to find good initial layouts for optimizing wind farms in complex terrain. The ability of the framework is demonstrated in a case study based on a real wind farm with 25 turbines in complex terrain. Results show that the framework can find a better design, with 2.70% higher net AEP than the original design, while keeping the occupied area and minimal distance between turbines at the same level. Comparison with two popular algorithms (Particle Swarm Optimization and Genetic Algorithm) also shows the superiority of the Random Search algorithm.

Keywords: wind farm; layout optimization; design; random search; complex terrain

---

1. Introduction

In the past two decades, the world has witnessed a remarkable growth of wind energy development. According to the latest statistics from the Global Wind Energy Council (GWEC), the global cumulative installed wind capacity has increased from 23.9 GW in 2001 to 439.1 GW in 2017, representing an average annual increase by 20.4% [1]. Although offshore wind is now attracting a lot of interests and has grown rapidly in recent years, especially in northern Europe, today it represents only 4.3% of the global installed capacity of wind energy [2]. The dominance of onshore wind is mainly due to its longer development history, lower cost, and easier deployment.

As a result of the rapid growth of onshore wind, a lot of onshore wind farms have been built worldwide. Since many suitable sites in flat terrain have already been developed with wind farms, more and more wind farms are going to be built in complex terrain, especially for countries featured by a large percentage of topography covered with mountains, such as China.

Compared with a wind farm on flat terrain, a wind farm built on complex terrain benefits from the possible richer wind resource at certain locations (brought by the speed-up effect due to the terrain topography change), but it is also more likely to be exposed to more complex flow conditions, higher fatigue loads, more expensive installation, operation and maintenance costs, and other disadvantages [3].

Designing wind farms in complex terrain is not a trivial task, mainly due to the complex interactions of the boundary layer flow with the complex terrain and wind turbine wakes, and also due to the multi-disciplinary nature of the wind farm design problem. This problem typically involves different design and engineering tasks, which may come from technical, logistical, environmental, economical, legal, and/or even social considerations.

For most industrial practitioners, the design of wind farms conventionally concerns mainly the micro-siting of wind turbines based on consideration of wind resource, i.e., determining the exact position for each turbine at a selected wind farm site [4]. Typically, a wind farm developer/designer uses a wind resource assessment tool, such as WAsP [5] (the industry-standard PC software for wind resource assessment, siting, and energy yield calculations for wind turbines and wind farms), to assess the wind resource of the wind farm site based on wind measurement data during a period [6]. Then, the final design of the wind farm, mainly the layout of turbines, is determined for maximizing the annual energy production (AEP) while considering certain constraints, such as the proximity of wind turbines. This final layout is usually obtained by either manual adjustments or using some optimization techniques [7]. After finding the turbine layout, other design tasks related to foundations, access roads, electrical system, and other components or systems are handled separately, often by specialized engineering consulting companies.

The general problem of modelling the wind flow over complex terrain has been an important research field for a very long time. Seventy years ago, Queney published a review of theoretical models of inviscid flow over hills and mountains [8]. More recently, Wood gave a historical review of studies (until 2000) on wind flow over complex terrain [9]. Similarly, in the wind energy community, the problem of wind resource assessment in complex terrain was investigated by many researchers over a long period [10].

In contrast, wind farm design optimization has only become a hot research area quite recently and the majority of the published studies in this field deal with wind farms in flat terrain or offshore [11]. Few studies investigated the wind farm design optimization problems in complex terrain. Song et al. [12] proposed a bionic method for optimizing the turbine layout in complex terrain. In this study the objective was to maximize the power output, and the wake flow was simulated by using the virtual particle wake model. They also developed a new greedy algorithm for this problem in a later study [13]. Feng and Shen [14] considered the layout optimization problem for a wind farm on a 2D Gaussian hill, using computational fluid dynamics (CFD) simulations for the background flow field and an adapted Jensen wake model for the wake effect. They used the random search algorithm [15,16] to optimize the layout for maximizing the total power. Kuo et al. [17] proposed an algorithm that couples CFD with mixed-integer programming (MIP) to optimize the layouts in complex terrain. In this study, the wind farm domain was discretized into cells to use the MIP method.

All the studies mentioned above aimed to maximize either the total power [12–14] or the total kinetic energy [17] of wind farm, with constraints on wind farm boundary and minimal distance requirements.

In this paper, we present an optimization framework for wind farm design in complex terrain, which was developed in the Sino-Danish research cooperation project, FarmOpt [18]. This framework combines the state-of-the-art flow field solver, a fast engineering wake model, and an advanced optimization algorithm to solve the design optimization problem of wind farms in complex terrain. Various realistic objectives, constraints, and requirements that one might encounter in real life wind farm developments can be considered in this framework.

We will first introduce the modelling methodology in Section 2. Then, the optimization framework is presented in Section 3. Results of a real wind farm case study will be described and discussed in Section 4. Section 5 compares the performance of the Random Search (RS) algorithm with two popular algorithms, i.e., Particle Swarm Optimization (PSO) and the Genetic Algorithm (GA). Finally, conclusions are given in Section 6.

2. Wind Farm Modelling

A reliable estimation of AEP is crucial for wind farm design, since AEP represents the amount of electricity that a given wind farm can generate in a year, which in turn determines the annual income that the wind farm's owner can obtain. This makes AEP either the objective function or a critical component in the objective function for any wind farm design optimization problem. For example, the levelized cost of energy (LCOE), which is a popular choice for evaluating wind farm designs [19], can be calculated by considering AEP, together with capital expenditure (CAPEX), operational expenditure (OPEX), and some financial parameters, such as the interest rate. In this section, we present the aspects of wind farm modelling related to the AEP estimation.

2.1. Site Condition

Due to terrain effects, i.e., the effect of topographic changes on the flow filed over a complex terrain [9], different locations at a complex terrain site will experience different wind conditions. Using onsite wind measurements and standard wind resource assessment tools, such as WAsP [5] and WindPRO [20], one could get the essential information on location-specific site conditions, i.e., wind resource and terrain effects, for any given inflow wind direction sector.

For flat or moderately complex terrain, linear models, such as the conventional WAsP [6], provide reasonable results. For more complex terrain, nonlinear models, such as WAsP CFD [21], are needed. In the current framework, we assume that the location-specific values of wind resource and terrain effects have been obtained by using a wind resource assessment tool.

Considering a rectangle area covering the interested area for a wind farm design, we first discretize the area into a number of grids, defined by a range of x coordinates $[x_1, x_2, \ldots, x_{N_x}]$ and a range of y coordinates $\left[y_1, y_2, \ldots, y_{N_y}\right]$. For a certain number of wind direction sectors defined by a range of far field inflow wind directions $\left[\theta_1^\infty, \theta_2^\infty, \ldots, \theta_{N_\theta}^\infty\right]$, we can then obtain the relevant sector wise values of wind resource and terrain effects variables, such as Weibull-A, Weibull-k, frequency, speed up factor, terrain induced turning angle, and mean wind speed, at a given height above the ground. Note that Weibull-A is the scale parameter and Weibull-k is the shape parameter of Weibull distribution, which is the most commonly used distribution for wind speed [4]. Values of variables that do not depend on inflow wind direction, such as elevation and overall mean wind speed, are also obtained.

Using linear interpolation, we can then easily estimate the values of these variables for any arbitrary position (x, y) in this area, under any far field inflow wind direction, θ^∞, if wind direction dependent variables are concerned. The variables important for estimating AEP are listed in Table 1.

Table 1. Site condition variables important for AEP (annual energy production) estimation.

Variable	Unit	Notation
Elevation	m	$z(x, y)$
Weibull-A	m/s	$A(x, y, \theta^\infty)$
Weibull-k	-	$k(x, y, \theta^\infty)$
Frequency	-/°	$freq(x, y, \theta^\infty)$
Speed up	-	$sp(x, y, \theta^\infty)$
Turning angle	°	$tu(x, y, \theta^\infty)$

In Table 1, speed up is defined as the ratio between the local wind speed, v, at location (x, y) and the far field wind speed, v^∞, at the same height above the ground for a given far field inflow wind direction, θ^{∞}, while turning angle, tu, denotes the difference between the local wind direction, θ, and the far field inflow wind direction, θ^{∞}. Thus, they can be computed as:

$$sp(x, y, \theta^{\infty}) = v(x, y, \theta^{\infty})/v^{\infty}(\theta^{\infty}), \tag{1}$$

$$tu(x, y, \theta^{\infty}) = \theta(x, y, \theta^{\infty}) - \theta^{\infty}. \tag{2}$$

As a demonstration, four interpolated site condition variables are shown in Figure 1 for a real wind farm site in complex terrain. Note that the results on wind resource and terrain effects are imported from WAsP CFD simulations and the wind farm on this site will be used in Sections 4 and 5 as a test case.

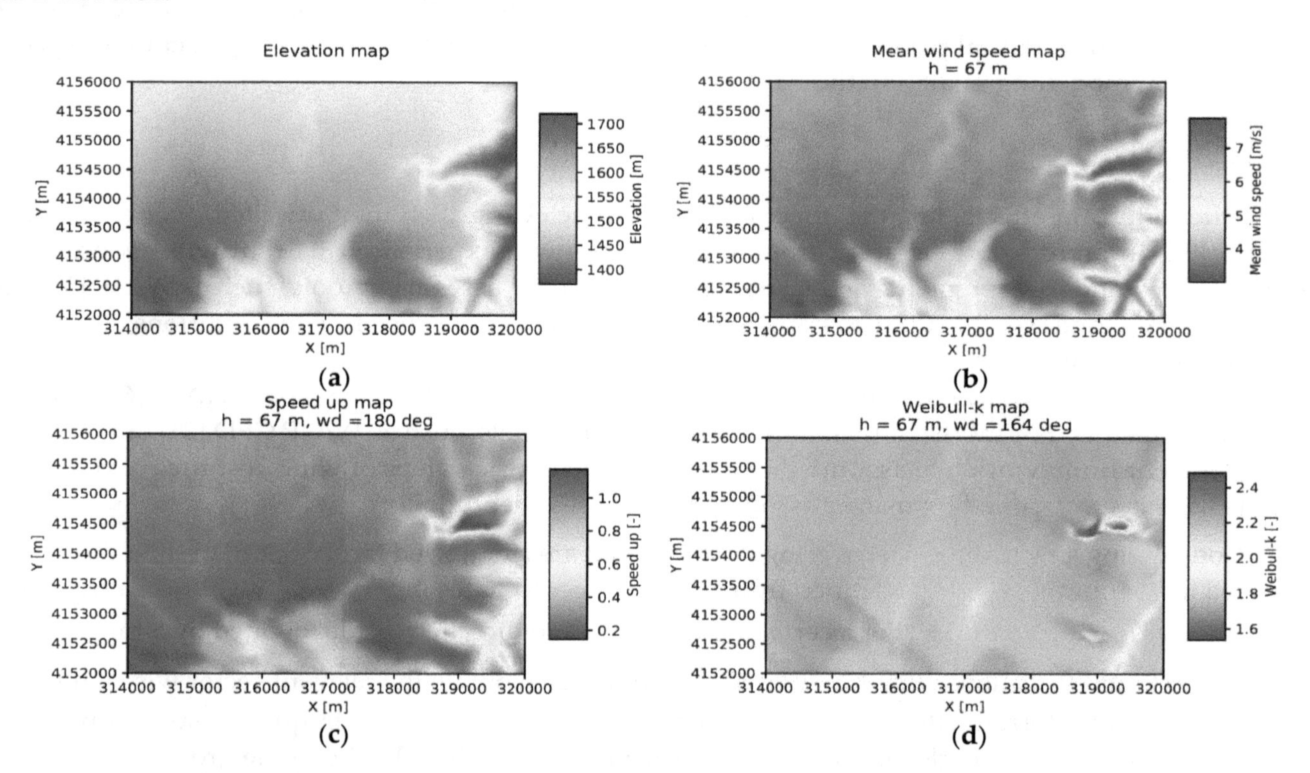

Figure 1. Interpolated site condition variables for a wind farm site in complex terrain: (**a**) elevation; (**b**) mean wind speed at 67 m height above the ground; (**c**) speed up at 67 m height under the far field inflow wind direction of 180°; (**d**) Weibull-k at 67 m height under the far field inflow wind direction of 164°.

2.2. *Wake Model*

Modelling the wake effects of wind turbines is essential for AEP estimation, but quite challenging for wind farms in complex terrain due to the complex interplay between terrain and wake flow [22]. While the CFD can model the wake flow field of a wind farm in complex terrain with a reasonable accuracy [23], its high computational cost makes it unsuitable for wind farm design optimization, since typical optimization algorithms require a large number of design evaluations.

Several studies have proposed a few fast wake models for wind farms in complex terrain: Song et al. proposed a virtual particle wake model [24] and applied it in wind farm layout optimizations [12,13]; Feng and Shen developed an adapted Jensen wake model in [14]; Kuo et al. [25] proposed a wake model by solving a simplified variation of the Navies-Stokes equations, which yields results with a reasonable accuracy and has a much less computational cost when compared with full CFD simulations.

In the current framework, the wake flow field of a wind farm in complex terrain is modelled by coupling the adapted Jensen wake model with the terrain flows obtained by CFD simulations, i.e., flow fields of the wind farm site without wind turbines, such as those imported from WAsP CFD simulations. The adapted Jensen wake model assumes the wake centerline of a wind turbine's wake follows the terrain at the same height above the ground along the local inflow wind direction, while its wake zone expands linearly and its wake deficit develops following the same rule as the original Jensen wake model [14]. The schematic of this wake model is shown in Figure 2.

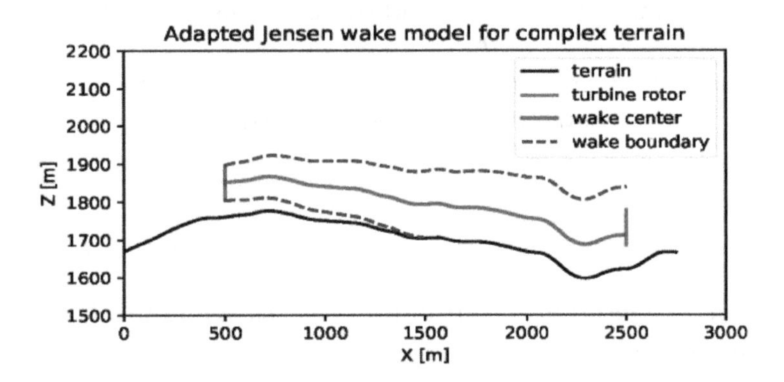

Figure 2. Schematic of the adapted Jensen wake model for a wind turbine in complex terrain.

Considering the wake effect between an upwind turbine, i, located at (x_i, y_i) and a downwind turbine, j, located at (x_j, y_j) for a given far field inflow condition: v^∞ and θ^∞, we can calculate the downwind distance, $d_{ij}^{down}(\theta^\infty)$, and the crosswind distance, $d_{ij}^{cross}(\theta^\infty)$, from the upwind turbine to the location of the downwind turbine. The downwind distance can be computed by integration following the wake centerline (the red line in Figure 2) based on the elevation, $z(x, y)$, and the local inflow wind direction, $\theta_i(\theta^\infty) = \theta^\infty + tu(x_i, y_i, \theta^\infty)$, while the crosswind distance can be easily obtained by considering the coordinates of these two turbines and the local inflow wind direction of the upwind turbine. Then, the local inflow wind speed at turbine i and the effective wake deficit turbine i caused on turbine j can be calculated by:

$$v_i(v^\infty, \theta^\infty) = sp(x_i, y_i, \theta^\infty) \cdot v^\infty, \tag{3}$$

$$\Delta v_{ij}(v^\infty, \theta^\infty) = \frac{A_{ij}^{ol}(\theta^\infty)}{A_j^{rotor}} \cdot v_i(v^\infty, \theta^\infty) \cdot \left(\frac{1 - \sqrt{1 - C_T(v_i(v^\infty, \theta^\infty))}}{1 + \alpha \cdot d_{ij}^{down}(\theta^\infty)/R_i} \right), \tag{4}$$

in which $A_{ij}^{ol}(\theta^\infty)$ is the overlapping area between the wake zone of turbine i and the rotor area of turbine j; $A_j^{rotor} = \pi R_j^2$ is the rotor area of turbine j; $C_T(\cdot)$ represents the thrust coefficient function of the wind turbine; α denotes the wake decay coefficient, and R_i represents the rotor radius of turbine i. Note that $A_{ij}^{ol}(\theta^\infty)$ can be calculated based on the wake zone radius, $R_{wake} = R_i + \alpha \cdot d_{ij}^{down}(\theta^\infty)$, the rotor radius, R_j, and the relevant crosswind distance, $d_{ij}^{cross}(\theta^\infty)$, using the method described in [14]. Additionally, $\alpha = 0.075$ is used in this study.

Denoting the set of indexes of all the turbines upwind of turbine j as I^{up}, and using the energy deficit balance assumption [14], we can then calculate the effective wind speed of turbine j under the far field inflow wind condition, v^∞ and θ^∞, as:

$$\overline{v}_j(v^\infty, \theta^\infty) = v_j(v^\infty, \theta^\infty) - \sqrt{\sum_{i \in I^{up}} (\Delta v_{ij}(v^\infty, \theta^\infty))^2}. \tag{5}$$

A more detailed description of the wake model governed by Equations (3)–(5) is referred to in [14].

2.3. AEP Estimation

After the local wind speeds with and without wake effects of each wind turbine are obtained by using Equations (3)–(5), it is then easy to calculate the corresponding power outputs using the power curve of the wind turbine.

Note that the site condition described in Section 2.1 provides wind resource parameters at each turbine site for a given far field inflow wind direction, i.e.,: $A_i(\theta^\infty) = A(x_i, y_i, \theta^\infty)$, $k_i(\theta^\infty) = k(x_i, y_i, \theta^\infty)$ and $freq_i(\theta^\infty) = freq(x_i, y_i, \theta^\infty)$. Based on these parameters, the probability of each turbine site's local wind condition, i.e., $v_i(v^\infty, \theta^\infty)$ and $\theta_i(\theta^\infty)$, with respect to the far field inflow wind condition (v^∞ and θ^∞) can be estimated by using the joint distribution proposed in [26] as:

$$\begin{aligned} pdf_i(v^\infty, \theta^\infty) = & \quad pdf(v_i(v^\infty, \theta^\infty),\ \theta_i(\theta^\infty)) \\ & = \frac{k_i(\theta^\infty)}{A_i(\theta^\infty)} \cdot \left(\frac{v_i(v^\infty, \theta^\infty)}{A_i(\theta^\infty)}\right)^{k_i(\theta^\infty)-1} \cdot \exp\left(-\left(\frac{v_i(v^\infty, \theta^\infty)}{A_i(\theta^\infty)}\right)^{k_i(\theta^\infty)}\right) \cdot freq_i(\theta^\infty) \end{aligned} \tag{6}$$

Thus, for a wind farm composed of N_{wt} turbines, its gross and net AEP, i.e., AEP with and without wake effects, can be estimated as:

$$\text{AEP}_{gross} = \sum_{i=1}^{N_{wt}} 8760 \cdot \eta_i \cdot \iint P(v_i(v^\infty, \theta^\infty)) \cdot pdf_i(v^\infty, \theta^\infty) dv^\infty \theta^\infty, \tag{7}$$

$$\text{AEP}_{net} = \sum_{i=1}^{N_{wt}} 8760 \cdot \eta_i \cdot \iint P(\overline{v}_i(v^\infty, \theta^\infty)) \cdot pdf_i(v^\infty, \theta^\infty) dv^\infty \theta^\infty, \tag{8}$$

in which 8760 is the total number of hours in a year, η_i denotes the availability factor of turbine i, and $P(\cdot)$ represents the power curve of the wind turbine. The integrations in Equations (7) and (8) are done numerically after properly discretizing the interested ranges of v^∞ and θ^∞. In this study, the availability factors of all the turbines are assumed to be 1.0 and the discretization used in numerical integration is $\Delta v^\infty = 1$ m/s, and $\Delta\theta^\infty = 5°$.

3. Optimization Framework

3.1. Problem Formulation

For a wind farm with N_{wt} turbines, its design can be specified by the turbines' locations, i.e., $\mathbf{X} = [x_1, x_2, \ldots, x_{N_{wt}}]$ and $\mathbf{Y} = [y_1, y_2, \ldots, y_{N_{wt}}]$. Then, a general optimization problem of the wind farm design can be formulated as:

$$\left.\begin{array}{lll} \text{min} & f_m(\mathbf{X}, \mathbf{Y}), & m = 1, 2, \ldots, M, \\ \text{subject to :} & g_k(\mathbf{X}, \mathbf{Y}) \geq 0, & k = 1, 2, \ldots, K; \\ & \mathbf{X}^{(L)} \leq \mathbf{X} \leq \mathbf{X}^{(U)}; & \\ & \mathbf{Y}^{(L)} \leq \mathbf{Y} \leq \mathbf{Y}^{(U)}. & \end{array}\right\} \tag{9}$$

where f_m is the mth objective function, g_k is the kth inequality constraint function, and $\mathbf{X}^{(L)}$, $\mathbf{X}^{(U)}$, $\mathbf{Y}^{(L)}$, and $\mathbf{Y}^{(U)}$ denote the lower and upper bounds.

Common objective functions for wind farm design optimization include AEP, LCOE, profit, noise emission, and so on. Typical constraints that can be modelled as inequality constraint functions include wind farm boundary, exclusive zones, and wind turbine proximity. In the current framework, the default objective function is AEP and the considered constraints are summarized in the next subsection.

3.2. Constraints

Wind farm design is subject to various constraints, which may come from technical, logistical, environmental, economical, legal, and/or even social considerations. In the current framework, three types of constraints are considered: (1) inclusive and exclusive boundaries; (2) minimal distance requirements; and (3) bounds on certain site condition variables.

Inclusive and exclusive boundaries denote the feasible and infeasible area for placing turbines, which may come from limitations and requirements on leased land, existing roads and properties, soil conditions, and so on. As a general method, polygons can be used to model inclusive and exclusive zones defined by the inclusive and exclusive boundaries. Thus, for a wind farm with N_{inc} inclusive and N_{exc} exclusive boundaries, the individual inclusive and exclusive zones and the overall feasible zone can be modeled as:

$$S^i_{inc} = \left\{ (x,\ y) \middle| \ a^i_k x + b^i_k y \leq c^i_k, \quad k = 1, 2,\ \ldots,\ m^i \ \right\}, \quad i = 1, 2,\ \ldots,\ N_{inc} \tag{10}$$

$$S^j_{exc} = \left\{ (x,\ y) \middle| \ a^j_l x + b^j_l y \leq c^j_l, \quad l = 1, 2,\ \ldots,\ n^j \ \right\}, \quad j = 1, 2,\ \ldots,\ N_{exc} \tag{11}$$

$$S_{\text{feasible}} = \left\{ (x,\ y) \middle| \ (x,\ y) \in \cup_{i=1}^{N_{\text{inc}}} S^i_{inc}, \quad (x,\ y) \notin \cup_{j=1}^{N_{\text{exc}}} S^j_{exc} \ \right\} \tag{12}$$

where m^i and n^j are the polygon edge numbers of the ith inclusive and jth exclusive boundaries, respectively.

Thus, the constraints on inclusive and exclusive boundaries of the wind farm design can be written as:

$$(x_i,\ y_i) \in S_{\text{feasible}}, \quad \text{for } i = 1, 2,\ \ldots,\ N_{\text{wt}}. \tag{13}$$

The second type of constraint is also typical in the engineering practice, which requires the distance between any two turbines larger than a minimal value. This is because a shorter distance between turbines will give a larger wake loss and a higher turbulence intensity, which results in a higher level of fatigue loads and maintenance cost, and even a shorter lifetime of turbines. Also, a minimal distance is required to make sure two turbines' blades never contact each other and one turbine never falls on the other turbine. In the current framework, a fixed minimal distance requirement, $Dist_{\text{min}}$, is assumed for any two turbines. Thus, the constraints on the minimal distance requirement are governed by:

$$\sqrt{(x_i - x_j)^2 + (y_i - y_j)^2} - Dist_{\text{min}} \geq 0 \quad \text{for } i, j = 1, 2,\ \ldots,\ N_{\text{wt}} \text{ and } i \neq j. \tag{14}$$

The third type of constraints exposes requirements on certain site condition variables to rule out certain unfavorable sites, such as those with too low a mean wind speed, too high a turbulence intensity, or too rugged terrain. When applied properly, this type of constraint can help the optimizer to focus on the more favorable design space and speed up the convergence process, as found in [18].

To measure the degree of the terrain ruggedness, a terrain ruggedness index (TRI) is introduced. This index was first proposed by Riley et al. [27] as a quantitative measure of topographic heterogeneity. A dimensionless version of this index is developed in the current framework, which can be calculated based on the elevation data of grids as defined in Section 2.1. Given the grid sizes along the x and y directions are Δx and Δy, the TRI at grid $(x_i,\ y_j)$ can be computed based on its elevation and the elevation values of the eight surrounding grids as:

$$\mathrm{TRI}(x_i, y_j) = \frac{1}{8}\Bigg\{ \frac{[z(x_i, y_j) - z(x_{i+1}, y_j)]^2}{\Delta x^2} + \frac{[z(x_i, y_j) - z(x_{i-1}, y_j)]^2}{\Delta x^2} + \frac{[z(x_i, y_j) - z(x_i, y_{j+1})]^2}{\Delta y^2} + \frac{[z(x_i, y_j) - z(x_i, y_{j-1})]^2}{\Delta y^2} + \frac{[z(x_i, y_j) - z(x_{i+1}, y_{j+1})]^2}{\Delta x^2 + \Delta y^2} + \frac{[z(x_i, y_j) - z(x_{i+1}, y_{j-1})]^2}{\Delta x^2 + \Delta y^2} + \frac{[z(x_i, y_j) - z(x_{i-1}, y_{j+1})]^2}{\Delta x^2 + \Delta y^2} + \frac{[z(x_i, y_j) - z(x_{i-1}, y_{j-1})]^2}{\Delta x^2 + \Delta y^2} \Bigg\}^{1/2} \tag{15}$$

For the wind farm site in Figure 1a, the TRI map calculated in Equation (15) is shown in Figure 3a. Note that the grid sizes are chosen as $\Delta x = \Delta y = 25$ m in this study according to the default setting of WAsP, which makes the total number of grids in the rectangle area shown in Figure 1 as 38,801. In the current framework, the constraints on TRI and mean wind speed, v^{mean}, are considered. Note that the mean wind speed values on the grids, $v^{mean}(x_i, y_j)$, can be imported from wind resource assessment tools or calculated based on the sector wise wind resource parameters, i.e., Weibull-A, Weibull-k, and frequency. For an arbitrary site, linear interpolation is used to find the value of TRI and v^{mean}. Thus, constraints on TRI and mean wind speed can be written as:

$$\mathrm{TRI}_{max} - \mathrm{TRI}(x_i, y_i) \geq 0, \quad \text{for } i = 1, 2, \ldots, N_{\mathrm{wt}}, \tag{16}$$

$$v^{mean}(x_i, y_i) - v^{mean}_{min} \geq 0, \quad \text{for } i = 1, 2, \ldots, N_{\mathrm{wt}}. \tag{17}$$

where TRI_{max} is the allowed maximal TRI and v^{mean}_{min} is the allowed minimal mean wind speed.

For the wind farm site in Figure 1, the feasibility of each grid, (x_i, y_j), can then be estimated as $F(x_i, y_j)$ based on the first and third types of constraints, i.e., constraints defined in Equations (13), (16) and (17). Note that $F(x_i, y_j)$, with a value of 1 (true) or 0 (false), means whether the given site, (x_i, y_j), is a feasible site to install a turbine, i.e., a site satisfies the constraints on inclusive/exclusive boundaries and bounds of certain site condition variables. The map of $F(x_i, y_j)$ is shown in Figure 3b. Note that the black area in this figure denotes the feasible area. Compared with the feasible area defined by Equations (13) and (14), the bounds on TRI and v^{mean} help to reduce the feasible grids from 12,150 to 9218, representing a 24.13% reduction in the feasible design space. This reduction can help certain optimization algorithms, such as RS, to limit the search space and converge faster.

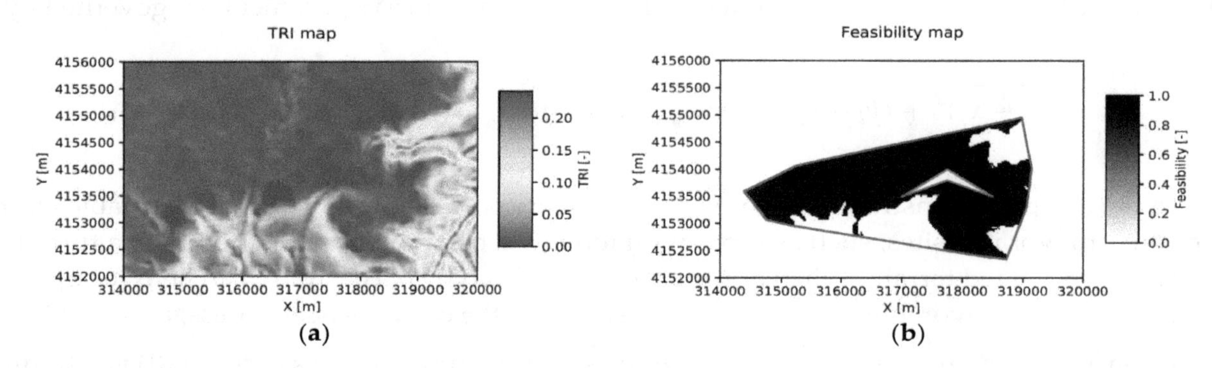

Figure 3. Constraint modelling: (**a**) TRI (terrain ruggedness index) map; (**b**) feasibility map with one inclusive (red line) and one exclusive (green line) boundary, and bounds on TRI ($\mathrm{TRI}_{max} = 0.05$) and v^{mean} ($v^{mean}_{min} = 6.5$ m/s).

3.3. Optimization Algorithm

To solve the optimization problem, the Random Search (RS) algorithm is used in the current framework. This algorithm was first proposed by Feng and Shen in [15] and improved in [16]. Recently, they also extended the algorithm to cover multiple objective cases [28] and overall design cases with multiple types of turbines [19], and to maximize the robustness of wind farm power production under wind condition uncertainties [29]. It has shown a superior performance when compared to other popular algorithms, such as GA (genetic algorithm) [16], NSGA-II (Non-dominated Sorting Genetic Algorithm II) [28], and mixed-discrete PSO (Particle Swarm Optimization) [19].

RS is a single solution search method. At each step, a new feasible design, i.e., a design satisfies all the constraints and bounds, is generated by slightly changing the current design, i.e., randomly choosing a turbine and moving it to a random position inside the feasible area. This new design is then compared with the current design with respect to the objective function. If the new design is better, it becomes the current design. If not, the current design remains unchanged. This step is iteratively repeated until a given stop condition is met. Normally, the stop condition can be set as a maximal number of steps (equivalent to the number of objective function evaluations). The procedure of this algorithm is shown as a flowchart in Figure 4.

There are two parameters controlling the RS algorithm: maximal step size, $M_{\max}$, which controls how far away a chosen turbine can be possibly moved from its current position; and maximal number of evaluations, $E_{\max}$, which defines when to terminate the optimization process. $M_{\max}$ should be set large enough to allow the algorithm to explore the design space more thoroughly and escape local minima more likely. When no trials and experimentations are first applied, $M_{\max}$ could be set in the order of the maximal distance between any two points in the feasible wind farm area and $E_{\max}$ should be in the range of 1000 to 10,000.

Due to the randomness involved in the optimization process, meta-heuristics, such as RS, exhibits a stochastic nature, meaning different runs typically obtain different results. This also means that the quality of the initial solution(s) can have profound influences on the final optimization results and the convergence speed [30]. Saavedra-Moreno et al. [31] tackled this problem by seeding an evolutionary algorithm with initial solutions obtained by a greedy heuristic algorithm, and applied this method in wind farm layout optimization.

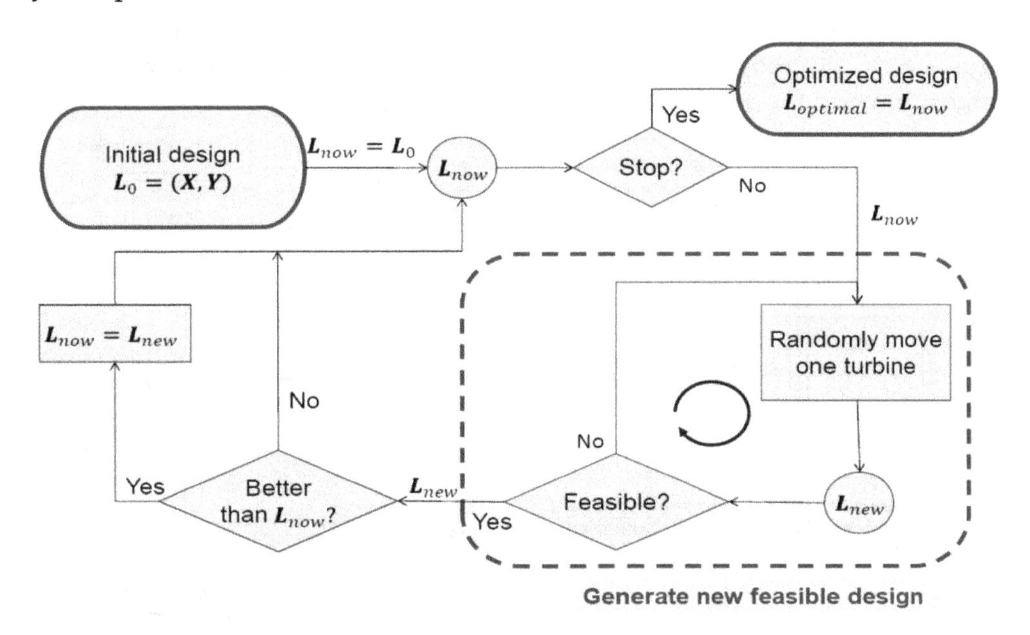

Figure 4. Flowchart of the Random Search algorithm.

In this study, a simple algorithm called Heuristic Fill is proposed to find a good initial design, based solely on site conditions and constraints. As described in Section 3.2, the feasibility of each grid, (x_i, y_j), can be calculated as $F(x_i, y_j)$ based on constraints defined in Equations (13), (16), and (17). The mean wind speed of each grid, $v^{mean}(x_i, y_j)$, can also be obtained from site conditions. The algorithm then tries to fill the feasible grids with N_{wt} wind turbines one by one according to the corresponding mean wind speed from high to low, while respecting the minimal distance constraints in Equation (14). This algorithm can be summarized in the pseudo code shown in Algorithm 1.

Algorithm 1. Pseudo code of the Heuristic Fill algorithm.

(1) Find the set of feasible grid points:
$G_{feasible} = \left\{(x_i, y_j) \middle| F(x_i, y_j) = 1, \text{ for } i = 1, 2, \ldots, N_x, j = 1, 2, \ldots, N_y\right\}$;

(2) Sort the feasible grid points according to mean wind speed, i.e., find a list of feasible grid points as $\left[(x_{i^1}, y_{j^1}), (x_{i^2}, y_{j^2}), \ldots, (x_{i^m}, y_{j^m}),\right]$, where $(x_{i^k}, y_{j^k}) \in G_{feasible}$ with $k = 1, 2, \ldots, m$ and $v^{mean}(x_{i^k}, y_{j^k}) \geq vv^{mean}(x_{i^{k+1}}, y_{j^{k+1}})$ with $k = 1, 2, \ldots, m-1$;

(3) Heuristically fill with N_{wt} turbine sites:

(a) Set $w = 1, g = 1$;

(b) **While** $1 < w \leq N_{\text{wt}}$ and $\sqrt{(x_i - x_{i^g})^2 + \left(y_i - y_{j^g}\right)^2}^2 < Dist_{\text{min}}$ holds for $i = 1, \ldots, w-1$:
$g = g + 1$
End While
Set $x_w = x_{i^g}, y_{\text{w}} = y_{j^g}, g = g + 1, w = w + 1$;

(c) Set the initial layout as $\boldsymbol{X} = [x_{\text{w}} | w = 1, 2, \ldots, N_{\text{wt}}], \boldsymbol{Y} = [y_{\text{w}} | w = 1, 2, \ldots, N_{\text{wt}}]$.

3.4. Framework

Putting things together, we can describe the framework in the architecture diagram shown in Figure 5. Note that WAsP (or another wind resource assessment code) needs to be used only once to get the relevant results for defining the site conditions, since the wake modelling and AEP estimation, as well as the modelling of constraints and optimization process, are all implemented by the modules inside the framework as introduced in the previous subsections. Thus, this framework can work as a stand-alone tool for wind farm design optimization. If objective functions other than AEP and more constraints are considered, the relevant modules can also be extended accordingly, while the optimization algorithm and the architecture remain unchanged.

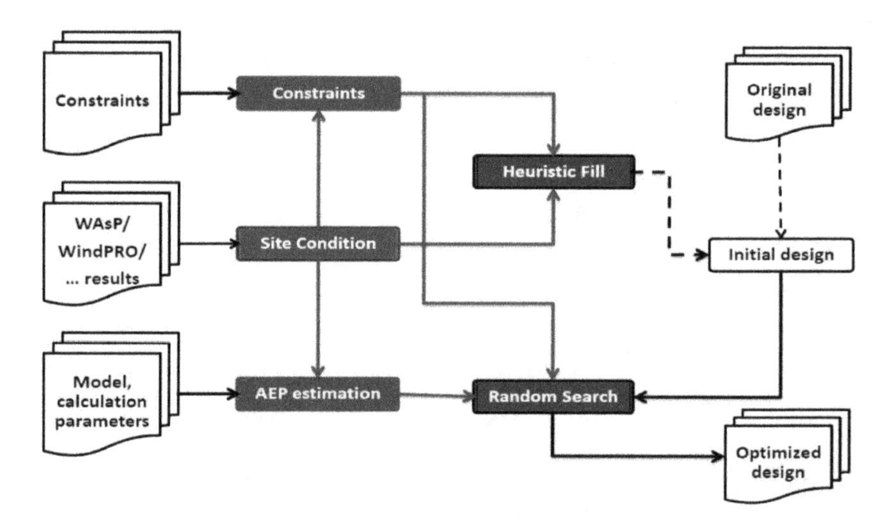

Figure 5. Architecture of the framework.

4. Case Study and Results

As a demonstration, a case study for an anonymous wind farm in complex terrain using the proposed framework is presented here. This wind farm is located in Northwest China, and has a total capacity of 50 MW with 25 turbines. This wind farm has been studied in [23,32] to investigate the wake flow simulation in complex terrain using CFD, and the preliminary results of the layout optimization study of this wind farm have been presented in [18]. Some of this wind farm's site conditions have already been shown in Figure 1. Its original layout is shown in Figure 6 and the turbine characteristics are shown in Figure 7. This type of turbine is variable speed, pitch regulated and has a rotor diameter of $D = 93$ m and a hub height of $H = 67$ m. The cut-in, rated, and cut-out wind speeds are 3 m/s, 12 m/s, and 25 m/s, respectively.

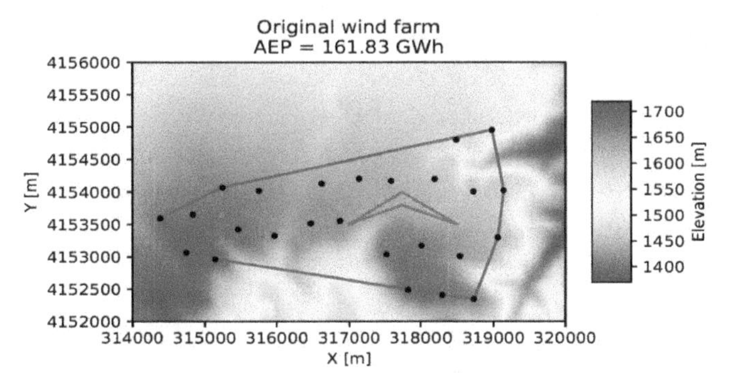

Figure 6. Original design of the anonymous wind farm in complex terrain.

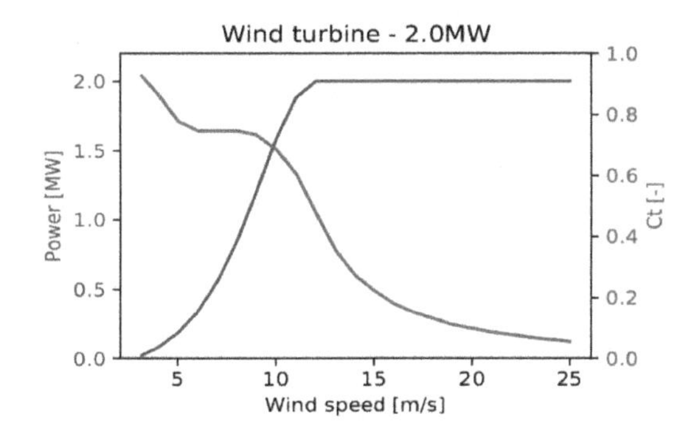

Figure 7. Turbine characteristics.

Note that constraints on inclusive and exclusive boundaries and bounds on parameters that are considered in this case study are shown in Figure 3. Additionally, the minimal distance requirement used in Equation (14) is chosen as the minimal distance between the turbines in the original layout, which is $D_{min} = 403.8$ m. The inclusive boundary shown with red line in Figure 6 is set as the minimal convex polygon that covers all the turbines, and the exclusive boundary (the green line) is arbitrarily set to test the modelling capacity of the exclusive boundary of the framework. The choice of the inclusive boundary is to make sure the optimized wind farm occupies the same or a smaller area and the required cost of electrical cables and access roads that connect all turbines stays at the same level, so that the comparison of AEP with the original wind farm makes sense.

To find a good feasible initial design, the Heuristic Fill algorithm is applied in this wind farm, which obtains an initial design with AEP = 164.61 GWh, as shown in Figure 8.

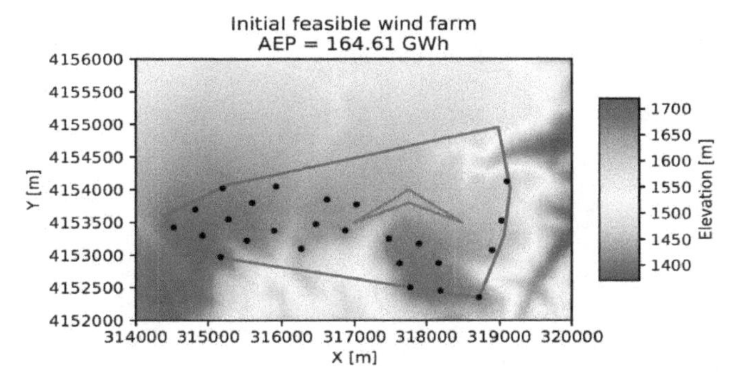

Figure 8. Initial design obtained by the Heuristic Fill algorithm.

Compared to the original wind farm, the design found by the Heuristic Fill algorithm obtains an impressive 1.72% increase of AEP (from 161.83 GWh to 164.61 GWh). Note that this increase is achieved by a deterministic filling process without any searching process, thus can be done in several seconds.

To show the performance of the proposed framework, multiple runs of Random Search with different settings of the optimization algorithm are carried out, i.e., optimization runs with different numbers of total evaluation number, E_{max}, and maximal step size, M_{max}. For each of these parameter combinations, 10 optimization runs are done to obtain the statistics. Cases starting from the original design (as shown in Figure 6) and the initial design found by the Heuristic Fill algorithm (as shown in Figure 8) are both tried. The results are summarized in Table 2. Note that the percentages of AEP that increase in the optimized wind farm design compared to the original wind farm (161.83 GWh) are marked with underlines inside the parentheses, and the CPU time shown here is the mean value per run for 10 runs by a Python 3.6 implementation of the framework on a laptop with an Intel® i5-2520M CPU @2.50 GHz.

Table 2. Performance of the framework in multiple runs with different settings.

Opt. Setting		AEP [GWh] (Increase [%])				CPU Time [s]
E_{max} [-]	M_{max} [m]	Initial	Opt. min	Opt. mean	Opt. max	Mean
1000	5000	161.83	164.62 (1.72)	165.03 (1.98)	165.67 (2.37)	2137.5
1000	50	164.61	165.18 (2.07)	165.27 (2.12)	165.30 (2.14)	1824.6
1000	500	164.61	165.17 (2.06)	165.28 (2.13)	165.41 (2.21)	1923.7
1000	5000	164.61	165.17 (2.06)	165.45 (2.24)	165.79 (2.45)	2272.3
5000	5000	164.61	165.76 (2.42)	166.02 (2.58)	166.17 (2.68)	12,042.6
10,000	5000	164.61	165.95 (2.55)	166.11 (2.64)	**166.21 (2.70)**	24,613.0

As shown in Table 2, allowing the optimization process to run longer, i.e., with a larger number of evaluations, generally obtains better results (see the last three rows in Table 2). The other advantage for an optimization process is starting from a better initial design, as the results from the initial design obtained by the Heuristic Fill algorithm are much better than those from the original design using the same optimization setting (see the first and fourth rows of the results in Table 2).

The maximal step size also has an effect on the results, as it controls the degree of change that can be made to the current best design in each step. As discussed in Section 3.3, this parameter should be set large enough to explore the design space more thoroughly and thus find a better solution after a sufficient large number of evaluations. This is supported by the comparison of optimization runs with the same E_{max}, but different M_{max}, as shown in the second to fourth rows of Table 2.

The best optimized design found by all the optimization runs summarized in Table 2 is under the optimization setting: $E_{max} = 10,000$, $M_{max} = 5000$ m, with the optimized AEP as 166.21 GWh, representing a 2.70% increase to the original wind farm design. This design is found by the Random Search algorithm starting from the initial design obtained by the Heuristic Fill algorithm, and shown in Figure 9.

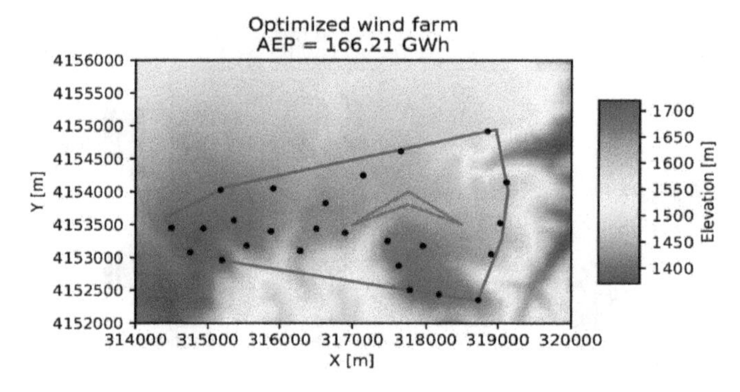

Figure 9. Best optimized design found by the framework with $E_{max} = 10,000$, $M_{max} = 5000$ m.

To better compare the performance of the original, initial, and optimized designs as shown in Figures 6, 8 and 9, their gross and net AEP values are listed in Table 3.

Table 3. AEP values comparison of the original, initial, and optimized wind farms.

Wind Farm	Net AEP [GWh]	Gross AEP [GWh]	Wake Loss [%]
Original (Figure 6)	161.83	169.75	4.67
Initial (Figure 8)	164.61	175.42	6.16
Optimized (Figure 9)	166.21	174.91	4.97

As Table 3 shows, the initial design found by the Heuristic Fill algorithm has both higher gross AEP and higher net AEP than the original design. Although the wake loss is higher for the initial design than the original one (6.16% versus 4.67%), the larger increase (3.34%) of gross AEP thanks to the higher mean wind speed of the sites found by the Heuristic Fill algorithm compensates the higher wake loss and still yields a substantial increase (1.72%) of net AEP. Note that the higher wake loss of the initial design can also be seen from the layout shown in Figure 8, as the turbines are placed much closer to each other than the original layout shown in Figure 6.

Since the optimized design in Figure 9 is obtained by the optimization run starting from the initial design in Figure 8, we can also compare these two designs. Apparently, the optimization algorithm makes a small sacrifice in gross AEP, i.e., allowing the gross AEP to decrease by 0.28% from 175.42 GWh to 174.91 GWh, while decreasing the wake loss percentage from 6.16% to 4.97%. The combined effect allows the optimized design to gain a 0.97% increase of net AEP to the initial design (from 164.61 GWh to 166.21 GWh), representing a 2.70% increase of net AEP when compared to the original design (from 161.83 GWh to 166.21 GWh).

Examining the layout of the initial design in Figure 8 and the layout of the optimized design in Figure 9, we can see that the turbines are scattered away from each other while most of the turbines are still located at high elevation sites. This shows that the optimization algorithm tries to lower wake effects while not sacrifice too much on the gross AEP, as suggested by the net and gross AEP comparison. To show the performance of the framework in multiple runs, the evolution histories of the 10 runs under the optimization setting: $E_{\max} = 10,000$, $M_{\max} = 5000$ m are shown in Figure 10, which demonstrate the random nature of the RS algorithm.

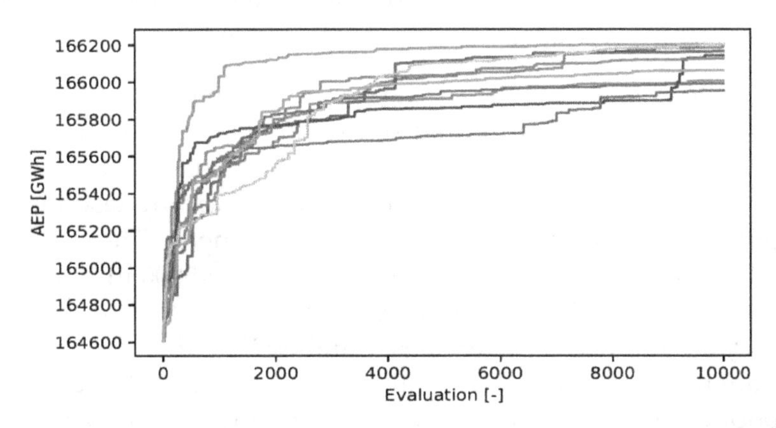

Figure 10. Evolution histories of the 10 Random Search runs under optimization setting: $E_{\max} = 10,000$, $M_{\max} = 5000$ m.

As Figure 10 shows, different runs of the optimization process yield different results. Thus, if it is possible, it is always advised to have multiple runs (10 or more) of the optimization process and use the best result. Also, it is shown for these 10 runs that the largest portion of the AEP increase is achieved in the first 2000 evaluations.

5. Comparison with Other Algorithms

To better show the effectiveness of the RS algorithm for wind farm design in complex terrain, we compare its performance with two widely used algorithms: Particle Swarm Optimization (PSO) and Genetic Algorithm (GA) [33].

PSO is a population based meta-heuristic algorithm inspired by bird flocking, which maintains a population of solutions (called particles) and moves each particle in the design space according to the global best solution in the population and the local best solution found by each particle. In this study, the version of PSO used by Chowdhury et al. in [34] is implemented in Python with the algorithm parameters set as recommended in Table 4 of [34]. Details of this algorithm are referred to in the paper [34].

GA is also a nature inspired population based meta-heuristic algorithm for optimization problems. Various versions of GA have been applied to solve the wind farm layout optimization problem, most of which are binary coded [33]. As a binary-coded GA makes restrictions on where to place turbines and thus limits the design space, we choose to implement a real-coded GA (RCGA) recently proposed by Chuang et al. [35], which has no such restrictions. This version of RCGA was developed for constrained optimization based on three specially designed evolutionary operators, i.e., ranking selection, direction-based crossover, and dynamic random mutation, and showed a better performance than traditional versions of RCGA for a variety of benchmark constrained optimization problems [35]. A Python version of this algorithm is implemented in this study using the parameters recommended by the authors of [35]. Details of this algorithm can be found in [35].

For the sake of simplicity, we consider the design optimization problem for the same wind farm as in the case study with only bounds on the design variables, i.e., x and y coordinates of wind turbines. No constraints on wind resource, terrain parameters, and minimal distances are considered. The bounds are set as the boundary of the rectangle area shown in Figure 6, i.e., $314,000 \text{ m} \leq x \leq 320,000 \text{ m}$ and $4,152,000 \text{ m} \leq y \leq 4,156,000 \text{ m}$. Following the recommendations in [34,35], the size of the population is set as 250 for both PSO and RCGA.

Two scenarios for setting the initial solution(s) are tested. In the random scenario, 250 initial layouts are generated randomly by putting 25 turbines randomly inside the bounds. These initial layouts are used in PSO and RCGA as the initial population, while the worst performing initial layout, i.e., the layout with lowest AEP, is used as the initial solution in RS.

As the initial solution(s) has influences on the optimization results obtained by meta-heuristics, there has been an effort on seeding good initial solution(s) for meta-heuristics in the field of wind farm layout optimization [31]. We address this issue in the heuristic fill scenario. In this scenario, an initial layout is first obtained by using the Heuristic Fill algorithm described in Section 3 with the minimal distance requirement set as four times of the rotor diameter. Then, a group of 124 layouts are generated by running RS 124 times with $E_{\max} = 100$ and $M_{\max} = 5000$ m from the initial layout found by Heuristic Fill. Thus, these 124 initial layouts have a higher or the same AEP as the one found by the Heuristic Fill, as they are improved better ones found by RS. After this process, we have 125 good initial layouts, which are then seeded in the initial populations of PSO and GA, while the remaining halves of the initial populations are filled with random layouts. In this way, the initial populations for both PSO and GA have some good quality initial solutions and some random initial solutions, thus they have a good degree of randomness, which is essential for the diversification. For RS, the worst performing layout among the 125 good initial layouts is used as the initial solution.

The performance of these three algorithms with a different number of evaluations ($E_{\max}$) for both scenarios are summarized in Table 4.

Table 4. Performance comparison of three algorithms: PSO (particle swarm optimization), RCGA (real-coded genetic algorithm) and RS (random search).

Initial Scenario	Algorithm	AEP [GWh]			E_{max} [-]
		Worst Initial	Best Initial	Optimized	
Random	PSO	119.12	152.44	160.76	5000
	RCGA	119.12	152.44	156.68	5000
	RS	119.12	119.12	173.58	5000
	PSO	119.12	152.44	161.66	50,000
	RCGA	119.12	152.44	165.43	50,000
	RS	119.12	119.12	**174.26**	10,000
Heuristic fill	PSO	170.27	171.18	171.74	10,000
	RCGA	170.27	171.18	171.96	10,000
	RS	170.27	170.27	174.10	10,000
	PSO	170.27	171.18	171.84	50,000
	RCGA	170.27	171.18	172.58	50,000
	RS	170.27	170.27	173.80	5000

Note: 'Worst/best initial' represents the worst/best performing initial solution among the 250 random initial layouts for the random scenario and the 125 seeded good initial layouts for the heuristic fill scenario. These two values are the same for RS as it maintains only one solution. 'Optimized' denotes the global best solution found by PSO/RCGA.

It is worth to note that in both scenarios, RS starts from the worst initial solution among the solutions seeded in the initial population for PSO and GA, thus with a disadvantage in terms of the quality of the initial solution(s). Despite this disadvantage, RS shows a much better performance than PSO and RCGA. As Table 4 shows, RS with 10,000 evaluations obtains the best optimized layout in both scenarios. Compared with PSO and RCGA, RS with only 5000 evaluations largely outperforms PSO and RCGA with 50,000 evaluations (equivalent to 200 generations) in both scenarios, which clearly demonstrates the effectiveness of RS. To better visualize the difference between algorithms, the evolution histories of the best solutions in the different runs of PSO, RCGA, and RS are shown in Figure 11.

Figure 11 shows that: (1) starting from good initial solutions makes PSO and RCGA converge to much better optimized solutions, while its benefit for RS is not so profound; (2) RCGA obtains better results than PSO after a sufficient number of evaluations (20,000); and (3) RS converges faster to a better solution than PSO and RCGA in both scenarios. Based on the results described above, it is natural to conclude that RS used in the current framework performs much better than PSO and RCGA for this case.

The best optimized design, which is found by RS with 10,000 evaluations in the random scenario, is shown in Figure 12.

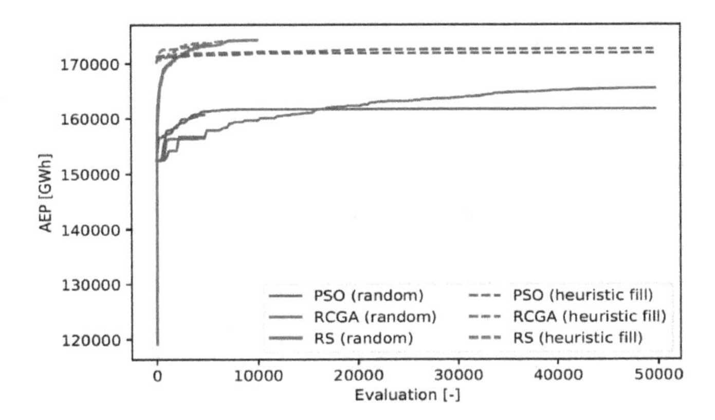

Figure 11. Evolution histories of the best solutions in different runs of PSO (particle swarm optimization), RCGA (real-coded genetic algorithm), and RS (random search).

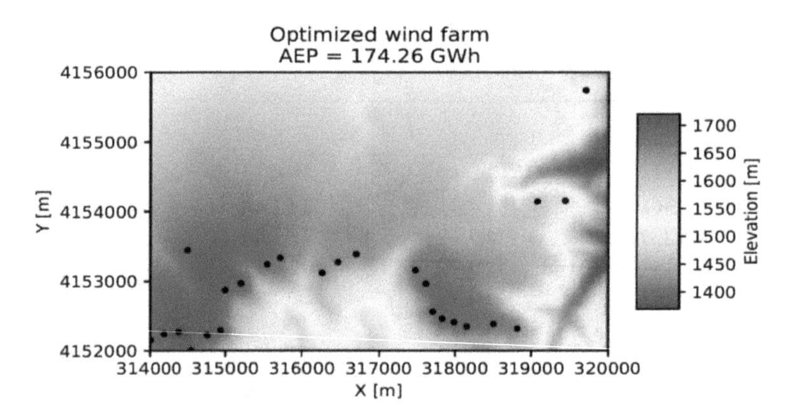

Figure 12. Best optimized design for the wind farm without any constraints (found by Random Search with $E_{\max} = 10,000$ and $M_{\max} = 5000$ m in the random scenario)

Without the limitation of inclusive and exclusive boundaries, the turbines in the optimized design in Figure 12 are spread out in the whole rectangle area and mostly occupying high elevation sites. Note that these sites also have high mean wind speeds as shown in Figure 1b. To place more turbines on the limited high elevation sites, some of the turbines are placed quite close (down to 170.44 m), which could cause problems on fatigue loads for certain turbines. Although the optimized wind farm has a much higher AEP (174.26 GWh), representing a 7.68% increase over the original wind farm, it should be interpreted with caution, as in one aspect, it occupies a much larger area and thus requires more investments on electrical cables and access roads, and in the other aspect, it places several turbines too close, thus it will have difficulty satisfying the requirements on fatigue loads.

6. Conclusions

Wind farm design in complex terrain is a crucial yet challenging task. In this work, we present an overall design optimization framework for tackling this task. This framework combines results from state-of-the-art wind resource assessment tools, such as WAsP CFD, a fast engineering wake model, and an advanced optimization algorithm, to solve the wind farm design optimization problem in complex terrain under various realistic constraints and requirements. This makes the framework a valuable tool that can be used by designers/developers who need to optimize wind farm designs in complex terrain under a real-life scenario.

While the framework in its default setting considers maximizing the total net AEP, its modular architecture makes it easy to be extended to consider other objective function(s). More constraints, requirements, and other optimization algorithms can also be easily implemented and added.

Other contributions of this work include: a terrain ruggedness index (TRI) to characterize the terrain feature and model constraints on the terrain ruggedness; and a fast and simple algorithm called Heuristic Fill to find good initial designs for wind farms in complex terrain, by using a deterministic process based on wind resource considerations and constraints.

The case study of a real wind farm in complex terrain demonstrates the effectiveness of the framework and the performance of proposed algorithms. An impressive increase of net AEP (up to 2.70%) is achieved by using the framework, while respecting realistic constraints on inclusive and exclusive boundaries, minimal mean wind speed, and maximal TRI, and minimal distance constraints between turbines. In comparison with two widely used algorithms (PSO and real-coded GA) for the same wind farm without any constraints, the Random Search algorithm also shows a much better performance in terms of convergence speed and optimization results. This of course demonstrates the advantages of the developed framework and proves the usefulness for wind farm designers/developers.

Although the wind farm design considered in the current framework concerns only the turbine layout, other design variables, such as number, hub height(s), and type(s) of turbines, foundations, access roads, and electrical systems, also play important roles for an overall design optimization of a wind farm. Our future works will investigate these problems and extend the framework in this direction.

Author Contributions: J.F. and W.Z.S. conceived and designed the study; J.F. wrote the code and performed the computation; J.F., W.Z.S. and Y.L. analyzed the results and wrote the paper.

Acknowledgments: The authors wish to give special thanks to the Danish and Chinese partners (DTU Wind Energy, EMD International A/S, North West Survey and Design Institute of Hydro China Consultant Corporation, and HoHai University) for their active collaborations.

References

1. GWEC: Global Cumulative Installed Capacity 2001–2017. Available online: http://gwec.net/global-figures/graphs/ (accessed on 7 July 2018).
2. GWEC: Global Cumulative and Annual Offshore Wind Capacity End 2017. Available online: http://gwec.net/global-figures/graphs/ (accessed on 7 July 2018).
3. Alfredsson, P.H.; Segalini, A. Wind farms in complex terrains: An introduction. *Phil. Trans. R. Soc. A* **2017**, *375*, 20160096. [CrossRef] [PubMed]
4. Manwell, J.; McGowan, J.; Rogers, A. *Wind Energy Explained: Theory, Design, and Application*, 2nd ed.; John Wiely & Sons: Hoboken, NJ, USA, 2009.
5. DTU Wind Energy: WAsP. Available online: http://www.wasp.dk/wasp (accessed on 7 July 2018).
6. Mortensen, N.G. *Wind Resource Assessment Using the WAsP Software*; DTU Wind Energy E-0135; Technical University of Denmark: Roskilde, Denmark, 2016.
7. Burton, T.; Jenkins, N.; Sharpe, D.; Bossanyi, E. *Wind Energy Handbook*, 2nd ed.; John Wiely & Sons: Hoboken, NJ, USA, 2011.
8. Queney, P. The problem of airflow over mountains: A summary of theoretical studies. *Bull. Am. Meteorol. Soc.* **1948**, *29*, 16–26. [CrossRef]
9. Wood, N. Wind flow over complex terrain: A historical perspective and the prospect for large-eddy modelling. *Bound.-Lay. Meteorol.* **2000**, *96*, 11–32. [CrossRef]
10. Landberg, L.; Myllerup, L.; Rathmann, O.; Petersen, E.L.; Jørgensen, B.H.; Badger, J.; Mortensen, N.G. Wind resource estimation—An overview. *Wind Energ.* **2003**, *6*, 261–271. [CrossRef]
11. González, J.S.; Payán, M.B.; Santos, J.M.R.; González-Longatt, F. A review and recent developments in the optimal wind-turbine micro-siting problem. *Renew. Sust. Energ. Rev.* **2014**, *30*, 133–144. [CrossRef]
12. Song, M.X.; Chen, K.; He, Z.Y.; Zhang, X. Bionic optimization for micro-siting of wind farm on complex terrain. *Renew. Energ.* **2013**, *50*, 551–557. [CrossRef]
13. Song, M.X.; Chen, K.; He, Z.Y.; Zhang, X. Optimization of wind farm micro-siting for complex terrain using greedy algorithm. *Energy* **2014**, *67*, 454–459. [CrossRef]
14. Feng, J.; Shen, W.Z. Wind farm layout optimization in complex terrain: A preliminary study on a Gaussian hill. *J. Phys. Conf. Ser.* **2014**, *524*, 012146. [CrossRef]
15. Feng, J.; Shen, W.Z. Optimization of wind farm layout: A refinement method by random search. In Proceedings of the 2013 International Conference on aerodynamics of Offshore Wind Energy Systems and wakes (ICOWES 2013), Lygnby, Denmark, 17–19 June 2013.
16. Feng, J.; Shen, W.Z. Solving the wind farm layout optimization problem using random search algorithm. *Renew. Energ.* **2015**, *78*, 182–192. [CrossRef]
17. Kuo, J.Y.; Romero, D.A.; Beck, J.C.; Amon, C.H. Wind farm layout optimization on complex terrains—Integrating a CFD wake model with mixed-integer programming. *Appl. Energ.* **2016**, *178*, 404–414. [CrossRef]
18. Feng, J.; Shen, W.Z.; Hansen, K.S.; Vignaroli, A.; Bechmann, A.; Zhu, W.J.; Larsen, G.C.; Ott, S.; Nielsen, M.; Jogararu, M.M.; et al. Wind farm design in complex terrain: The FarmOpt methodology. In Proceedings of the China Wind Power 2017, Beijing, China, 17–19 October 2017.
19. Feng, J.; Shen, W.Z. Design optimization of offshore wind farms with multiple types of wind turbines. *Appl. Energ.* **2017**, *205*, 1283–1297. [CrossRef]
20. EMD: WindPRO. Available online: https://www.emd.dk/windpro/ (accessed on 7 July 2018).

21. DTU Wind Energy: WAsP CFD. Available online: http://www.wasp.dk/waspcfd (accessed on 7 July 2018).
22. Politis, E.S.; Prospathopoulos, J.; Cabezon, D.; Hansen, K.S.; Chaviaropoulos, P.K.; Barthelmie, R.J. Modeling wake effects in large wind farms in complex terrain: The problem, the methods and the issues. *Wind Energ.* **2012**, *15*, 161–182. [CrossRef]
23. Sessarego, M.; Shen, W.Z.; van der Laan, M.P.; Hansen, K.S.; Zhu, W.J. CFD Simulations of flows in a wind farm in complex terrain and comparisons to measurements. *Appl. Sci.* **2018**, *8*, 788. [CrossRef]
24. Song, M.X.; Chen, K.; He, Z.Y.; Zhang, X. Wake flow model of wind turbine using particle simulation. *Renew. Energ.* **2012**, *41*, 185–190. [CrossRef]
25. Kuo, J.; Rehman, D.; Romero, D.A.; Amon, C.H. A novel wake model for wind farm design on complex terrains. *J. Wind Eng. Ind. Aerod.* **2018**, *174*, 94–102. [CrossRef]
26. Feng, J.; Shen, W.Z. Modelling wind for wind farm layout optimization using joint distribution of wind speed and wind direction. *Energies* **2015**, *8*, 3075–3092. [CrossRef]
27. Riley, S.J.; DeGloria, S.D.; Elliot, R. A terrain ruggedness index that quantifies topographic heterogeneity. *Intermt. J. Sci.* **1999**, *5*, 23–27.
28. Feng, J.; Shen, W.Z.; Xu, C. Multi-objective random search algorithm for simultaneously optimizing wind farm layout and number of turbines. *J. Phys. Conf. Ser.* **2016**, *753*, 032011. [CrossRef]
29. Feng, J.; Shen, W.Z. Wind farm power production in the changing wind: Robustness quantification and layout optimization. *Energ. Convers. Manag.* **2017**, *148*, 905–914. [CrossRef]
30. Blum, C.; Roli, A. Metaheuristics in combinatorial optimization: Overview and conceptual comparison. *ACM Comput. Surv.* **2003**, *35*, 268–308. [CrossRef]
31. Saavedra-Moreno, B.; Salcedo-Sanz, S.; Paniagua-Tineo, A.; Prieto, L.; Portilla-Figueras, A. Seeding evolutionary algorithms with heuristics for optimal wind turbines positioning in wind farms. *Renew. Energ.* **2011**, *36*, 2838–2844. [CrossRef]
32. Han, X.; Liu, D.; Xu, C.; Shen, W.Z. Atmospheric stability and topography effects on wind turbine performance and wake properties in complex terrain. *Renew. Energ.* **2018**, *126*, 640–651. [CrossRef]
33. Khan, S.A.; Rehman, S. Iterative non-deterministic algorithms in on-shore wind farm design: A brief survey. *Renew. Sust. Energ. Rev.* **2013**, *19*, 370–384. [CrossRef]
34. Chowdhury, S.; Zhang, J.; Messac, A.; Castillo, L. Unrestricted wind farm layout optimization (UWFLO): Investigating key factors influencing the maximum power generation. *Renew. Energ.* **2012**, *38*, 16–30. [CrossRef]
35. Chuang, Y.C.; Chen, C.T.; Hwang, C. A simple and efficient real-coded genetic algorithm for constrained optimization. *Appl. Soft Comput.* **2016**, *38*, 87–105. [CrossRef]

7

Measurements of High-Frequency Atmospheric Turbulence and its Impact on the Boundary Layer of Wind Turbine Blades

Alois Peter Schaffarczyk * and Andreas Jeromin

Mechanical Engineering Department, Kiel University of Applied Sciences, D-24149 Kiel, Germany; andreas.jeromin@fh-kiel.de

* Correspondence: Alois.Schaffarczyk@FH-Kiel.de

Abstract: To gain insight into the differences between onshore and offshore atmospheric turbulence, pressure fluctuations were measured for offshore wind under different environmental conditions. A durable piezo-electric sensor was used to sample turbulent pressure data at 50 kHz. Offshore measurements were performed at a height of 100 m on Germany's FINO3 offshore platform in the German Bight together with additional meteorological data provided by Deutscher Wetterdienst (DWD). The statistical evaluation revealed that the stability state in the atmospheric boundary does not seem to depend on simple properties like the Reynolds number, wind speed, wind direction, or turbulence level. Therefore, we used higher statistical properties (described by so-called shape factors) to relate them to the stability state. Data was classified to be either within an unstable, neutral, or stable stratification. We found that, in case of stable stratification, the shape factor was mostly close to zero, indicating that a thermally stable environment produces closer-to Gaussian distributions. Non-Gaussian distributions were found in unstable and neutral boundary layer states, and an occurrence probability was estimated. Possible impacts on the laminar-turbulent transition on the blade are discussed with the application of so-called laminar airfoils on wind turbine blades.

Keywords: turbulence; super-statistics; piezo-electric flow sensor; ABL stability; laminar-turbulent transition

---

1. Introduction

The use of wind energy has been very successful during the last decades [1], reaching a nearly stable annual investment corresponding to 50 GW of rated power word-wide. This was in connection and in parallel to an impressive development in Wind Turbine Aerodynamics [2] and even a special branch of *wind energy meteorology* has been established [3].

Due to the increased number of annual new installations, site assessment for wind farms has become more and more important and sophisticated, even under offshore conditions, with application to tailored turbine design as well. For a selection of important references on properties of the atmospheric boundary layer on- and off-shore see, for example, reference [4–7]. In most cases, turbulence has been treated in the context of loads, and frequency ranges higher than a few Hertz have generally been considered to be of no importance. However, this high frequency turbulence plays a significant role in laminar-to-turbulent transition inside the boundary layer of blades for airplanes [8,9] and wind turbines [10,11]. Due to the much higher drag of turbulent parts, this may give rise lower wind turbine efficiency (in terms of c_P, the number $0 \leq c_P \leq 0.596$ measures the fraction of extracted power from the wind) as desired, and the proper choice of airfoils is crucial.

In this paper, these high-frequency turbulent statistics were studied with respect to higher order statistical moments in some detail using the method of super-statistics [12]. The shape factor from reference [13]—which gives estimates of the extent to which turbulent fluctuations diverge from Gaussian behavior—was of special interest.

Unlike earlier investigations, the non-Gaussianity of turbulent pressure fluctuations could not be linked to one of the more popular atmospheric parameters like the Reynolds number based on Taylor's micro-scale [14], wind speed, or others. Therefore, we propose characterizing our findings according to the stability of the atmospheric boundary layer in terms of Richardson's number. This paper is organized as follows: Firstly, the measurement setup is described together with the locations at which they were performed. Then, we describe the procedure of data analysis. After that, we present and discuss our findings and finally, draw some conclusions.

Parts of the material have been presented earlier in unpublished proceedings of ICOWES2013 [15].

2. Measurements

Our first set of high-resolution measurements (during the years 2009 until 2011) was performed using piezoelectric pressure sensors from PCB Piezotronics (Figure 1) that were connected to an imc Meßsystem GmbH CS-1208 data logger. The diameter of the sensing element was 15 mm. The minimal pressure resolution was 0.13 Pa, and the possible temporal resolution ranged from 2.5 Hz to 80,000 Hz.

Reference [16] discusses how the **length**, which is usually more than 100 times larger than the diameter of a hot-wire, influences the spatially resolution.

In our measurements, the pressure data was sampled at 50 kHz with a total duration of 100 s. The wind speed and temperature were sampled with a 1 Hz temporal resolution.

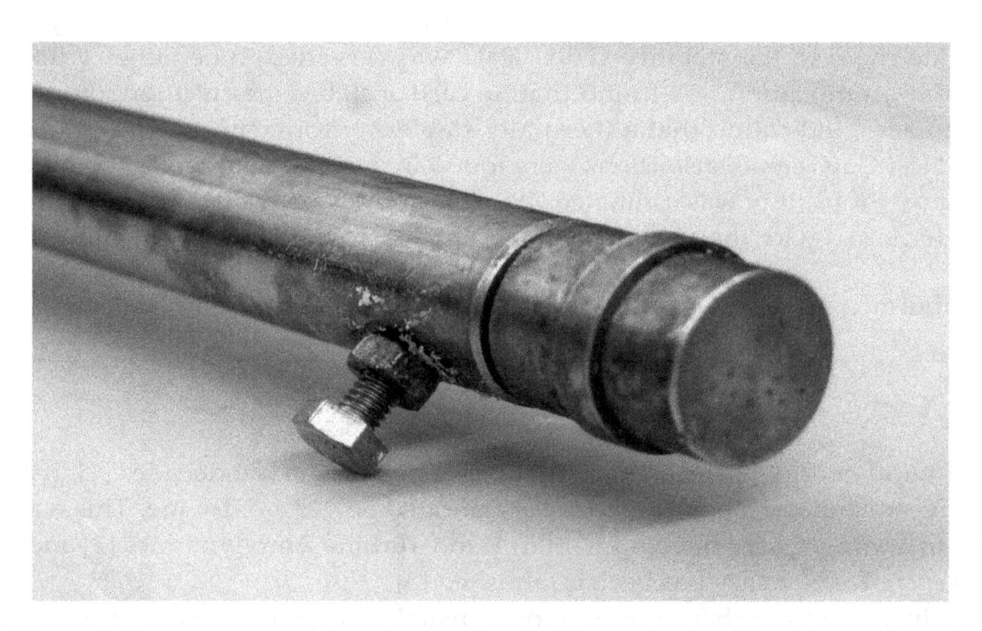

Figure 1. Pressure sensor after six months of offshore service (diameter = 12 mm).

The piezoelectric pressure sensor was calibrated against a hot wire anemometer in a wind tunnel at the University of Oldenburg. The turbulent wind speed from the hot wire anemometer and the variation in turbulent pressure showed the same statistical properties up to 3 kHz (see [17]).

Onshore measurements were conducted at the Kaiser–Wilhelm–Koog test site of Germanischer Lloyd/Garrad Hassan (see Figure 2). A lattice tower of 60 m height provided booms to mount the pressure sensor on and other measurement equipment. The pressure sensor was mounted at a height of 55 m. The wind speed and wind direction were recorded at a height of 55 m by calibrated cup anemometers and wind vanes, respectively. The temperature was recorded at height of 53 m by a resistor-type thermometer.

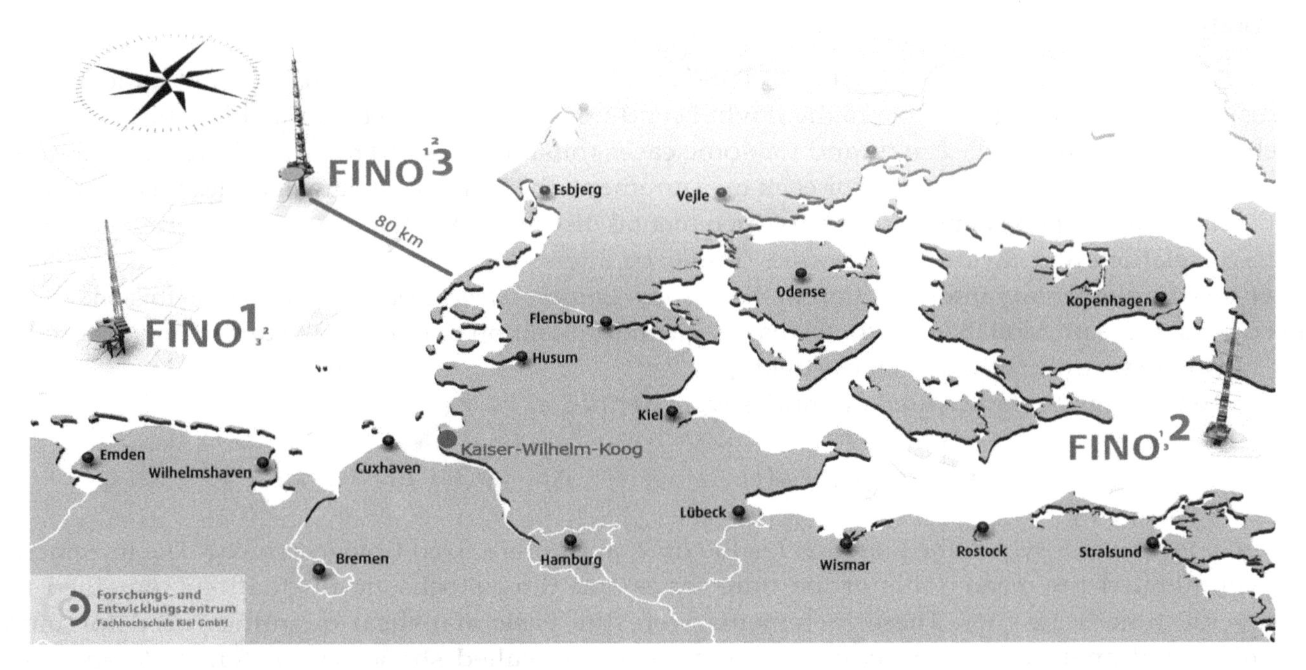

Figure 2. Locations of onshore (Kaiser–Wilhelm–Koog) and offshore (FINO3 platform) test sites, 80 km west of the island of Sylt. ©FEZ FH Kiel GmbH, Graphics: Bastian Barton.

FEZ Kiel's platform FINO3 was the location for the offshore measurements. About 80 km west of the island of Sylt (see Figure 2) the platform was constructed close to (at that time not) operational wind farms like DanTysk and Sandbank 24. The tower is a lattice tower type with booms of sufficient length for undisturbed measurements. Two pressure sensors (Figure 3) were mounted at a height of about 100 m above the mean sea level for parallel operation. The data acquisition equipment for meteorological signals was similar to that used onshore. The wind speed and wind direction were measured at a height of 100 m, and the temperature was measured at a height of 95 m above the mean sea level.

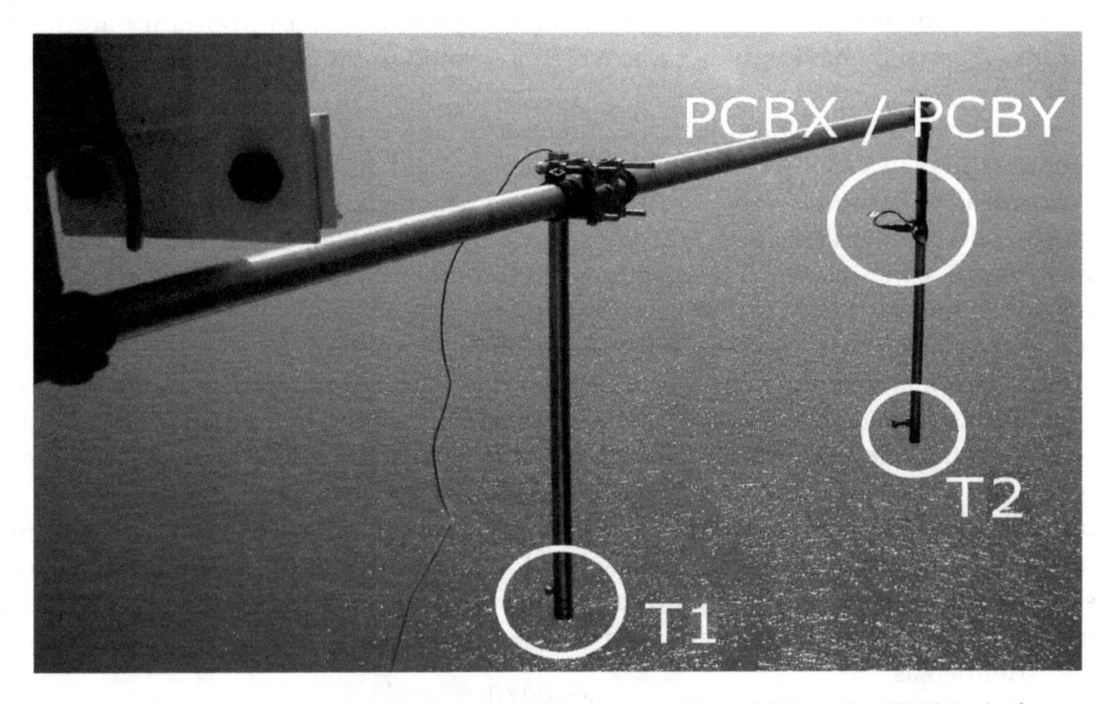

Figure 3. Installation of two parallel pressure sensors (T1 and T2) at the FINO3 platform.

3. Analysis

Our recorded data sets were put into wind speed classes starting at 6 m/s, 12 m/s, and 16 m/s. Measurements were triggered manually if wind conditions were regarded as suitable. The bin size for each class was generally ± 2 m/s, and for some cases (offshore class 10 m/s, offshore class 12 m/s), it was less (lower limit: -1 m/s). For each measurement, the statistical characteristics were analyzed with respect to the power-spectral density, incremental distributions, shape factors, structure functions, auto-correlations and Taylor's micro-scale. (It may be interesting to note that Taylor's micro-scale may have an complementary interpretation to the *Markov–Einstein coherence length* [18]. To use this one seems to be much more physical than the somewhat vague interpretation of the average turbulent vortex size.) We now explain our methods in more detail.

The increments of measured quantities were defined as

$$\Delta u = u(t + \Delta t) - u(t) \qquad \Leftrightarrow \qquad \Delta p = p(t + \Delta t) - p(t), \tag{1}$$

where u is the velocity, p is the pressure, t is the time, and Δt is a fixed time increment. The increments also eliminated the mean value of the time series and thus, stochastic fluctuations remained for a specific time scale (Δt). These increments were the basic statistical quantities used for more sophisticated analyses like structure functions or the so-called shape parameter. Reference [13] suggested that the distribution of increments may be described as a superposition of two distributions, one of them being lognormal. The shape factor then may be regarded as a measure of the level of intermittency. In accordance with Beck's approach from reference [12], the shape parameter can be calculated by the 2nd and 4th order moments of the distribution:

$$s_u^2 = \ln\left(\frac{1}{3}\frac{\langle \Delta u^4 \rangle}{\langle \Delta u^2 \rangle^2}\right) \qquad \Rightarrow \qquad s_p^2 = \ln\left(\frac{1}{3}\frac{\langle \Delta p^4 \rangle}{\langle \Delta p^2 \rangle^2}\right), \tag{2}$$

where s_u^2 is the shape factor for the velocity and s_p^2 is the shape factor for the pressure. The brackets $\langle x \rangle$ define the mean value of a quantity (x). Further properties are given in reference [12]. A value of $s^2 \approx 0$ indicates a normal distribution. It is worth noting that another important parameter for the deviation from Gaussian behavior, skewness, which is related to third-order moments, was found to be small [17].

It has well-known, at least since the work of reference [19], that turbulent velocities and pressures obey different scaling laws ($\sim k^{-5/3}$ and $\sim k^{-7/5}$, respectively), although they are seemingly related by $p \sim v^2$ according to Bernoulli's law. A direct comparison of a hot-wire with piezo-electric pressure sensor data showed comparable power density spectra up to 4 kHz (see [17], unpublished), however. Theoretical as well as experimental investigations of reference [20] are still incomplete, so that there is no clear answer so far about how s_u^2 and s_p^2 are related to each other.

When heat transfer occurs, such as in the atmospheric boundary layer if the sea is warmer than the air (autumn), thermal stratification becomes more important. Simple static stability of the boundary layer may be used to characterize levels of turbulence. As was shown by reference [21], the turbulence intensity differs remarkably for stable and unstable flow when used to describe fatigue loads of wind turbines.

A simple characterization of a stratified boundary layer exposed to heat transfer was introduced by reference [4] as being stable, neutral, or unstable. Different calculation methods have been used to distinguish these states ([4]) depending on what is known in terms of input. In our case, a Richardson's number approach was used, in accordance with common practice in wind energy meteorology [22–24].

This was defined as

$$Ri = \frac{g}{T}\frac{\partial \theta / \partial z}{(\partial v / \partial z)^2}, \tag{3}$$

where g is the acceleration of gravity, T is the mean absolute temperature, z is the height normal to the surface, and $\partial v/\partial z$ is the mean velocity the gradient. The z-direction was assumed to be vertical. The potential temperature (θ) is defined as

$$\theta = T\left(\frac{p_0}{p}\right)^{R/c_p}, \tag{4}$$

where p is the pressure, and $p_0 = 1000$ hPa is the reference pressure, $R = 289\ \frac{\mathrm{J}}{\mathrm{kg\,K}}$ is the specific gas constant, and $c_p = 1005\ \frac{\mathrm{J}}{\mathrm{kg\,K}}$ is the specific heat capacity of air. It has to be noted that we checked for the influence of humidity and found that the influence on density as well as specific humidity [5] was below 1%. We therefore neglected all moisture corrections to Richardson's Number (Ri).

In Equation (3), the quantity $\frac{g}{T}\partial\theta/\partial z$ describes the forces introduced by heat transfer in the boundary layer. The term $(\partial v/\partial z)^2$ represents the momentum forces in the boundary layer. If we consider two points in the boundary layer at heights z_1 and z_2 where $z_1 < z_2$ and $\Delta z = z_2 - z_1$, Ri from (3) can be rewritten with the differences between the two locations as

$$Ri = \frac{g}{T}\frac{\Delta\theta/\Delta z}{(\Delta v/\Delta z)^2} = \frac{g}{T}\frac{(\theta_2-\theta_1)/\Delta z}{((v_2-v_1)/\Delta z)^2}. \tag{5}$$

We always assumed the presence of positive velocity gradients so only the convective part of Equation (5) remained. Three typical situations were then distinguished as follows:

$\theta_2 > \theta_1$ The surface is colder than the fluid and the gradient becomes $\partial\theta/\partial z > 0$ and therefore, $Ri > 0$. Heat is transported by conduction only, and a convection flow does not occur. In this case, the stratification is strong, and turbulence gets damped. The boundary condition is stable for $Ri > 0$.

$\theta_2 = \theta_1$ The temperature gradient is zero and therefore, $Ri = 0$. There is no temperature gradient and therefore, no conduction nor convection. This condition is called neutral.

$\theta_2 < \theta_1$ The surface is warmer than the fluid and the gradient becomes $\partial\theta/\partial z < 0$ and therefore, $Ri < 0$. Heat is transported by conduction and by convection from the surface to the fluid. The convection results in a vertical, upward component of the flow that interacts with the horizontal velocity component. This leads to the production of turbulence in the boundary layer and therefore, is called unstable.

From our measurements we obtained the necessary values for p, T_{air} at a height of 100 m , wind speed v_{air}, and wind direction as well as the sensor signal of the piezoelectric microphone. By assuming the air was a perfect gas, the density ρ was calculated by the perfect gas law.

For the derivatives $\partial\theta/\partial z$ and $\partial v/\partial z$, the temperature and wind speed at a second height needed to be known. Unfortunately, these data were not available during most of our measurements. Therefore, these gradients had to be approximated in a different way.

The ground temperature (T_{gnd}) was estimated using data from the Deutscher Wetterdienst (DWD). For the onshore measurements, the ground temperature at a nearby location was available at hourly samples. The temperature of the sea for the offshore measurements was interpolated from measurement stations on the island of Helgoland and at List on the island of Sylt.

The sign of Ri mostly depends on the temperature gradient, so the velocity gradient can be regarded as of lower importance. Therefore, as a rough estimate, $v_{gnd} \approx 0$ was used to calculate the velocity gradient. This mostly affected the absolute value of Ri but not the heat transfer conditions in the atmospheric boundary layer.

With this procedure, estimates for Ri were possible. The identification of stable or unstable states was simple. However, the neutral state required a value of exactly $Ri = 0$ which was difficult to attain precisely within our set of approximations. To indicate a nearly neutral state, a bandwidth of $|Ri| < 0.02$ was used instead.

4. Results

The computation of the shape parameter was straightforward, and the results are shown in Figure 4. The scale for the shape factor is shown on the left. Colors indicate locations, and symbols represent velocity classes. Time series with $s_p^2 < 2$ were grayed out, and will not be considered further in the following analysis. For purposes of comparison, shape factors from a particular velocity measurement ([25], $\bar{U} = 7.6$ m/s, $u' = 1.36$ m/s) were included (black line) as well.

The general behavior of both shape parameter functions for velocity and pressure was similar, whereas absolute values differed significantly. This behavior was also observed in reference [26]. Surprisingly, most of the pressure measurements resulted in s_p^2 close to zero.

We noted differences between the measurements of velocity and pressure concerning sampling rates and length of the time series.

In Figure 4, offshore curves are distinguished with numbers in brackets. They were all recorded on the same day with a short delay at the following times:

(1) 19 October 2010, 8:44 a.m.
(2) 19 October 2010, 8:54 a.m.
(3) 19 October 2010, 8:55 a.m.
(4) 19 October 2010, 8:57 a.m.

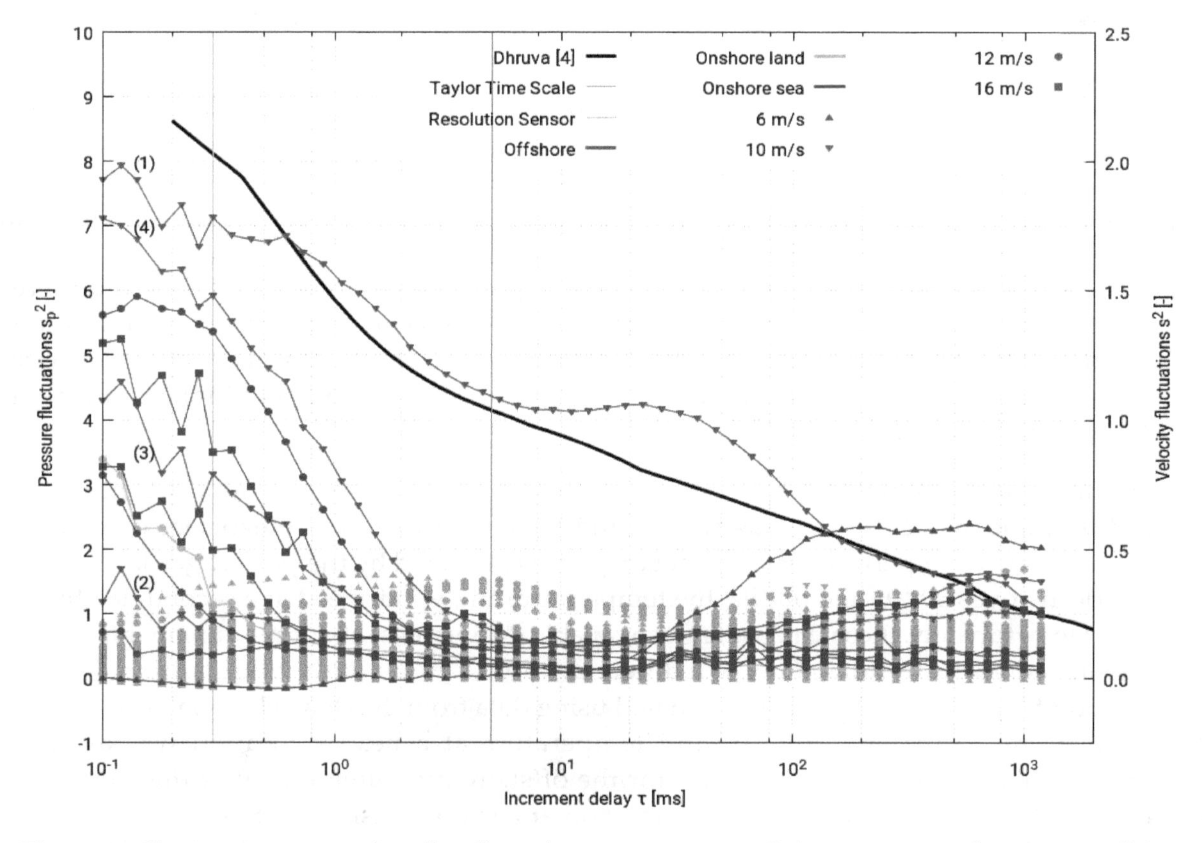

Figure 4. Shape parameters for all collected measurements. Colors represent locations: offshore, onshore with wind coming from land or sea side. The symbols represent the velocity classes for the wind speed. Measurements with $max(s_p^2) < 2$ are shown in grey. For reasons of comparison, Taylor's and Kolmogorov's time scales are given in 5 msec and 0.05 msec, respectively. The sensor's resolution goes down to approximately 0.3 msec only.

Pressure time series were recorded for a period of 100 s. For the first data set (1) s_p^2 was high. However, then, the pressure data resulted in $s_p^2 \approx 0$ until the values increased eight minutes later at

(2). The time series (2), (3), and (4) were consecutive, so it can be said that the shape parameter for the turbulent pressure fluctuations did not remain constant in time for this interval. When all time series from 8:44 a.m. to 8:57 a.m.were concentrated into one and evaluated according to a common s_p^2, a curve similar to the curve for velocity fluctuations was obtained. This has already been presented in reference [14].

We now try to relate these different statistical behaviors to the atmospheric boundary layer state. With our estimation of the Richardson number, a stability state was determined. Selected data from all our 119 measured data sets are shown in Tables 1 and 2. Measurements with $s_p^2 > 0$ are marked with a star in the last column.

As can be seen from Table 1, high s_p^2 ($s_p^2 > 0$) behavior corresponded to unstable to neutral Atmospheric Boundary Layer (ABL) conditions: $-0.13 \leq Ri \leq 0.01$. However, a significant correlation with the Richardson number was not found.

Table 1. Summary of selected Atmospheric Boundary Layer (ABL) stability cases: wind speeds (v_W) and turbulence intensities (Ti) from cup anemometers, Ri numbers and estimated ABL states.

Date & Time	Location	v_W (m/s)	Ti (%)	Ri (-)	Boundary Layer State	
21 October 2010, 7:56 a.m.	Offshore	6.6	8.2	−0.54	Unstable	
28 April 2008, 9:35 a.m.	Onshore	6.1	11.9	−0.15	Unstable	
19 October 2010, 8:57 a.m.	Offshore	10.8	2.2	−0.13	Unstable	*
19 October 2010, 8:44 a.m.	Offshore	10.2	3.1	−0.12	Unstable	*
19 August 2010, 8:13 a.m.	Offshore	12.7	2.8	−0.06	Unstable	
25 March 2008, 2:55 p.m.	Onshore	11.3	10.8	−0.03	Unstable	*
29 March 2008, 11:44 a.m.	Onshore	15.3	10.6	−0.01	Neutral	
30 March 2008, 6:04 p.m.	Onshore	15.7	5.6	0.00	Neutral	*
28 April 2008, 2:20 p.m.	Onshore	5.8	2.1	0.01	Neutral	*
1 May 2008, 2:55 a.m.	Onshore	6.0	6.0	0.10	Stable	
12 April 2008, 7:24 p.m.	Onshore	5.2	11.0	0.28	Stable	

4.1. Time Development of a Sample Time Series

Therefore, especially for the measurement on 19 October 2010, a possible transient behavior of the shape factor was investigated in more detail. The time development of measured data is presented in Table 2.

Table 2. Development of Time Series on October 19. Potential temperatures (θ), wind speeds (v_W) and turbulence intensities (Ti) from cup anemometers, Ri numbers, and estimated ABL states.

Time	θ_{air} (K)	θ_{gnd} (K)	v_W (m/s)	Ti (%)	Ri (-)	Boundary Layer State	$s_p^2 \neq 0$
8:44 a.m.	283.6	286.8	10.2	3.1	−0.12	Unstable	*
8:45 a.m.	283.5	286.8	10.2	3.9	−0.13	Unstable	
8:47 a.m.	283.4	286.8	10.0	7.5	−0.14	Unstable	
8:49 a.m.	283.4	286.8	10.1	7.3	−0.13	Unstable	
8:50 a.m.	283.5	286.8	9.3	6.7	−0.16	Unstable	
8:52 a.m.	283.5	286.8	11.2	6.0	−0.11	Unstable	
8:54 a.m.	283.5	286.8	10.8	7.7	−0.11	Unstable	*
8:55 a.m.	283.2	286.8	11.2	4.2	−0.12	Unstable	*
8:57 a.m.	283.1	286.8	10.8	2.2	−0.13	Unstable	*

The measurements started at 8:44 a.m. when s_p^2 was high and the turbulence intensity from cup anemometers was low. In the following 7 min, the time series were found to have $s_p^2 \approx 0$, with the mean velocity and potential temperature of the air remaining constant. Only the turbulence intensity increased from about 4 to about 8%.

At 8:54 a.m., the shape factor began to rise (see also curve (2) of Figure 4). Promptly, the turbulence intensity dropped more than 5.5 percentage points, and also, the potential temperature of the air decreased by 0.4 K with the rising shape factor.

Our interpretation is summarized as follows: the boundary layer, corresponding to time interval from 8:45 a.m. to 8:49 a.m. was stratified, and turbulent pressure fluctuations were distributed in a Gaussian way. At 8:50 a.m., a cluster of warm air rose up from the warmer sea surface to higher (and colder) regions ($\theta_{air} \approx$ consant, v_W fluctuating). This may have led to a stronger vertical shear, thereby disturbing the (Gaussian) turbulent structures.

4.2. Occurrence Probabilities

As was seen in our evaluations of boundary layer states and shape factors, the occurrence of high values for s_p^2 (a highly non-Gaussian behavior) stems from a non-linear, dynamic process with chaotic phases, for which the term *intermittency* was introduced [27]. The occurrence probabilities for all of our 119 measurements are listed in Table 3.

In the first row, the number of states is listed for all measurements. We see much more unstable (84) than neutral (23) or stable (12) conditions. The subsequent two rows list the numbers for offshore and onshore locations with their states. In the fourth row, the tallies for high shape factor events are listed for all measurements. The probability for the occurrence of a high-s_p^2 event at a specific boundary layer state is shown in the next line. The same was done for both locations, offshore and onshore.

To correctly interpret the data, one has to know the circumstances of the measurements. The onshore measurements took place in spring 2008 between March and May. During this time of year, the ground begins to warm up slowly while the air heats up much faster, yielding stable conditions. Project specific constraints limited the offshore measurement campaign to a period in late summer and autumn 2010. At this time of year, the sea is still heated up from the summer and warms the air near the surface, while the air temperature subsides, and many unstable conditions can be observed.

The focus was laid on the onshore measurements for the relation between the boundary layer state and the occurrence of high shape factors in the turbulent pressure fluctuations. With a joint probability of about 16.7%, a high-s_p^2 was found for unstable and neutral states (7 out of 42, see Table 3). However, in stable conditions, a high-s_p^2 was not observed. So, we propose that turbulent fluctuations in the pressure may deviate from a normal distribution much more frequently under unstable or neutral boundary conditions than under stable conditions.

Table 3. Correlation of the occurrence probability of a high shape factor to boundary layer state for all 119 measurements.

	Boundary Layer State			
	Unstable	**Neutral**	**Stable**	**Total**
All	84	23	12	119
Offshore	65	0	0	65
Onshore	19	23	12	54
All $s_p^2 \neq 0$	7	4	0	11
Probability (%)	8.3	17.4	0.0	9.2
Offshore $s_p^2 \neq 0$	4	0	0	4
Probability (%)	6.2	-/-	-/-	-/-
Onshore $s_p^2 \neq 0$	3	4	0	7
Probability (%)	15.8	17.4	0.0	-/-

4.3. Confidence Considerations

To check the confidence of our above assumption, possible sources for errors were investigated. Individual measurements at onshore locations were recorded at different hours and days, so they can be assumed to be independently and randomly sampled. The probability of occurrence of a high s_p^2 event under combined unstable or neutral conditions was estimated to be $\Phi_0 = 0.167$. The statistical test was based on a binomial distribution with the states *event occurred* and *no event occurred*.

To check the occurrence probability of an event in the onshore stable boundary layer states, for the null-hypothesis, an occurrence probability was assumed to be the same as for unstable/neutral states, and finding no event was just bad luck. The alternative hypothesis postulates a much lower probability of $\Phi_1 = 0.062$ for high-s_p^2 events under stable conditions (see Table 4).

Table 4. Error estimation for the occurrence of non-Gaussian turbulence for onshore stable conditions.

Hypotheses:	
H_0	Occurrence probability is $\Phi_0 = 0.167$ in stable conditions
H_1	Occurrence probability is $\Phi_1 < \Phi_0$ in stable conditions
	Error Type I
Number of samples	12
Occurrence probability	$\Phi_0 = 0.167$
Expected occurrences	$E_0 = 2.00$
Variance	$V_0 = 1.67$
Acceptance region for H_0	$\{1 \ldots 12\}$
Rejection region for H_0	$\{0\}$
Probability of error	11.2%

The number of expected events for H_0 in onshore stable conditions is given in Table 4, and $E_0 = 2$ and H_0 were accepted if at least one event was found. However, no event of a non-Gaussian distribution was found under onshore stable conditions, and H_0 was rejected. The type I error rate was about 11.2%. This is also the level of significance (α). To improve the statistical significance, the number of samples in stable conditions had to be raised up to 17 for $\alpha < 5\%$ (or 26 for $\alpha < 1\%$), but has not been done so far.

Since the offshore measurements were often consecutive within the same conditions, and no observations in neutral or stable conditions were acquired, the statistical confidence could not be estimated for the offshore turbulence. However, the calculated occurrence probability gave a good approximation to allow the quality of the type II error for the onshore unstable and neutral boundary layers to be checked. The setup for this test is presented in Table 5 and the errors were 11.4% for type I (H_0 was rejected despite it being true) and 14.9% for type II (H_0 was accepted when it was really false).

Table 5. Error estimation for the occurrence of non-Gaussian turbulence for the onshore **unstable/neutral** conditions.

Hypotheses:			
H_0	Occurrence probability is $\Phi_0 = 0.062$ in unstable/neutral conditions		
H_1	Occurrence probability is $\Phi_1 > \Phi_0$ in unstable/neutral conditions		
	Error Type I	**Error Type II**	
Number of samples			42
Occurrence probability	$\Phi_0 = 0.062$	$\Phi_1 = 0.167$	
Expected occurrences	$E_0 = 2.58$	$E_1 = 7.00$	
Variance	$V_0 = 2.43$	$V_1 = 5.83$	
Acceptance region for H_0			$\{0 \ldots 4\}$
Rejection region for H_0			$\{5 \ldots 42\}$
Probability of error	11.4%	14.9%	

These error estimates were acceptable with the small number of measurements and the occurrence probability onshore was reasonable.

5. Impact on Boundary Layer Transition on a Wind Turbine Blade

In the preceding sections, we have described that small-scale turbulence below a Taylor's length-scale of less than 50 mm are likely to show non-Gaussian behavior. We now briefly describe how this may be related to turbulent-laminar transitions on wind turbine blades. It has to be noted that more detailed investigations have been performed in the upper atmosphere for small aircraft [8,9]. Corresponding outdoor experiments on wind turbines were reported in reference [10,28]. It can be concluded that despite a high integral turbulence intensity of more that 10%, the energy content suitable for the receptivity of TS-waves is even smaller than in a wind-tunnel environment. More evidence for this **low-frequency cut-off** for aerodynamic important turbulence can be given by a simple argument to estimate an **upper** frequency $f^\star$ for a boundary layer (BL) responding to an oscillating outer flow. Stokes [29], pp 191 ff, showed that this outer flow (with ω) is damped out by viscosity according to $U = u_0 \cdot e^{-ky} \cdot cos(\omega t - ky)$ with $k = \sqrt{\omega/2\nu}$, ν being the kinematic viscosity and U being the main flow in the x-direction, with y being perpendicular to that. Now, if we introduce a *Stokes boundary layer thickness* by $\delta_S = 2\pi/k$, we get $f^\star = 4\pi\nu/\delta^2 \approx 200\ Hz$. Here, we have set δ_S to an approximate value of 1 mm, a typical value for laminar boundary layers on airfoils at Reynolds numbers of several million. Therefore, this high-frequency, non-Gaussian regime may play an important role in triggering the TS-type of (blade) boundary layer instability on the pathway towards fully developed turbulence.

6. Conclusions

High-frequency (above 100 Hz) resolved measurements of pressure fluctuations for different average wind velocities under onshore and offshore conditions were investigated with respect to higher order statistical properties and were related to atmospheric boundary layer stability. It was found that up to a frequency of 4 kHz, pressure and velocity fluctuations obey the same power spectrum, and further, out of 119 time series in total, high shape factors were shown to occur with a total probability of about 10%, independent of the Taylor-length based Reynolds number. Sixty-four percent of all high shape factor events were recorded during unstable thermal stratification, and the level of occurrence seemed to be almost doubled when compared to onshore cases. When a continuous sequence of time series was investigated, strong variations of the shape factor were sometimes observed even during short periods of about half an hour.

We argue for a separation of time scales to distinguish between instationary and turbulent flow at about 100 Hz, because it is clear that a severe reduction of total turbulence intensity takes place which makes the usage of laminar airfoils (NACA 63-215, for example) meaningful.

Author Contributions: A.P.S. and A.J. conceived and designed the experiments; A.J. performed the experiments; A.P.S. and A.J. analyzed the data; A.P.S. wrote the paper.

Acknowledgments: We would like to thank the Germanischer Lloyd/Garrad Hassan (formerly WINDTEST) for their support during the measurement periods, onshore as well as offshore. We thank the Department of Physics at the University of Oldenbourg for their beneficial assessments and testing of our pressure sensors. Our thanks go also to the Deutscher Wetterdienst (DWD) for providing additional local temperature data. Discussion with S. Emeis, Institute of Meteorology and Climate Research, Karlsruhe Institute of Technology, Germany, is gratefully acknowledged.

References

1. Schaffarczyk, A. (Ed.) *Understanding Wind Power Technology*; Wiley: Chichester, UK, 2014.
2. Schaffarczyk, A.P. *Introduction to Wind Turbine Aerodynamics*; Springer: Berlin, Germany, 2014.
3. Emeis, S. *Wind Energy Meteorology*; Springer: Heidelberg, Germany, 2013.

4. Obukhov, A.M. Turbulence in an atmosphere with a non-uniform temperature. *Bound.-Lay. Metereol.* **1971**, *2*, 7–29. [CrossRef]
5. Kaimal, J.; Finnigan, J. *Atmospheric Boundary Layer Flows: Their Structure and Measurements*; Oxford University Press: Oxford, UK, 1994.
6. Mahrt, L.; Vickers, D.; Howell, J.; Høstrup, J.; Wilzak, J.; Edson, J.; Hare, J. Sea surfuace drag coeficiecient in the Risøe Air Sea Experiment. *J. Geophys. Res.* **1996**, *14*, 327–335.
7. Grachev, A.; Leo, L.; Fernando, H.; Fairall, C.; Creegan, E.; Blomquist, B.W.; Christman, A.; Hocut, C. Air-sea/land interaction in the costal zone. *Bound.-Lay. Meteorol.* **2018**, *167*, 181–210.
8. Reeh, A.D.; Weissmüller, M.; Tropea, C. Free-Flight Investigation of Transition under Turbulent Conditions on a Laminar Wing Glove. In Proceedings of the 51st AIAA Aerospace Sciences Meeting, Grapevine, TX, USA, 7–10 January 2013.
9. Reeh, A.D.; Tropea, C. Behaviour of a natural laminar flow airfoil in flight through atmospheric turbulence. *J. Fluid Mech.* **2015**, *767*, 394–429. [CrossRef]
10. Schaffarczyk, A.; Schwab, D.; Breuer, M. Experimental detection of Laminar-Turbulent Transition on a rotating wind turbine blade in the free atmosphere. *Wind Energy* **2016**, *19*. [CrossRef]
11. Schwab, D. Aerodynamische Grenzschichtuntersuchungen an Einem Windturbinenblatt im Freifeld. Ph.D. Thesis, Helmut-Schmidt-Universität, Hamburg, Germany, 2018.
12. Beck, C. Superstatistics in hydrodynamic turbulence. *Physica D* **2004**, *193*, 195–207. [CrossRef]
13. Castaing, B.; Gagne, Y.; Hopfinger, E. Velocity probability density functions of high Reynolds number turbulence. *Physica D* **1990**, *64*, 177–200. [CrossRef]
14. Jeromin, A.; Schaffarczyk, A.P. Advanced statistical analysis of high-frequency turbulent pressure fluctuations for on- and off-shore wind. In Proceedings of the EUROMECH Colloquium 528: Wind Energy and the Impact of Turbulence on the Conversion Process, Oldenburg, Germany, 22–24 February 2012.
15. Jeromin, A.; Schaffarczyk, A.P. Relating high-frequency offshore turbulence statistics to boundary layer stability. In Proceedings of the 2013 International Conference on Aerodynamics of Offshore Wind Energy Systems and Wakes (ICOWES2013), Frankfurt, Germany, 19–21 November 2013; Shen, W.Z., Ed.; 2013; pp. 162–172.
16. Segalini, A.; Ramis, O.; Schlatter, P.; Henrik Alfredsson, P.; Rüedi, J.D.; Alessandro, T. A method to estimate turbulenFphysicace intensity and transvers Taylor microscale in turbulent flows from spatially averaged hot-wire data. *Exp. Fluids* **2011**, *51*, 693–700. [CrossRef]
17. Jeromin, A.; Schaffarczyk, A.P. *Statistische Auswertungen Turbulenter Druckfluktuationen auf der Off-Shore Messplattform FINO3*; Technical Report Unpublished Interal Report No. 78; University of Applied Sciences Kiel: Kiel, Germany, 2012.
18. Lück, S.; Renner, C.; Peinke, J.; Friedrich, R. The Markov-Einstein coherence length—An new meaning for the Taylor length in turbulence. *Phys. Lett.* **2006**, *359*, 335–338. [CrossRef]
19. Batchelor, G. Pressure fluctuations in isotropic turbulence. *Proc. Camb. Philos. Soc.* **1951**, *47*, 359–374. [CrossRef]
20. Xu, H.; Ouellette, T.; Vincenzi, D.; Bodenschatz, B. Acceleration correlations and pressure structure functions in high-reynolds number turbulence. *Phys. Rev. Lett.* **2007**, *99*, 204501. [CrossRef] [PubMed]
21. Sathe, M.; Mann, J.; Barlas, T.; Bierbooms, W.; van Bussel, G. Influence of atmospheric stability on wind turbine loads. *Wind Energy* **2013**, *16*, 1013–1032. [CrossRef]
22. Coelingh, J.; van Wijk, A.; Holtslag, A. Analysis of wind speed observations over the North Sea. *J. Wind Eng. Ind. Aerodyn.* **1996**, *61*, 51–69. [CrossRef]
23. Oost, W.; Jacobs, C.; van Oort, C. Stability effects on heat and moisture fluexes at sea. *Bound.-Lay. Meteorol.* **2000**, *95*, 271–302. [CrossRef]
24. Emeis, S. Upper limit for wind shear instable stratified conditions expressed in terms of a bulk Richardson number. *Meteorol. Z.* **2017**, *16*, 421–430. [CrossRef]
25. Dhruva, B. An Experimental Study of High Reynolds Number Turbulence in the Atmosphere. Ph.D. Thesis, Yale University, New Haven, CT, USA, 2000.
26. Jeromin, A.; Schaffarczyk, A.; Puczylowski, J.; Peinke, J.; Hoelling, M. Highy resolved measurements of atmospheric turbulence with the new 2D-atmospheric Laser Cantilever Anemometer. *J. Phys. Conf. Ser.* **2014**, *555*, doi:10.1088/1742-6596/555/1/012054. [CrossRef]
27. Lohse, D.; Grossmann, S. Intermittency in turbulence. *Physica A* **1993**, *194*, 519–531. [CrossRef]

28. Madsen, H.; Fuglsang, P.; Romblad, J.; Enevoldsen, P.; Laursen, J.; Jensen, L.; Bak, C.; Paulsen, U.S.; Gaunna, M.; Sorensen, N.N.; et al. The DAN-AERO MW experiments. *AIAA* **2010**, 645, doi:10.2514/6.2010-645. [CrossRef]
29. Batchelor, G. *An Introduction to Fluid Dynamics*; Cambridge University Press: Cambridge, UK, 1967.

8

The Influence of Tilt Angle on the Aerodynamic Performance of a Wind Turbine

Qiang Wang [1], Kangping Liao [1,*] and Qingwei Ma [2]

[1] College of Shipbuilding Engineering, Harbin Engineering University, Harbin 150001, China; wangqiang918@hrbeu.edu.cn
[2] School of Mathematics, Computer Sciences & Engineering, City, University of London, London EC1V 0HB, UK; q.ma@city.ac.uk
* Correspondence: liaokangping@hrbeu.edu.cn

Featured Application: This research has certain reference significance for improved wind turbine performance. It can also provide reference value for wind shear related study.

Abstract: Aerodynamic performance of a wind turbine at different tilt angles was studied based on the commercial CFD software STAR-CCM+. Tilt angles of 0, 4, 8 and 12° were investigated based on uniform wind speed and wind shear. In CFD simulation, the rotating motion of blade was based on a sliding mesh. The thrust, power, lift and drag of the blade section airfoil at different tilt angles have been widely investigated herein. Meanwhile, the tip vortices and velocity profiles at different tilt angles were physically observed. In addition, the influence of the wind shear exponents and the expected value of turbulence intensity on the aerodynamic performance of the wind turbine is also further discussed. The results indicate that the change in tilt angle changes the angle of attack of the airfoil section of the wind turbine blade, which affects the thrust and power of the wind turbine. The aerodynamic performance of the wind turbine is better when the tilt angle is about 4°. Wind shear will cause the thrust and power of the wind turbine to decrease, and the effect of the wind shear exponents on the aerodynamic performance of the wind turbine is significantly greater than the expected effect of the turbulence intensity. The main purpose of the paper was to study the effect of tilt angle on the aerodynamic performance of a fixed wind turbine.

Keywords: wind turbine; tilt angle; unsteady aerodynamics; computational fluid dynamics

---

1. Introduction

The use of wind energy has increased over the past few decades. Today, wind energy is the fastest growing renewable energy source in the world [1]. Despite the amazing growth in the installed capacity of wind turbines in recent years, engineering and science challenges still exist [2]. The main goals in wind turbine optimization are to improve wind turbine performance and to make them more competitive on the market. Studies have shown that the wind turbine tilt angle affects the shear force and bending moment at the tower top and the blade root [3], and the interaction between the blade and the tower also affects the aerodynamic performance of the wind turbine [4]. Therefore, it is necessary to study the effect of tilt angle on wind turbine performance and analyze the characteristics of blade–tower interaction, aiming to improve the wind turbine performance.

In recent years, more and more scholars have been paying attention to the interaction between the blades and towers of wind turbines. Kim et al. [4] studied the interaction between the blade and the tower using the nonlinear vortex correction method. They concluded that as the yaw angle and wind shear exponent increase, the interaction between the blade and the tower decreases. The influence of the tower diameter on the interaction between the blades and the tower is higher than that of the tower clearance. Meanwhile, this interaction may increase the total fatigue load at low wind speed. Guo et al. [5] used blade element moment (BEM) theory to study the interaction between the blade and tower. Their results show that the blade–tower interaction is much more significant than that of the wind shear. Wang et al. [6] researched the blade–tower interaction using computational fluid dynamics (CFD). Their research shows that the influence of the tower on the total aerodynamic performance of the upwind wind turbine is small, but the rotating blade will cause an obvious periodic drop in the front pressure of the tower. At the same time, we can see the strong interaction of blade tip vortices. Narayana et al. [7] researched the gyroscopic effect of small-scale wind turbines. Their findings show that changing the tilt angle can improve the aerodynamic performance of small-scale wind turbines. Recently, Zhao et al. [3] proposed a new wind turbine control method. In their control method, tilt angle increases as wind speed increases, with the purpose of reducing the blade loading and maintaining the power of the wind turbine at high wind speeds. Their research shows that the new control method can reduce the shear force at the top and bottom of the tower when compared with the yaw control strategy.

Many researchers have studied the effect of tilt angle on the structural performance of a wind turbine. For example, Zhao et al. [8] studied the structural performance of a two-blade downwind wind turbine at different tilt angles. However, there is little research on the effect of tilt angle on the aerodynamic performance of wind turbines. In this paper, aerodynamic performance of a wind turbine at different tilt angles is studied. All simulations are performed in CFD software STAR-CCM+ 12.02. Through a comparison of aerodynamic performance of the wind turbine at different tilt angles, the effects of tilt angle on the thrust, power and wake of the wind turbine are studied.

2. Numerical Modeling

2.1. Physical Model

In this study, the governing equation uses the unsteady Reynolds-averaged Navier–Stokes equation. The $SST\ k-\omega$ turbulence model was used in current simulations. A separated flow model was used to solve the flow equation. SIMPLE solution algorithm was used for pressure correction. Convection terms used the second-order upwind scheme. In the unsteady simulation, the time discretization used the second-order central difference scheme. In addition, due to the sliding mesh approach, no hole cutting was necessary, making the calculations more efficient than with the use of an overset mesh. Thus the sliding mesh technique was used to handle rotating motion of a blade [9].

2.2. Turbulence Model

The $SST\ k-\omega$ turbulence model can consider the complex flow of the adverse pressure gradient near the wall region and the flow in the free shear region. Thus, the $SST\ k-\omega$ turbulence model is suitable for simulating the rotational motion of the blade [10]. In addition, this turbulence model can accurately capture wind turbine wake [11,12].

In the Reynolds-averaged N-S equations, $\tau_{ij} = -\rho\overline{u_i' u_j'}$ refers to the Reynolds stress tensor. Reynolds stress tensor and mean strain rate tensor (S_{ij}) are related by the Boussinesq eddy viscosity assumption:

$$\tau_{ij} = 2\nu_t S_{ij} - \frac{2}{3}\rho k \delta_{ij} \tag{1}$$

where ν_t refers to the eddy viscosity, ρ refers to the density, k refers to the turbulence kinetic energy and δ_{ij} refers to the Kronecker delta function.

To provide closure equations, in the $SST\ k-\omega$ turbulence model, the turbulent kinetic energy (k) and specific dissipation of turbulent kinetic energy (ω) also need governing transport equations, which are given as follows:

$$\frac{D\rho k}{Dt} = \tau_{ij}\frac{\partial u_i}{\partial x_j} - \beta^* \rho \omega k + \frac{\partial}{\partial x_j}\left[(\mu + \sigma_k \mu_t)\frac{\partial k}{\partial x_j}\right] \tag{2}$$

$$\frac{D\rho\omega}{Dt} = \frac{\gamma}{\nu_t}\tau_{ij}\frac{\partial u_i}{\partial x_j} - \beta\rho\omega^2 + \frac{\partial}{\partial x_j}\left[(\mu+\sigma_\omega\mu_t)\frac{\partial\omega}{\partial x_j}\right] + 2(1-F_1)\rho\sigma_{\omega 2}\frac{1}{\omega}\frac{\partial k}{\partial x_i}\frac{\partial\omega}{\partial x_j} \tag{3}$$

In the formulas above, the model coefficients are defined as follows:

$$\beta^* = F_1\beta_1{}^* + (1-F_1)\beta_2{}^* \tag{4}$$

$$\beta = F_1\beta_1 + (1-F_1)\beta_2 \tag{5}$$

$$\gamma = F_1\gamma_1 + (1-F_1)\gamma_2 \tag{6}$$

$$\sigma_k = F_1\sigma_{k1} + (1-F_1)\sigma_{k2} \tag{7}$$

$$\sigma_\omega = F_1\sigma_{\omega 1} + (1-F_1)\sigma_{\omega 2} \tag{8}$$

The blending function F_1 is defined as follows:

$$F_1 = \tanh\left\{\left\{\min\left[\max\left(\frac{\sqrt{k}}{\beta^*\omega y}, \frac{500\upsilon_\infty}{y^2\omega}\right), \frac{4\rho\sigma_{\omega 2}k}{CD_{k\omega}y^2}\right]\right\}^4\right\} \tag{9}$$

where $CD_{k\omega}$ refers to the cross-diffusion term, y refers to the distance to the nearest wall and υ refers to the kinematic viscosity. F_1 is equal to zero in the region away from the wall ($k-\varepsilon$ turbulence model) and one in the region near the wall ($k-\omega$ turbulence model).

The eddy viscosity is

$$\nu_t = \frac{a_1 k}{\max(a_1\omega, \Omega F_2)} \tag{10}$$

where Ω is the absolute value of the vorticity and F_2 is the second blending function, defined as

$$F_2 = \tanh\left[\max\left(\frac{2\sqrt{k}}{\beta^*\omega y}, \frac{500\upsilon}{y^2\omega}\right)^2\right] \tag{11}$$

A more detailed description of the *SST* $k-\omega$ turbulence model is provided in [10]. In this study, the parameters for the *SST* $k-\omega$ turbulence model are as follows:

$\sigma_{k1} = 0.85$ $\sigma_{\omega 1} = 0.5$ $\beta_1 = 0.075$ $a_1 = 0.31$ $\beta^* = 0.09$ $k = 0.41$ $\sigma_{k2} = 1$
$\sigma_{\omega 2} = 0.856$ $\beta_2 = 0.0828$ $\gamma_1 = \frac{\beta_1}{\beta^*} - \frac{\sigma_{\omega 1}k^2}{\sqrt{\beta^*}}$ $\gamma_2 = \frac{\beta_2}{\beta^*} - \frac{\sigma_{\omega 2}k^2}{\sqrt{\beta^*}}$

2.3. Computational Domain

The computational domain was divided into the rotating and outer domains, as shown in Figure 1. The size of the entire outer domain was 12D(x) × 5D(y) × 4D(z). The distance from the wind turbine to the velocity inlet was 3D, and the distance to the pressure outlet was 9D, where D is the diameter of the wind turbine. Due to the complex geometry of the blades, we used the trimmed cell mesh technology to generate high-quality meshes. In order to capture the complex flow around the blade, a fine mesh was used around the blade. A 10-layer boundary layer mesh was generated near the blade and the hub. The total thickness of the boundary layer was 0.03 m, and the growth rate was 1.2. A six-layer boundary layer mesh was generated near the tower and the nacelle. The total thickness of the boundary layer was 0.1 m, and the growth rate was 1.2. Figure 2b shows the refined sliding mesh regions around the blade. Figure 2c,d shows a close-up view of the blades and nacelle tower.

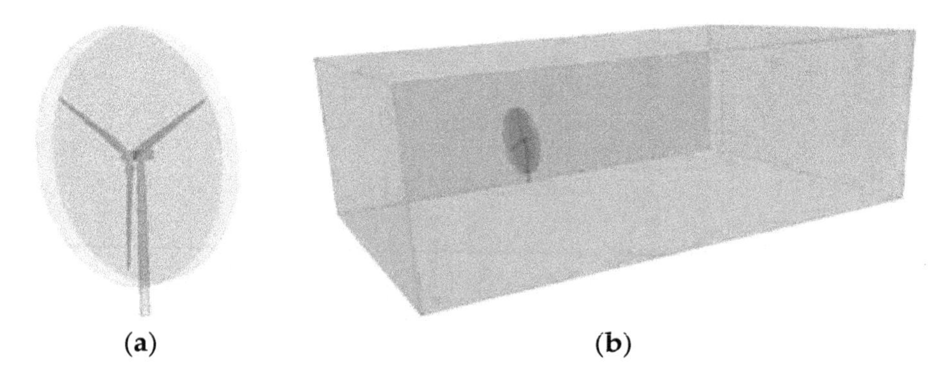

Figure 1. Rotating and outer domain: (**a**) rotation domain for wind turbine simulation; (**b**) entire computational domain for numerical simulation.

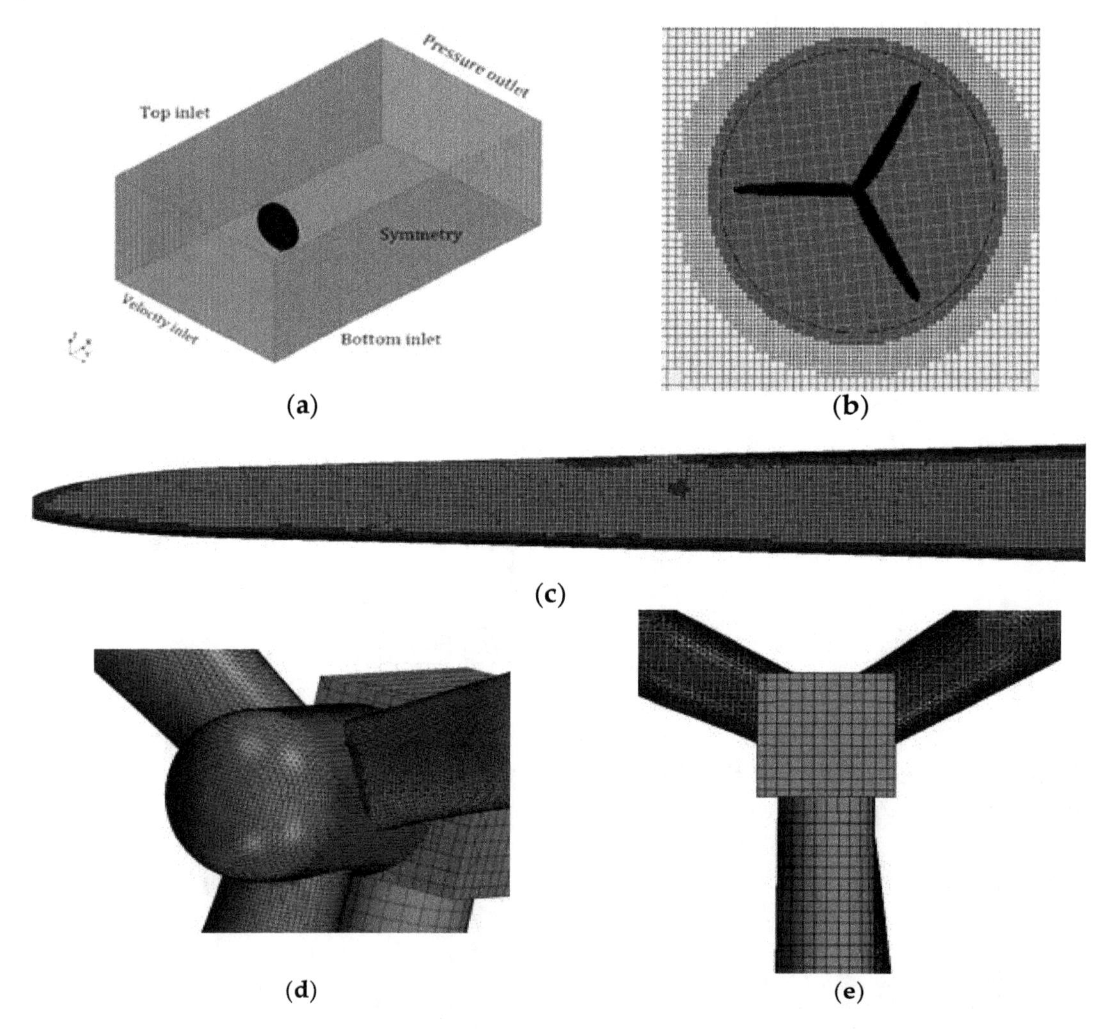

Figure 2. The computational mesh domain for the wind turbine: (**a**) full grid domain, (**b**) sliding mesh regions, (**c**) close-up view of the blade surface, (**d**) close-up view of the hub surface mesh and (**e**) close-up view of nacelle and tower.

2.4. Boundary Conditions

Figure 2a illustrates the setting of the boundary conditions in this study. In the computational domain, the inlet boundary, bottom and top surfaces were set as velocity inlets. The pressure outlet was set at the outlet boundary. The sides of the computational domain were set to the plane of symmetry. In this simulation, all of the y+ wall treatment of near-wall modeling was applied. In order to reduce the convergence order and improve the solution accuracy, the maximum internal iterations within each time-step was 10 [13].

3. Results and Discussion

3.1. Validations

The 1/75 scale model of a DTU 10 MW reference wind turbine was used for the mesh independence test. In the numerical verification, the tilt angle of the wind turbine was not considered. The main parameters of the scale model are given in Table 1. A detailed introduction of the blade parameters at 40 different blade sections is provided by [14]. Figure 3 shows the wind turbine geometric model and the surface grid. After scaling according to the scale factor, the boundary layers near the blade and hub surface have five layers of refined grid with the total layer thickness of 0.004 m and a progression factor of 1.2.

Table 1. Principal dimensions of the scale model.

Specifications	DTU Down-Scaled
Number of Blades	3
Rotor Diameter (m)	2.37
Hub Diameter (m)	0.178
Rated Wind Speed (m/s)	5.53
Rated Rotor Speed (rpm)	330

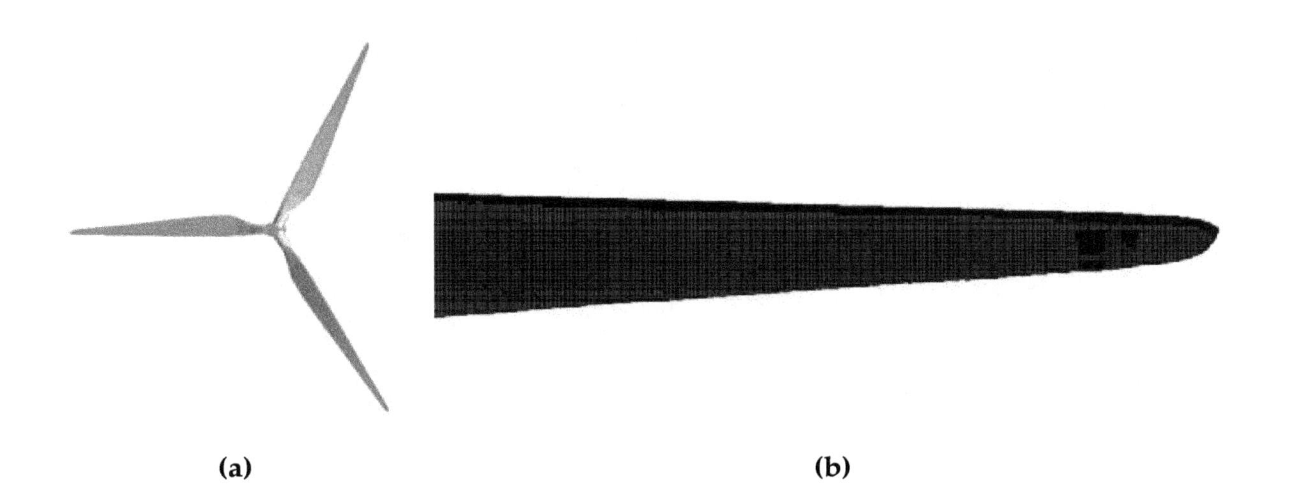

(a) **(b)**

Figure 3. Geometric model and surface grid: (**a**) the rotor geometric model; (**b**) the blade surface grid.

The blade surface mesh size includes the maximum mesh size and the minimum mesh size. The number of meshes corresponding to different mesh sizes is shown in Table 2. According to previous study, the time-step size corresponding to 1° increment of azimuth angle of the wind turbine per time-step was applied in all simulations [15]. Moreover, the simulation was run under unsteady conditions. The comparison of thrust and torque for different grid resolutions with the same wind speed of 5.53 m/s, rotor speed of 330 rpm and time-step size of 5×10^{-4} s is presented in Tables 3 and 4. It can be observed from Tables 3 and 4 that the grid resolution of Case 3 is sufficient to solve the unsteady aerodynamics of the wind turbine. Therefore, the grid resolution of Case 3 was used in subsequent simulations.

Table 2. Mesh size of blade surface.

CFD Mesh Type	Case 1	Case 2	Case 3	Case 4
Maximum Size (mm)	3.000	2.000	1.500	1.100
Minimum Size (mm)	0.500	0.350	0.250	0.180
Total Mesh Number (million)	1.850	3.240	4.630	9.400

Table 3. Comparison of thrust between experiment and CFD simulation at different grid densities.

CFD Mesh Type	LIFES50+ Wind Tunnel Data (N), [14]	Present Study (N)	Error (%)
Case 1	68.631	70.010	2.000
Case 2		69.660	1.500
Case 3		69.520	1.300
Case 4		69.500	1.300

Table 4. Comparison of torque between experiment and CFD simulation at different grid densities.

CFD Mesh Type	LIFES50+Wind Tunnel Data (*N·M*), [14]	Present Study (*N·M*)	Error (%)
Case 1	6.232	5.690	8.700
Case 2		5.850	6.100
Case 3		5.900	5.300
Case 4		5.920	5.000

Simulations at different wind speeds were performed, and the simulation results were compared with wind tunnel experiment data, as presented in Figure 4. In this paper, we always keep the pitch angle at 0°, so we have not considered the working conditions above the rated wind speed. When the wind speed is close to the rated wind speed, the thrust and torque of the CFD simulation are lower than those of the wind tunnel experiment, but the maximum error is not more than 10%. This means that STAR-CCM+ can accurately simulate the aerodynamic performance of the wind turbine under rotating motion.

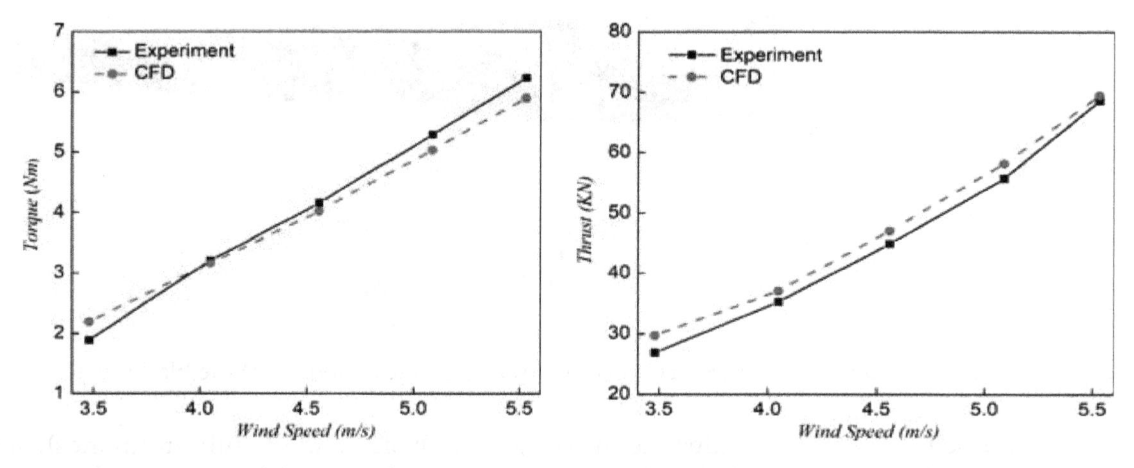

Figure 4. Comparison of thrust and torque between wind tunnel experiment and CFD simulation at different wind speeds (Case 3).

In order to ensure the reliability of the NREL 5 MW real-scale wind turbine simulation, the NREL 5 MW real-scale wind turbine was used for grid convergence analysis. Major properties of the NREL 5 MW reference wind turbine are given in Table 5 [16]. Figure 5 shows the blade alone geometric model and full configuration geometric model with the tower. The blade alone model was used for numerical verification, and the full configuration model was used to investigate the effect of tilt angle on the aerodynamic performance of the wind turbine. Near the wall surface of the blades and hub, the boundary layers have 10 layers of refined grid with the total layer thickness of 0.03 m and a progression factor of 1.2. The same wind speed of 11.4 m/s and rotor speed of 12.1 rpm were applied in all simulations. Meanwhile, in all simulations, the time step is the time taken by the wind turbine to increase the azimuth angle by 1°.

Table 5. Principal dimensions of the NREL 5 MW reference wind turbine.

Specifications	
Rated Power (MW)	5
Rotor Orientation, Configuration	Upwind, 3 blades
Rated Wind Speed (m/s)	11.4
Rated Rotor Speed (rpm)	12.1
Rotor Diameter (m)	126
Hub Diameter (m)	3
Hub Height (m)	90
Tower Base Diameter (m)	6
Tower Top Diameter (m)	3.87
Pre-cone (°)	2.5

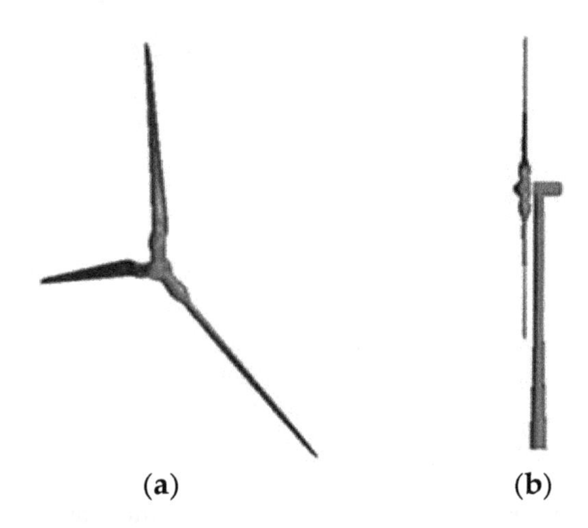

Figure 5. Geometric model of a 5 MW reference wind turbine: (**a**) the rotor geometric model; (**b**) the full configuration model.

The number of meshes corresponding to different mesh sizes is shown in Table 6. The comparison of power for different grid resolutions with the same wind speed of 11.4 m/s and rotor speed of 12.1 rpm is presented in Table 7. It can be observed from Table 7 that the grid resolution of Case 2 is sufficient to solve the unsteady aerodynamics of the wind turbine. Therefore, the grid of Case 2 was used for the simulation of NREL 5 MW real-scale wind turbines at different wind speeds.

Table 6. Mesh size of blade surface.

CFD Mesh Type	Case 1	Case 2	Case 3
Maximum Size (m)	0.20	0.10	0.05
Minimum Size (m)	0.04	0.02	0.01
Total Mesh Number (Million)	1.52	4.80	9.53

Table 7. Comparison of power between NREL data and CFD simulation at different grid densities.

CFD Mesh Type	NREL Data (MW), [16]	Present Study (N)	Error (%)
Case 1		4.767	4.700
Case 2	5.000	4.981	0.380
Case 3		5.020	0.400

Aerodynamic simulations of a wind turbine with various wind speeds were tested and compared with the FAST results. The obtained thrust and power were compared with the corresponding NREL data calculated by FAST V8, as presented in Figure 6. The power agrees well with the NREL data, but the thrust tends to be smaller than that from NREL data. The reason for the difference between the CFD method and the FAST can be summarized as follows: (a) the FAST does not consider the three-dimensional flow effects around blades; (b) in the BEM method, in order to calculate a rotor with a limited number of blades, a tip loss correction model needs to be added. The results obtained by different tip loss correction models are also quite different [17]. FAST uses a Prandtl tip loss correction model [16]. Therefore, the CFD result of the thrust is significantly lower than the FAST result. A similar phenomenon appeared in [18]. However, at the rated wind speed, compared with FAST data, the errors of the thrust and power obtained by CFD are less than 5% Through the above analysis, the grid of Case 2 can accurately simulate the aerodynamic performance of NREL 5 MW real-scale wind turbines. Therefore, the grid of Case 2 was used to simulate the effect of tilt angle on the aerodynamic performance of the wind turbine.

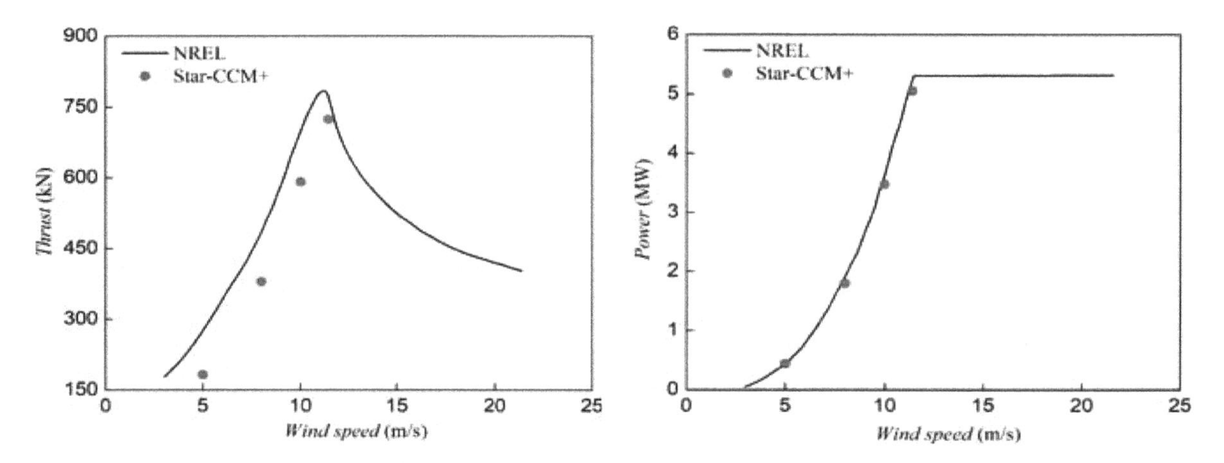

Figure 6. Comparisons of thrust and power.

3.2. The Effect of the Tilt Angle on the Aerodynamic Performance of the Wind Turbine

In this study, nacelle tilt angles of 0, 4, 8 and 12° were investigated. Figure 7 shows the structure of the wind turbine at different tilt angles. In the picture, β is the pre-coning angle, and γ is the shaft tilt angle. The azimuth of the rotor is defined as ψ, as presented in Figure 8. In Figure 8, the blue rotor is the initial position with the 0-azimuth angle. Subsequent analysis is based on the results after the wind turbine has stabilized. Under different tilt angles, the change in wind turbine thrust and power with the azimuth is shown in Figure 9. Comparing the no-tower curve with the other four curves, it can be seen that the thrust and power generate periodic fluctuations due to the influence of the tower. When the blades pass through the tower, the thrust and power will periodically decrease. This phenomenon is called the blade–tower interaction (BTI) [19]. The BTI effects begin at approximately 30° rotor azimuth and dissipate at approximately 100° rotor azimuth, as presented in Figure 10. This agrees with previous studies, which all show effects in approximately this same 70° range [19].

Figure 9 shows the difference between the thrust and power at approximately 60, 180 and 300° azimuth with the same nacelle tilt. This phenomenon is due to the interaction between the blade and the tower creating a random vortex. As the nacelle tilt increases, the blade and tower interactions gradually weaken. Therefore, this phenomenon becomes less important as the nacelle tilt increases. In Figure 10, when ψ is approximately 65°, the thrust and power of the wind turbine at 4 and 8° nacelle tilt are higher than 0 and 12° nacelle tilt.

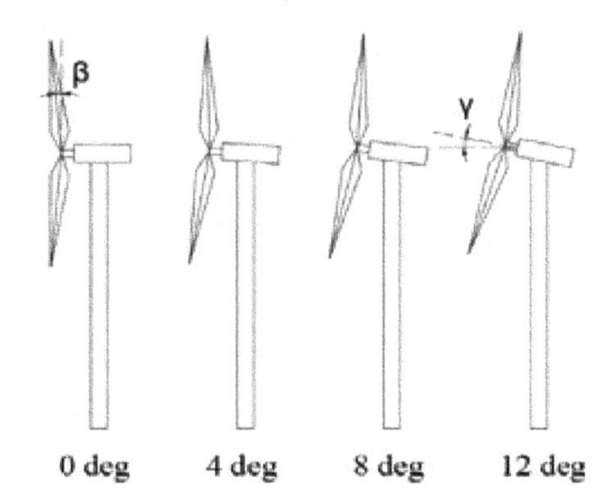

Figure 7. Structure of the wind turbine at different tilt angles.

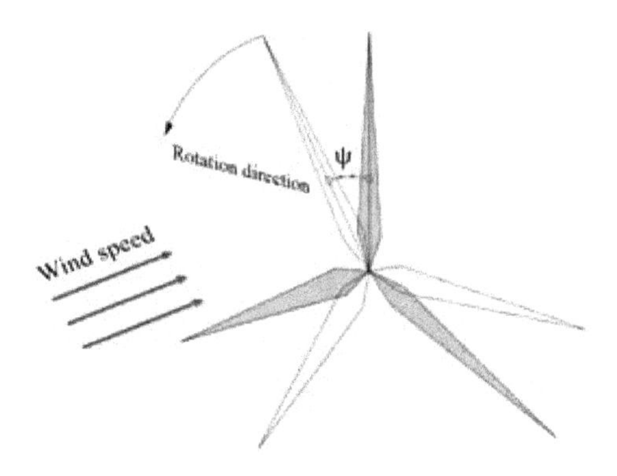

Figure 8. Definition of the azimuth.

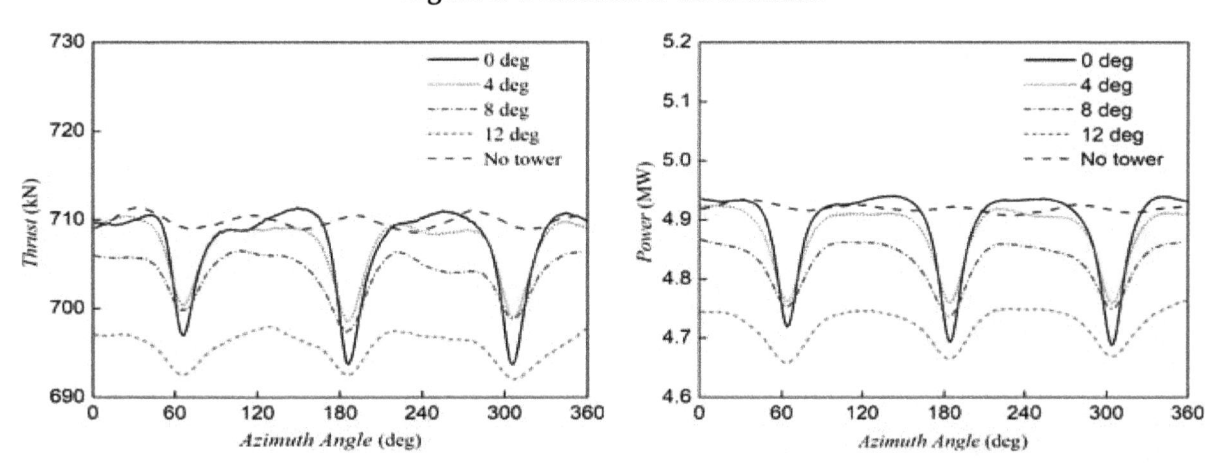

Figure 9. Comparison of thrust and power at different nacelle tilt angles.

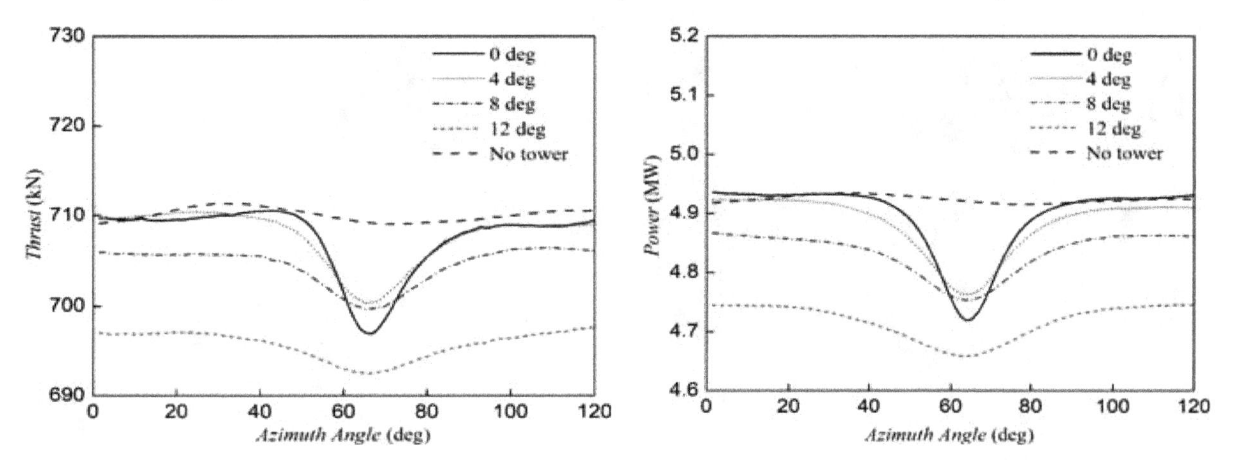

Figure 10. Comparison of thrust and power at different nacelle tilt angles (partial enlargement).

The position of the blade relative to the tower with the 60° azimuth is shown in Figure 11. Instantaneous pressure magnitude and streamlines at blade sections r/R = 0.5, r/R = 0.7 and r/R = 0.9 (Blade 1) of the wind turbine are presented in Figures 12–14. In the low span (r/R = 0.5) suction side, the flow separation phenomenon can be observed. However, the flow remains attached for higher radial sections (r/R = 0.7 and r/R = 0.9). In addition, with the increase of the nacelle tilt, the flow separation of the low span suction side is gradually weakened. The variation of the pressure distribution around different sections airfoil with the nacelle tilt can also be observed.

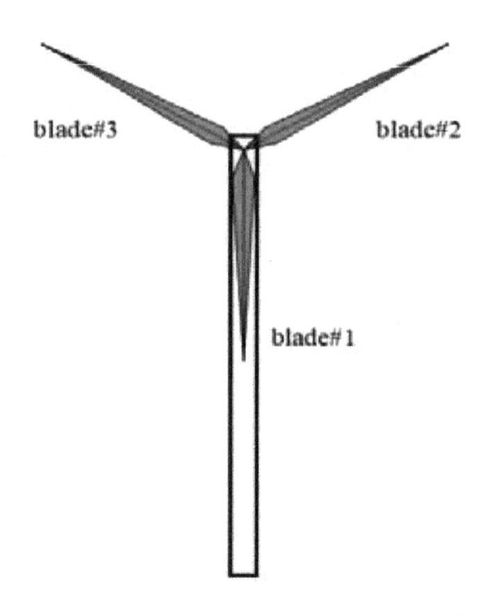

Figure 11. Blade position (ψ = 60°).

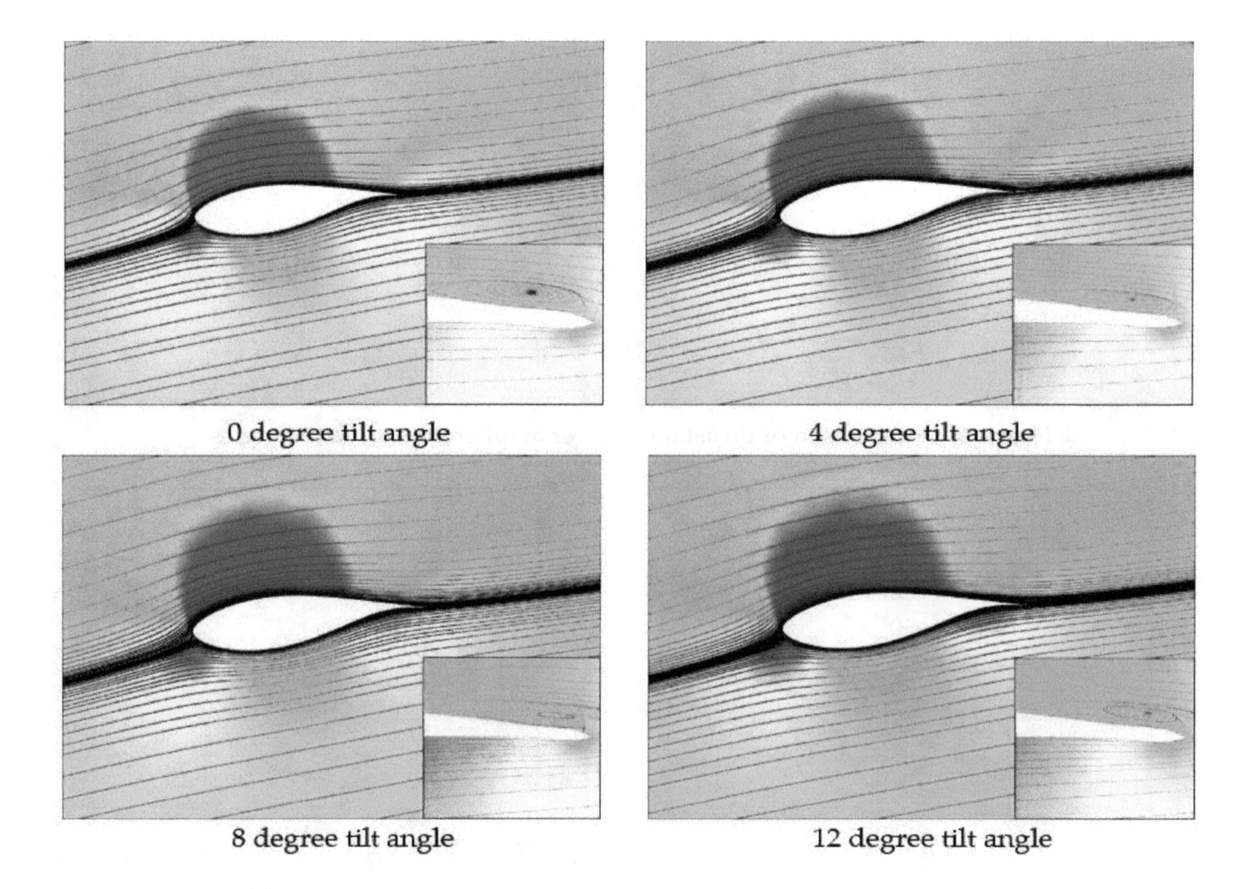

Figure 12. Instantaneous pressure magnitude and streamlines diagram (r/R = 0.5).

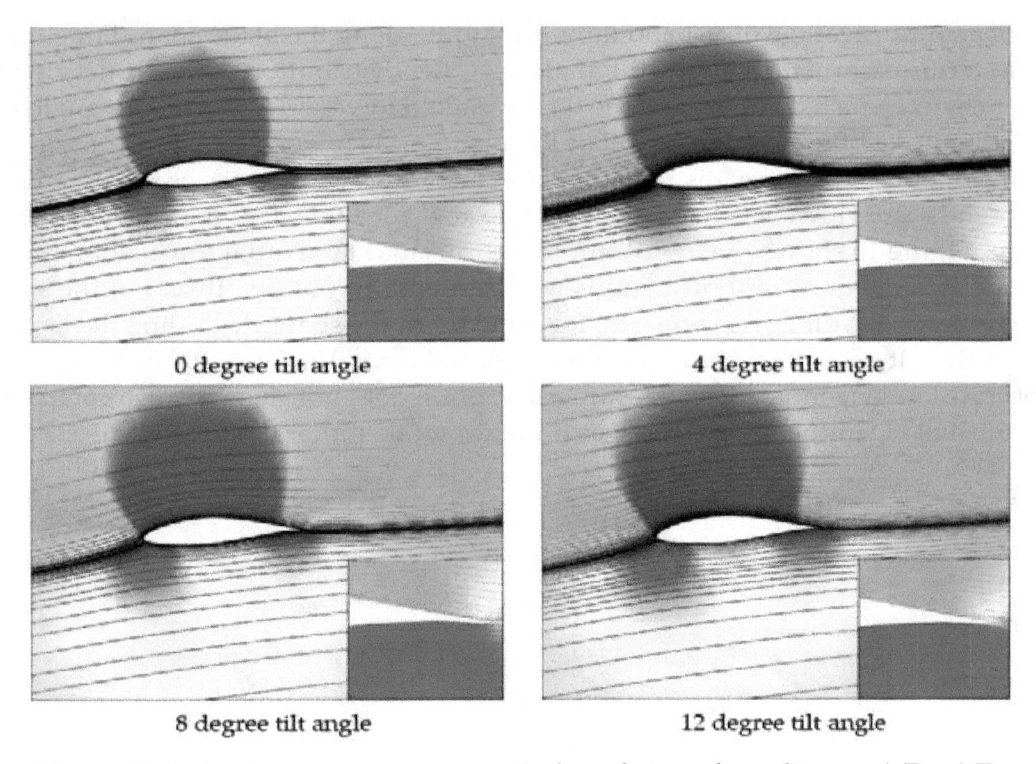

Figure 13. Instantaneous pressure magnitude and streamlines diagram (r/R = 0.7).

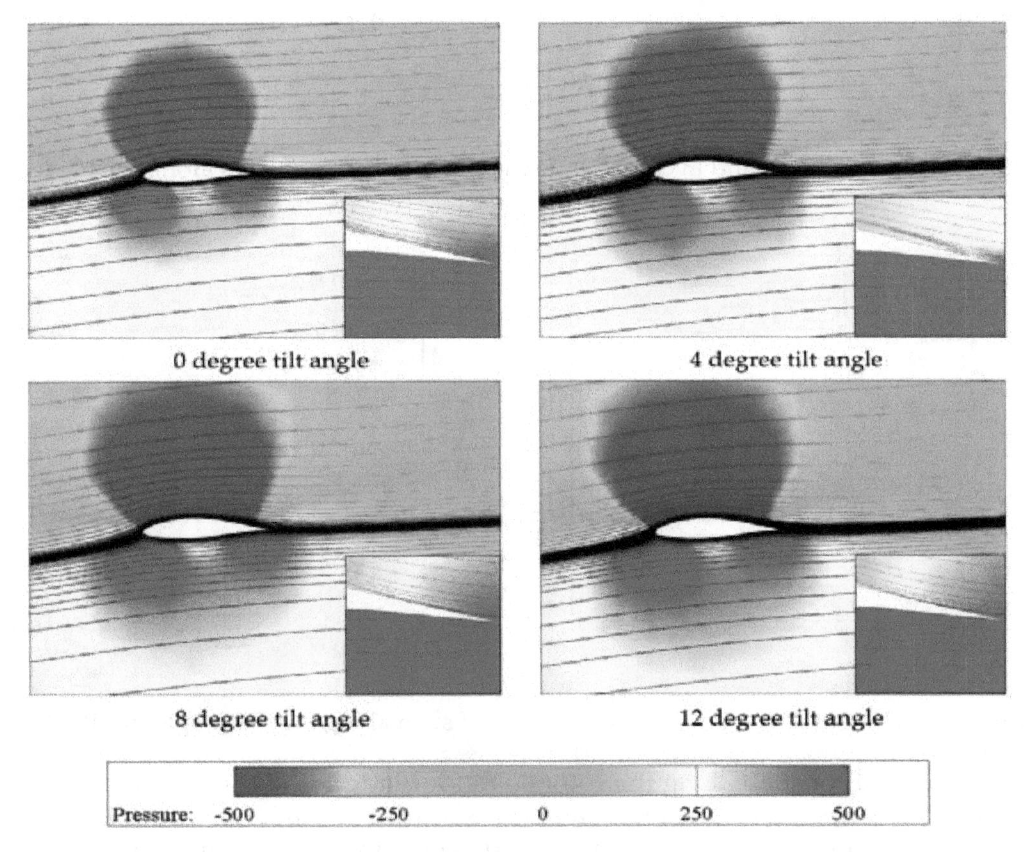

Figure 14. Instantaneous pressure magnitude and streamlines diagram (r/R = 0.9).

H. Rahimi et al. [20] studied different methods of calculating the angle of attack of the wind turbine section airfoil. However, in CFD, when considering the interaction between the blade and the tower, it is difficult to calculate the angle of attack of the blade section airfoil. Therefore, only the effect of tilt angle on the blade section airfoil load is considered in this paper. Figure 15 shows the distribution of azimuth average thrust and tangential force along the blade span. From the figure, we can see that in terms of thrust, when the tilt angle is 4°, the distribution of the thrust along the blade span does not change much compared to the 0° tilt angle. However, when the tilt angle is increased to 8 and 12°, the thrust of the section airfoil at the blade tip is lower than the values at 0 and 4° tilt angle. In terms of tangential force, the tangential force gradually decreases as the tilt angle increases, for up to 0.5 of the span. However, the tangential force at 4° tilt angle does not change much compared to 0° tilt angle. Figure 16 shows the distribution of thrust and tangential force along the blade span when the blade is located in front of the tower. In terms of thrust, the thrust of the section airfoil gradually increases as the tilt angle increases, for up to 0.7 of the span. Regarding the tangential force, the increase of the tilt angle also increases the tangential force, for up to 0.6 of the span. However, regardless of thrust or tangential force, the value at 8° of tilt does not change much compared to 12° of tilt. This means that the influence of the tower becomes weaker after the tilt angle exceeds 8°.

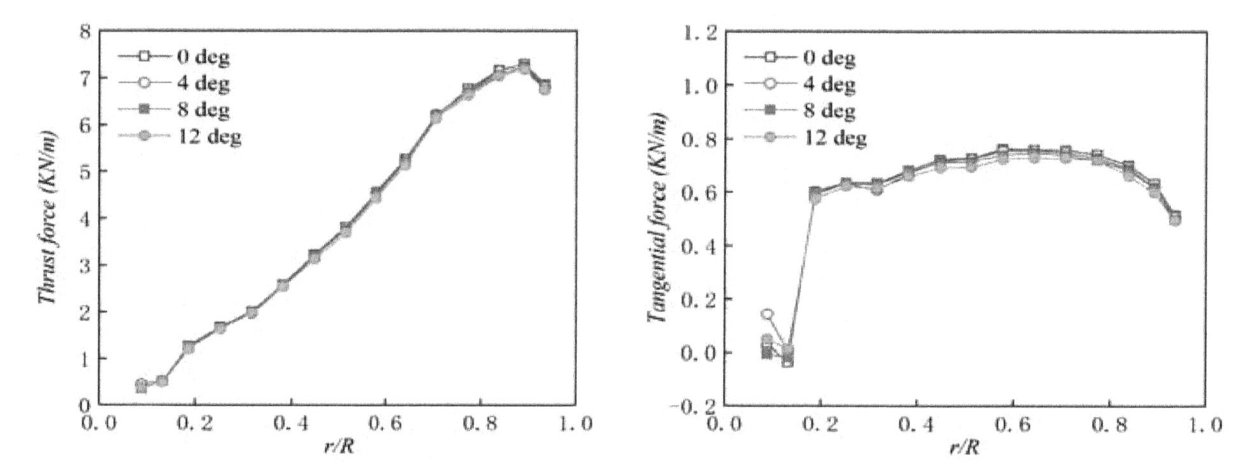

Figure 15. The average thrust and average tangential force per unit of span along the blade span for Blade 1.

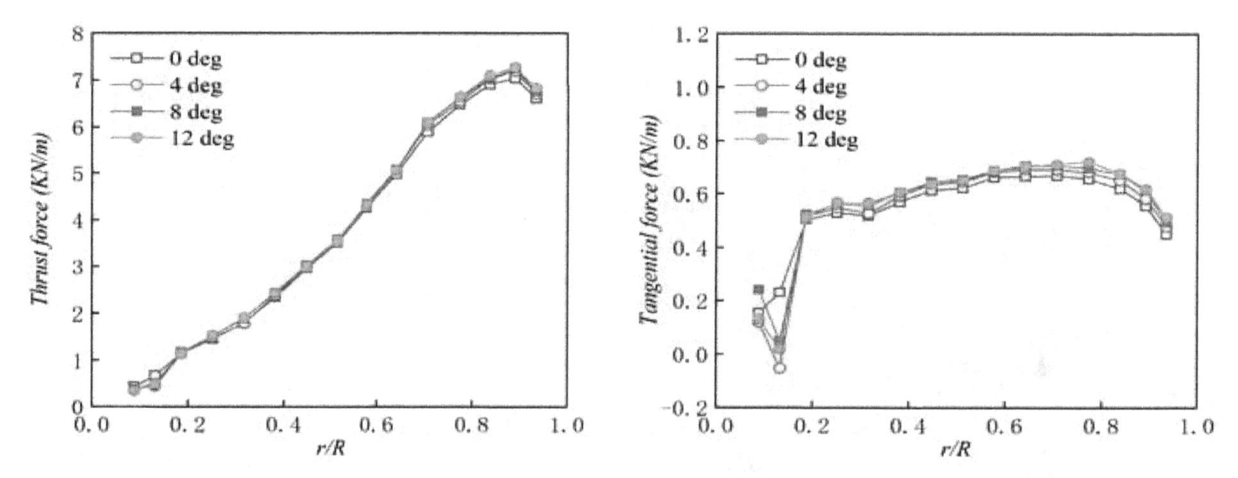

Figure 16. Thrust and tangential force per unit of span along the blade span for Blade 1.

Thrust force per unit of span along the rotor span for Blade 1 is shown in Figure 17. In the blade root, the thrust will fluctuate with the change of the azimuth angle, which is mainly caused by the three-dimensional flow of the blade root. In the middle of the blade, when the azimuth angle is 0-180°, the thrust is the largest at 4° tilt angle, and the thrust is the smallest at 12° tilt angle. When the azimuth angle is 180-360°, the thrust gradually decreases as the tilt angle increases. In the vicinity of the blade tip, when the azimuth angle is 0-180°, except for the tilt angle of 0°, the thrust has a change that increases first and then decreases with the change of the azimuth angle. When the azimuth angle is 180-360°, the thrust curve decreases first and then increases, and the thrust gradually decreases as the elevation angle increases. Figure 18 shows the tangential force per unit of span along the rotor span for Blade 1. We can see that in the middle of the blade and near the tip of the blade, the tangential force of the section airfoil in the 180-360° angle range is higher than the value in the 0–180° angle range, except for the case of the 0° tilt angle. At the same time, we found that in the middle of the blade and near the tip of the blade, when the azimuth angle is 180°, the thrust and tangential force at 0° tilt are the smallest, which is mainly due to the maximum interaction between the blade and the tower at 0° tilt angle.

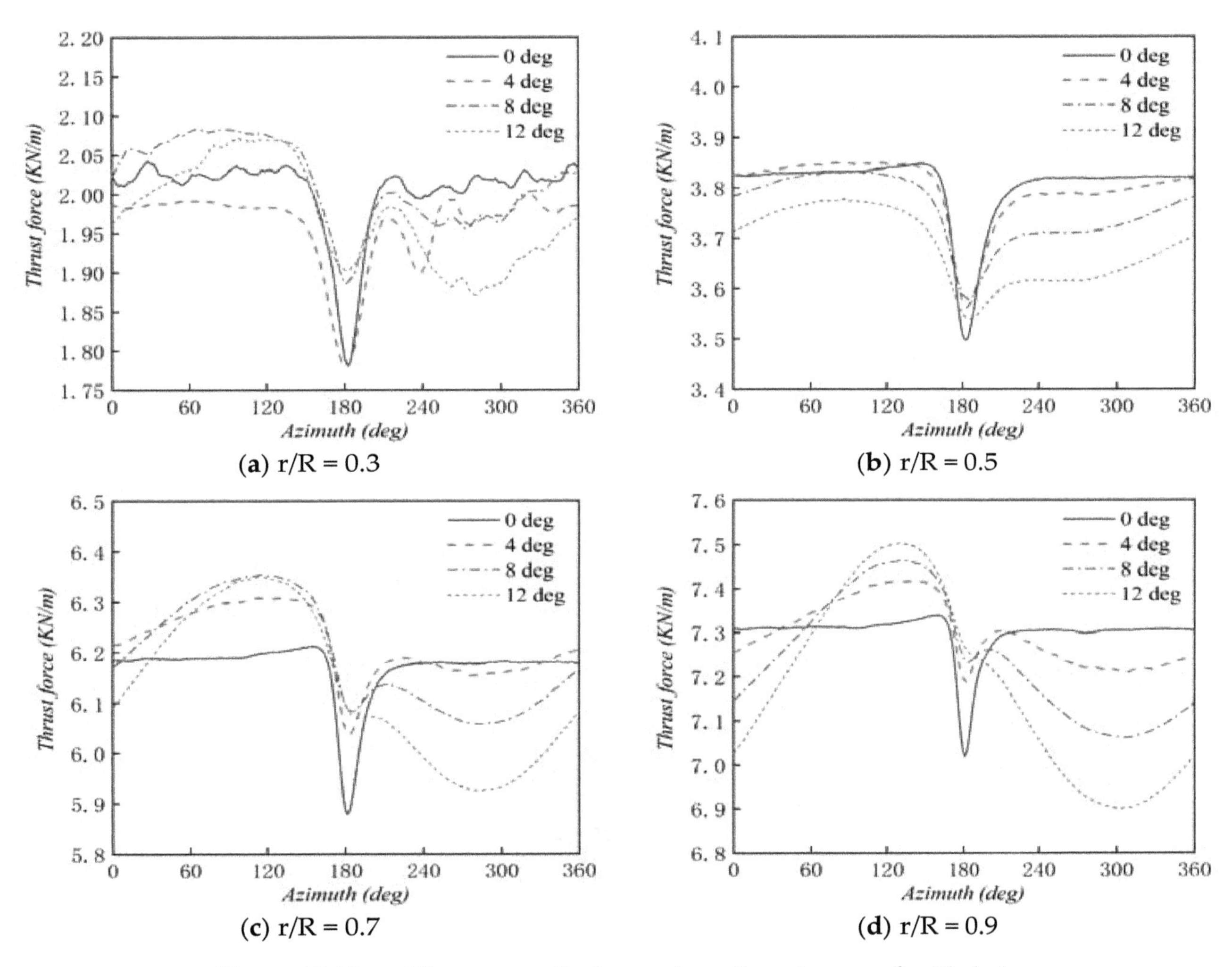

(**a**) r/R = 0.3 (**b**) r/R = 0.5

(**c**) r/R = 0.7 (**d**) r/R = 0.9

Figure 17. Thrust force per unit of span along the rotor span for Blade 1.

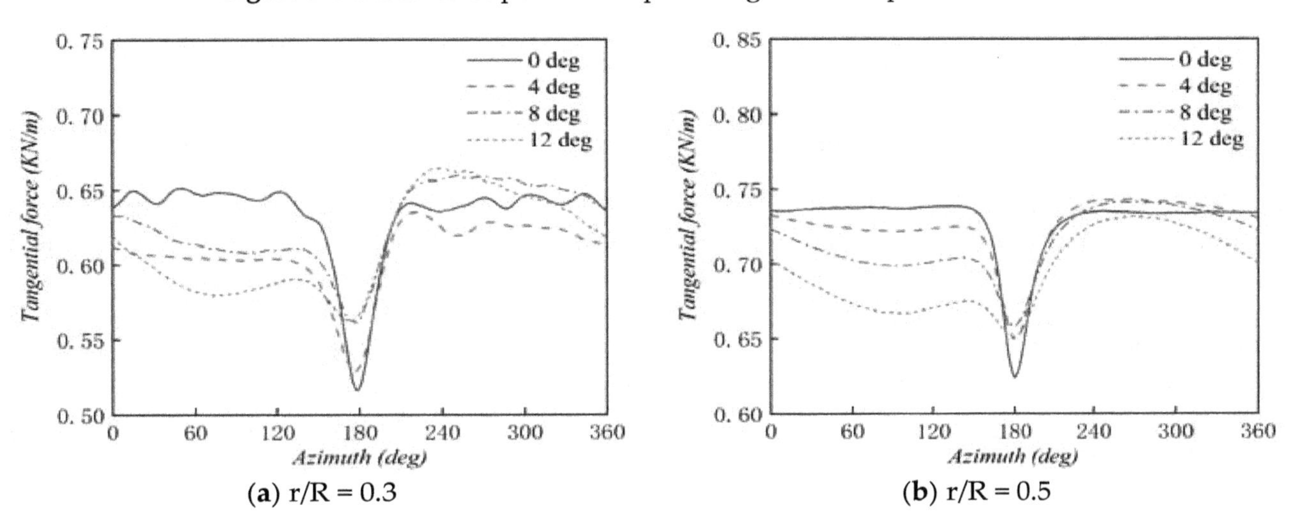

(**a**) r/R = 0.3 (**b**) r/R = 0.5

Figure 18. *Cont.*

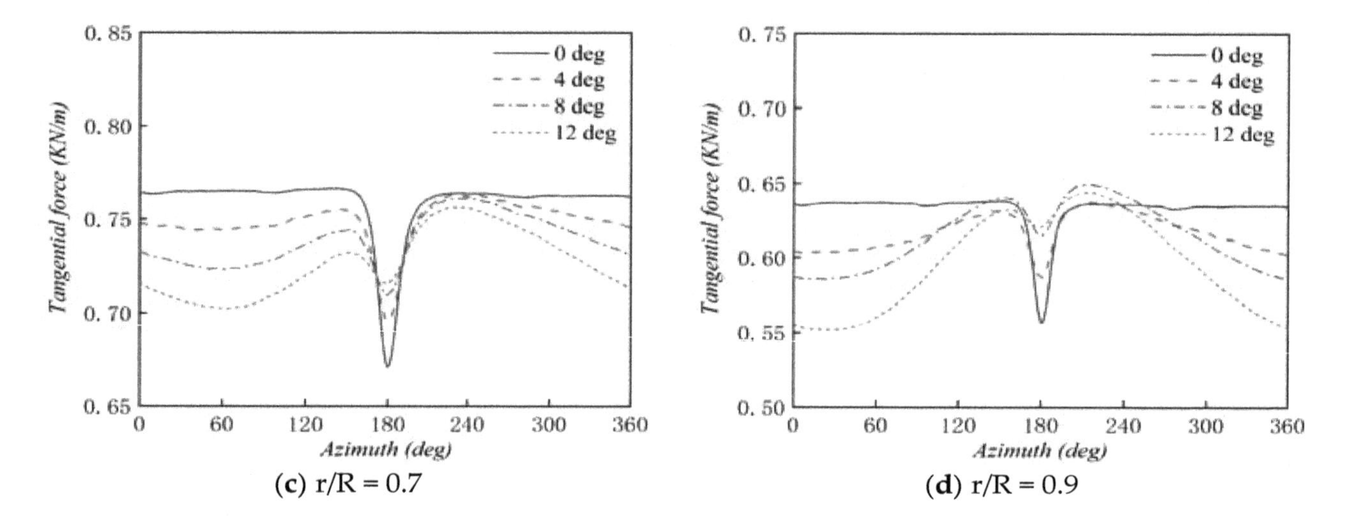

Figure 18. Tangential force per unit of span along the rotor span for Blade 1.

3.3. The Effect of the Tilt Angle on the Wind Turbine Wake

The instantaneous isovorticities occurring when the blade is in front of the tower are presented in Figure 19. One can clearly see that these instantaneous diagrams with nacelle tilt angle shows that there is a strong flow interaction between the wake generated by the blade root, hub and tower regions. Because of the existence of the tower, there are strong unsteady flow interactions between tower vortex and blade tip vortex during downstream propagation. This interaction caused the blade tip vortex to break behind the tower. In addition, an increase in tilt angle will cause the blade tip vortex tube to tilt.

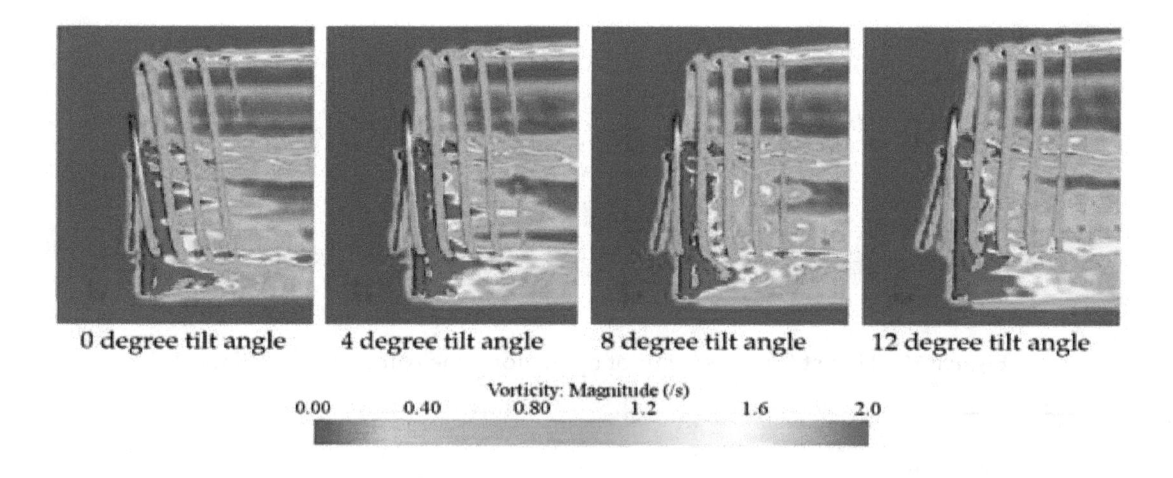

Figure 19. Side-view of instantaneous isovorticity contours for different nacelle tilt angles.

The instantaneous x-vorticities at different sections in four tilt angles are presented in Figure 20. We can observe that there is a clear difference in the blade tip vortex at different tilt angles. At 0 and 4° tilt angles, the blade tip vortex has only negative x-vorticities. When the tilt angle is changed to 8°, positive x-vorticity and negative x-vorticity appear in the right half of the blade tip vortex. When the tilt angle is changed to 12°, the left part of blade tip vortex is negative and right part is positive. At the same time, it can be seen that there are slight differences in the tower-generated vortexes of the four cases. By comparison, at the positions of x/D = 0.25 and x/D = 0.5, the vortex generated by the tower behind the rotor at the tilt angle of 4° is slightly less than other cases. When the tilt angle reaches 12°, the vortex generated by the tower is broken.

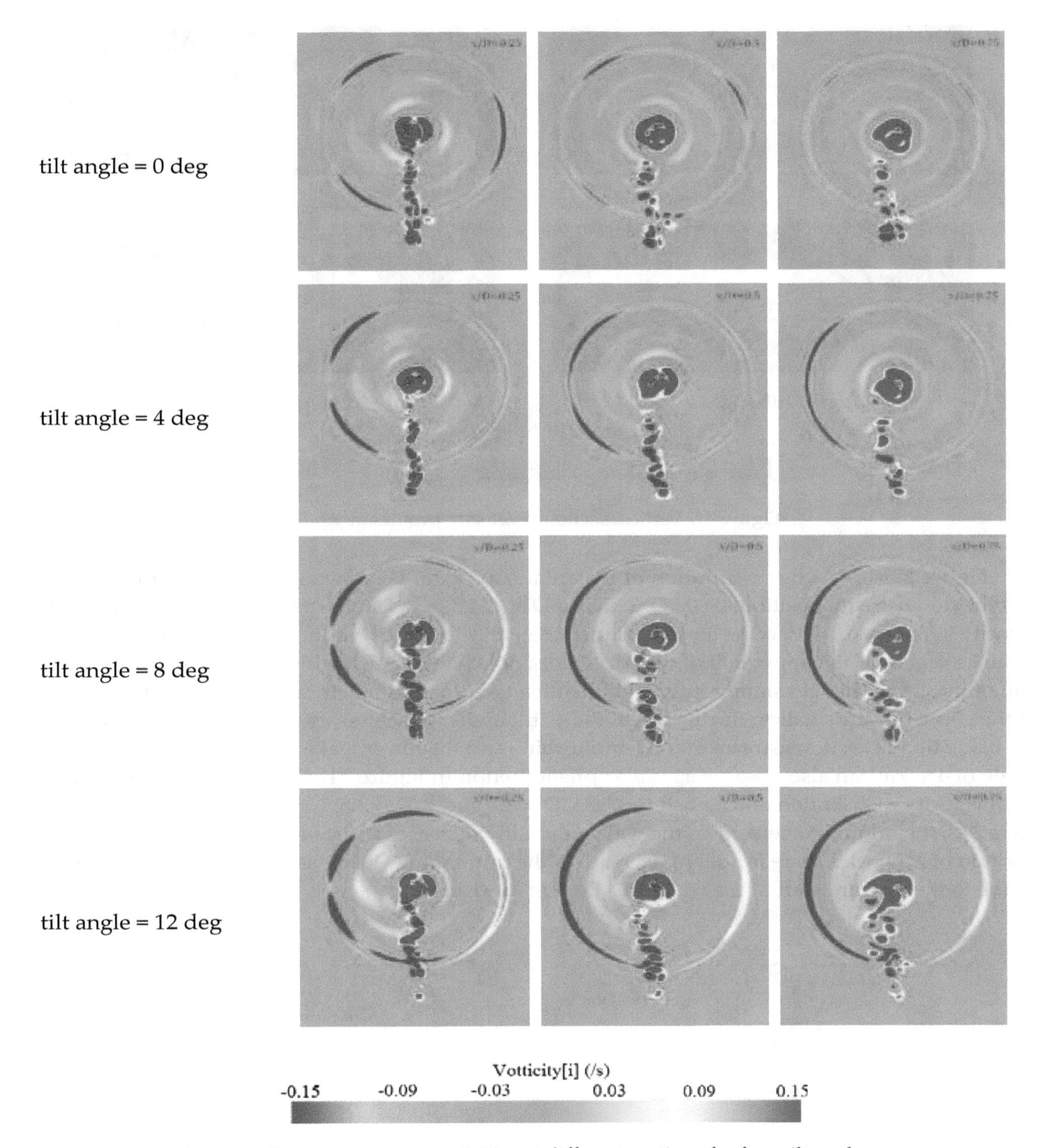

Figure 20. Instantaneous x-vorticities at different sections for four tilt angles.

The corresponding vertical x-velocity profiles are presented in Figure 21. When the tilt angle is 0°, as the downstream distance increases, the velocity field behind the wind turbine is approximately symmetrical about the centerline and keeps a circular shape. However, as the tilt angle increases, the velocity field behind the wind turbine shows asymmetry and gradually moves to the upper right. Meanwhile, the low-velocity region at the end of the wake gradually decreases with increasing tilt angle.

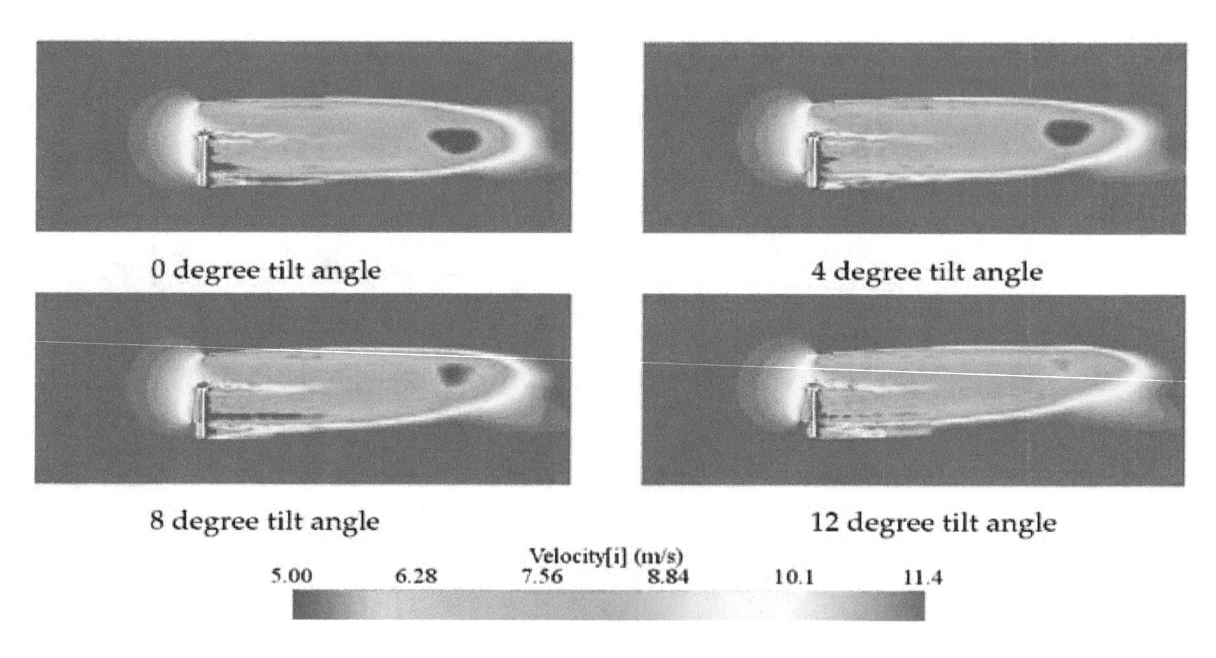

Figure 21. Vertical section x-velocity profiles at y = 0 m.

Figure 22 shows the distribution of instantaneous axial velocity along blade span at the wind turbine downstream positions of 0.5D, 2.5D, 3.5D and 4.5D, which represent the development of the velocity in the wake. Observing the instantaneous axial velocity distribution at the position of X/D = 0.5, it can be seen that the upper half of the curve does not change much with the tilt angle, but the lower half of the curve changes significantly with the tilt angle. In addition, it can be seen that the lower half of the curve has the smallest fluctuation at the 4° tilt angle, which means that the interaction between the blade tip vortex downstream of the wind turbine and the tower wake vortex is the weakest at a tilt angle of 4°. We can also observe a similar phenomenon in Figure 21. Observing the instantaneous axial velocity distribution at the positions of X/D = 2.5 and X/D = 3.5, we can see that as the tilt angle increases, the minimum velocity in the wake gradually increases and shifts upwards. However, at the position of X/D = 4.5, there is a slight decrease in the minimum velocity as the tilt angle increases. This is due to the upward shift of the wake-end deceleration zone.

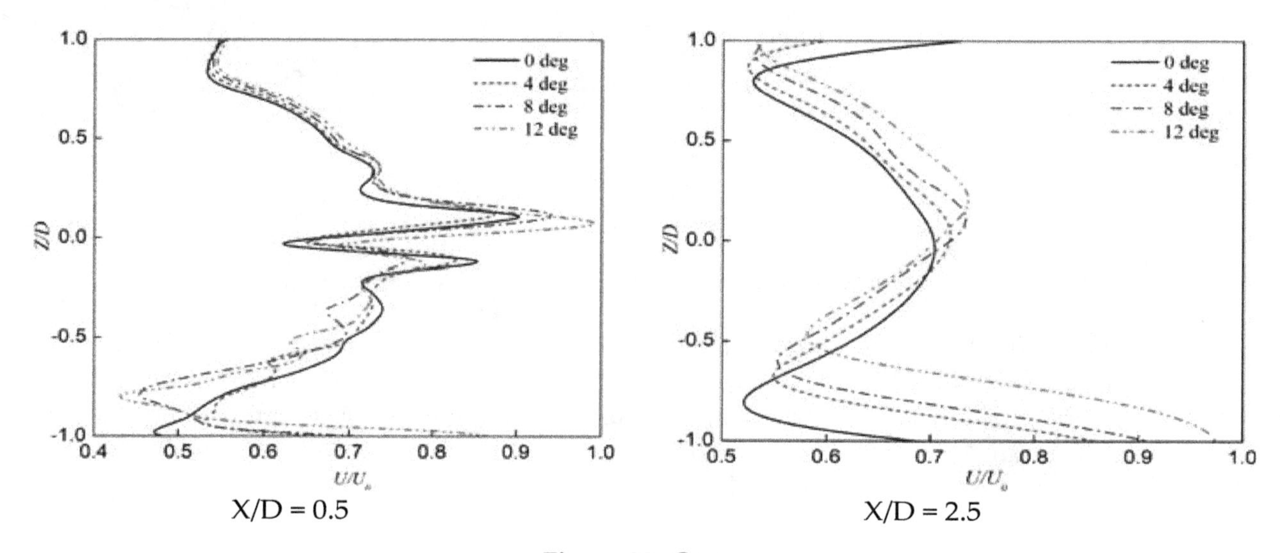

Figure 22. *Cont.*

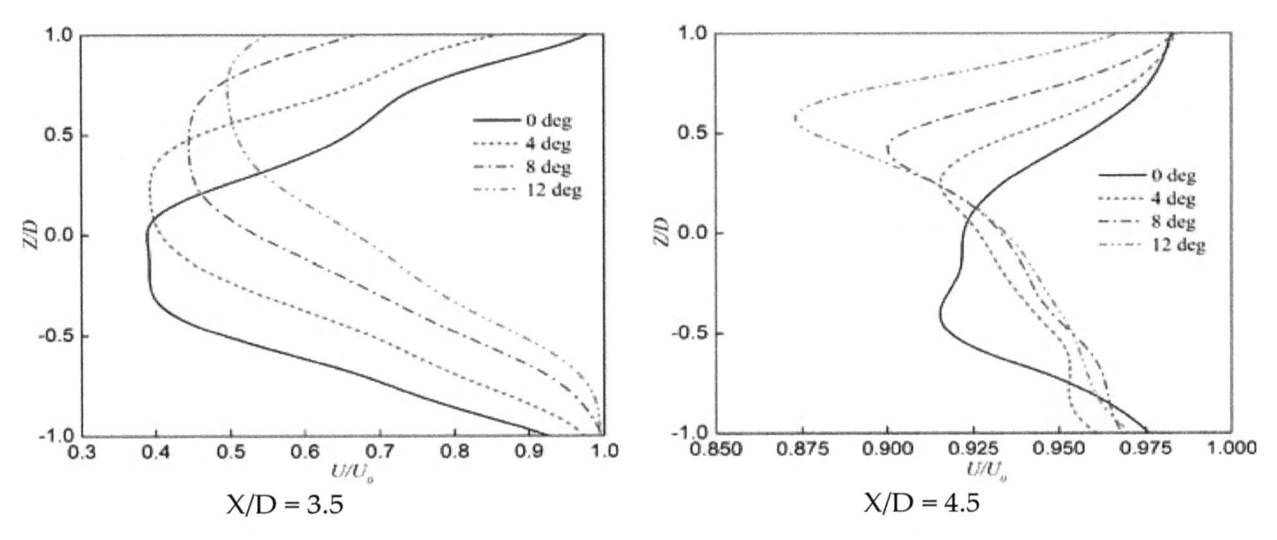

Figure 22. The distribution of instantaneous axial velocity along blade span at the wind turbine downstream positions of 0.5D, 2.5D, 3.5D and 4.5D.

3.4. Wind Shear

The change in wind speed with height was determined according to the power function given in International Electrotechnical Commission (IEC) 61400-1 [21] and presented as follows:

$$\frac{V_Z}{V_{Z_r}} = \left(\frac{Z}{Z_r}\right)^{\gamma} \tag{12}$$

where V_Z refers to the wind speed at height z, V_{Z_r} refers to the reference wind speed at height Z_r and γ refers to the wind shear exponent. Z_r refers to the hub height. In this study, the reference wind speed is 11.4 m/s. In this paper, wind shear exponents are 0.09, 0.2 and 0.41. The wind shear exponent of 0.09 indicates a very unstable atmospheric state, 0.20 represents a neutral state and 0.41 represents a very stable state [4].

The turbulence intensity was calculated according to the formula in IEC 61400-1 [21] and given as follows:

$$I_T = I_{ref}(0.75V_{hub} + 5.6)/V_{hub} \tag{13}$$

where I_T is the turbulence intensity, I_{ref} is the expected value of the turbulence intensity and V_{hub} is the reference velocity at the hub. In this paper, I_{ref} values are 0.12, 0.14 and 0.16. I_{ref} of 0.12 represents lower turbulence characteristics, 0.14 describes medium turbulence characteristics and 0.16 describes higher turbulence characteristics.

Table 8 shows the average power along one rotation of the wind turbine after it has stabilized. It can be seen from Table 8 that, compared with uniform wind, wind shear will cause the average power of the wind turbine to decrease by about 14%. At the same time, it can be found that the average power of the 4° tilt angle is close to that of the 0° tilt angle and is higher than the average power of the 8 and 12° tilt angles under uniform wind or wind shear conditions. The deviation ($|P_a - P_m|$) of the power relative to the average power at an azimuth angle of 180° gradually decreases as the tilt angle increases (see Figure 23, Table 8). When the tilt angle reaches 8° and continues to increase, $|P_a - P_m|$ will remain unchanged. This means that as the tilt angle increases, the interaction between the blade and the tower gradually weakens. When the tilt angle exceeds 4°, the influence of the tilt angle on the interaction between the blade and the tower can be ignored. However, when the tilt angle exceeds 4°, it will cause a significant decrease in the average power of the wind turbine. Therefore, considering the power of the wind turbine and the interaction between the blade and the tower, it is more appropriate to set the wind turbine tilt angle to about 4°.

Table 8. Power for uniform wind and wind shear flow conditions at $V_{hub} = 11.4$ m/s ($\gamma = 0.2$, $I_{ref} = 0.14$).

Tilt Angle (°)	Average Power Pa (MW)			Power Pm at 180° Azimuth Angle (MW)			$\lvert P_a - P_m \rvert$ (MW)	
	Uniform Wind	Wind Shear	Error (%)	Uniform Wind	Wind Shear	Error (%)	Uniform Wind	Wind Shear
0	4.92	4.20	14.63	4.73	4.08	13.74	0.19	0.12
4	4.91	4.21	14.26	4.79	4.15	13.36	0.12	0.06
8	4.85	4.18	13.81	4.78	4.14	13.39	0.07	0.04
12	4.75	4.12	13.26	4.68	4.08	12.82	0.07	0.04

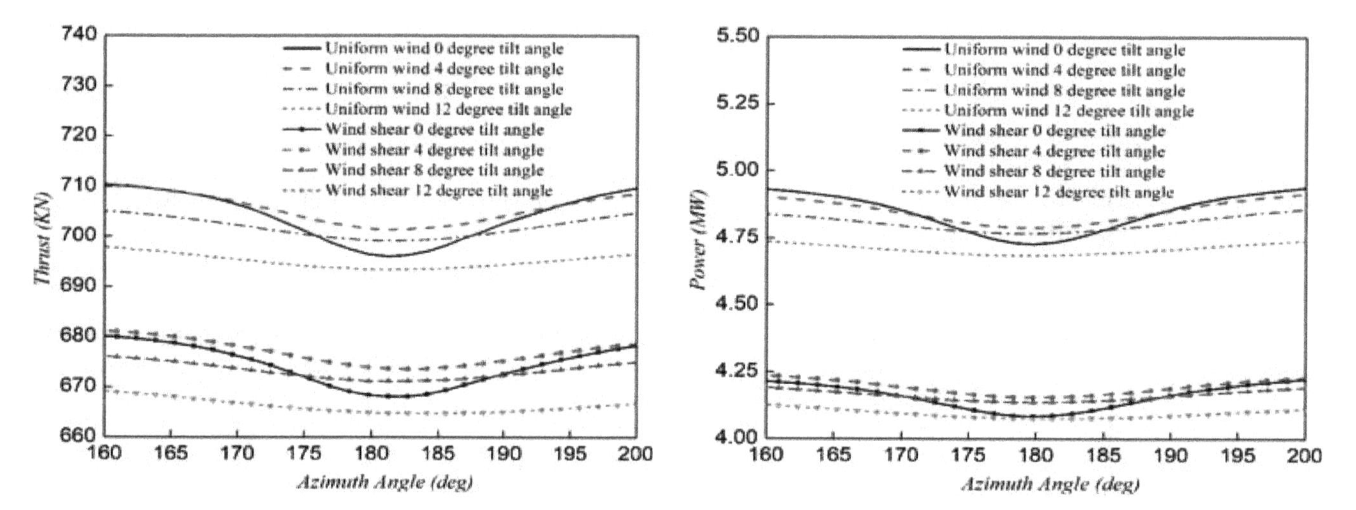

Figure 23. Thrust and power versus azimuth angle for various tilt angles at $V_{hub} = 11.4$ m/s ($\gamma = 0.2$, $I_{ref} = 0.14$).

Figure 24 describes the influence of wind shear exponents (γ) on the aerodynamic performance of the wind turbine. It can be seen from Figure 24 that the thrust and power of the wind turbine when the wind shear exponent is 0.41 are higher than the values when the wind shear exponents are 0.09 and 0.20. It can be found from Table 9 that the average thrust and power of the wind turbine under different wind shear exponents have the smallest error when the wind shear factor is 0.41 compared with the uniform wind, and the average thrust and power of the wind turbine are almost the same when the wind shear factors are 0.09 and 0.2. In Table 9, the wind shear exponent of 0.00 means uniform wind inlet conditions. In Figure 24, it can be seen that the fluctuation of the wind turbine thrust and power curve when the wind shear factor is 0.09 is significantly higher than the other two cases. This means that the wind shear exponent has an effect on the interaction between the blade and the tower.

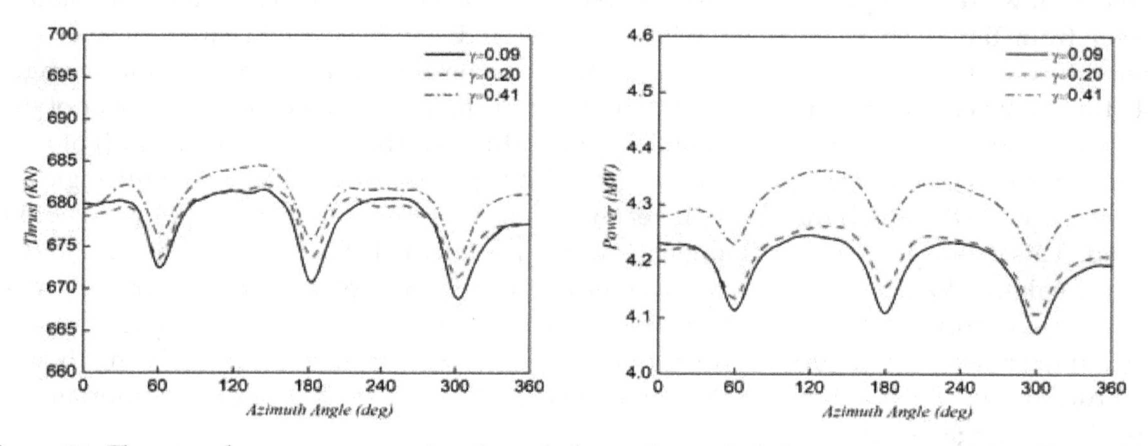

Figure 24. Thrust and power versus azimuth angle for various wind shear exponents (γ) at $V_{hub} = 11.4$ m/s ($I_{ref} = 0.14$, tilt angle = 4°).

Table 9. The power for various wind shear exponents (γ) at V_{hub} = 11.4 m/s (I_{ref} = 0.14, tilt angle = 4°).

Wind Shear Exponents	Average Power Pa (MW)		Average Thrust Ta (KN)	
	Power	Relatively Uniform Wind Error (%)	Thrust	Relatively Uniform Wind Error (%)
0.00	4.91	0.00	709.16	0.00
0.09	4.19	14.66	677.89	4.41
0.20	4.21	14.26	678.38	4.34
0.41	4.30	12.42	680.64	4.02

At the same time, as can be seen from Figure 25, at different turbulence intensity expectations, the thrust and power of the wind turbine are basically the same. This shows that the expected value of the turbulence intensity has little effect on the thrust and power of the wind turbine. Therefore, when using wind shear to simulate a wind turbine, it is necessary to focus on the size of the wind shear exponents according to simulated working conditions.

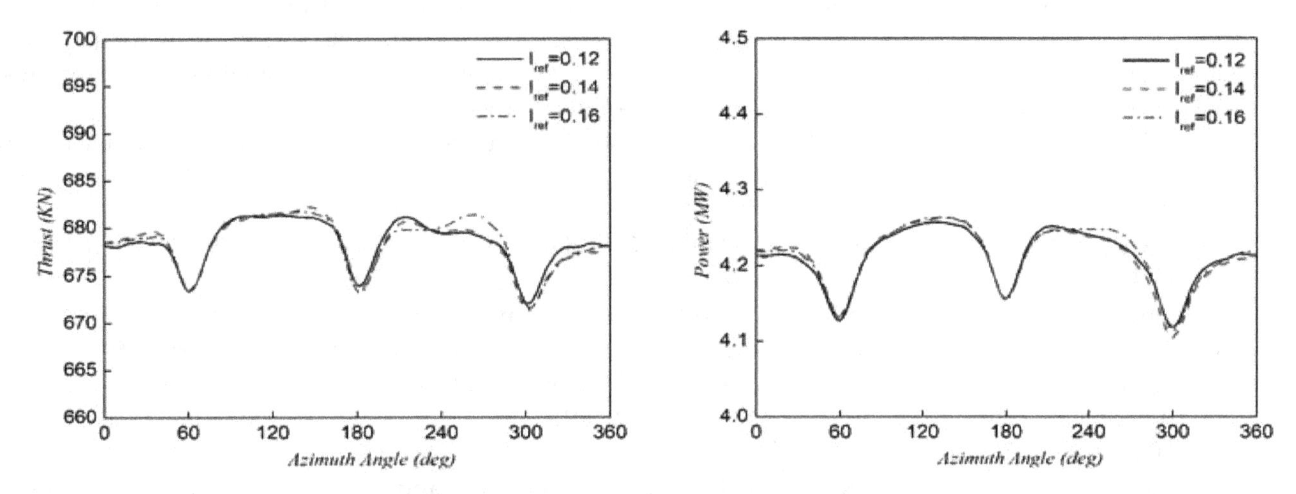

Figure 25. Thrust and power versus azimuth angle for various expected values of the turbulence intensity (I_{ref}) at V_{hub} = 11.4 m/s (γ = 0.2, tilt angle = 4°).

4. Conclusions

The computational fluid dynamics (CFD) method was used to simulate the aerodynamic performance of a fixed wind turbine with different tilt angles. By comparing the aerodynamic performance of a wind turbine at different tilt angles, it was found the aerodynamic performance of the wind turbine is better when the tilt angle is about 4°. The main purpose of the paper was to study the practical importance of effect of tilt angle on the aerodynamic performance of a wind turbine. The main conclusions of the paper are as follows:

1. In order to balance the power generation efficiency of the wind turbine and the interaction between the blade and the tower, the tilt angle of a wind turbine can be set at about 4° to obtain better aerodynamic performance.

2. The increase of the tilt angle will cause the load of the section airfoil to change, thus affecting the thrust and power of the wind turbine. When the blade is located in front of the tower, increasing the tilt angle will increase the load of the section airfoil. At the same time, after the tilt angle reaches 8°, the change in the load of the section airfoil with the tilt angle will not be obvious.

3. Wind shear will cause the thrust and power of the wind turbine to decrease, and the effect of the wind shear exponents on the aerodynamic performance of the wind turbine is significantly greater than the expected effect of the turbulence intensity. When performing wind turbine simulations, it is recommended to use a wind shear that is closer to that found in the real environment instead of uniform wind.

In summary, in order to ensure that a fixed wind turbine has an improved aerodynamic performance, the tilt angle of the wind turbine when installed should be about 4°. In reality, for a floating offshore wind turbine, a tilt angle of about 4° may not be appropriate, so the effect of tilt angle on a floating offshore wind turbine should be further studied in future works.

Author Contributions: Conceptualization, K.L.; Data curation, Q.W.; Formal analysis, Q.W.; Software, Q.W.; Supervision, Q.M.; Validation, Q.W.; Writing—original draft, Q.W.; Writing—review & editing, K.L. and Q.M. All authors have read and agreed to the published version of the manuscript.

References

1. Laks, J.H.; Pao, L.Y.; Wright, A.D. Control of wind turbines: Past, present, and future. In Proceedings of the 2009 American Control Conference, St. Louis, MO, USA, 10–12 June 2009.
2. Pao, L.Y.; Johnson, K.E. A tutorial on the dynamics and control of wind turbines and wind farms. In Proceedings of the 2009 American Control Conference, St. Louis, MO, USA, 10–12 June 2009.
3. Zhao, Q.; AlKhalifin, Y.; Li, X.; Sheng, C.; Afjeh, A. Comparative study of yaw and nacelle tilt control strategies for a two-bladed downwind wind turbine. *Fluid Mech. Res. Int. J.* **2018**, *2*, 85–97. [CrossRef]
4. Kim, H.; Lee, S.; Lee, S. Influence of blade-tower interaction in upwind-type horizontal axis wind turbines on aerodynamics. *J. Mech. Sci. Technol.* **2011**, *25*, 1351–1360. [CrossRef]
5. Guo, P. Influence analysis of wind shear and tower shadow on load and power based on blade element theory. In Proceedings of the 2011 Chinese Control and Decision Conference (CCDC), Mianyang, China, 23–25 May 2011.
6. Wang, Q.; Zhou, H.; Wan, D. Numerical simulation of wind turbine blade-tower interaction. *J. Mar. Sci. Appl.* **2012**, *11*, 321–327. [CrossRef]
7. Narayana, M. Gyroscopic Effect of Small Scale Tilt Up Horizontal Axis Wind Turbine. In *World Renewable Energy Congress VI 2000*; Elsevier: Amsterdam, The Netherlands, 2000; pp. 2312–2315. [CrossRef]
8. Zhao, Q.; Sheng, C.; Al-Khalifin, Y.; Afjeh, A. Aeromechanical analysis of two-bladed downwind turbine using a nacelle tilt control. In Proceedings of the ASME 2017 Fluid Division Summer Meeting, Waikoloa, HI, USA, 30 July–3 August 2017.
9. Abdulqadir, S.A.; Iacovides, H.; Nasser, A. The physical modelling and aerodynamics of turbulent flows around horizontal axis wind turbines. *Energy* **2017**, *119*, 767–799. [CrossRef]
10. Menter, F.R. Two-equation eddy-viscosity turbulence models for engineering applications. *AIAA J.* **1994**, *32*, 1598–1605. [CrossRef]
11. Vermeer, N.-J.; Sørensen, J.N.; Crespo, A. Wind turbine wake aerodynamics. *Prog. Aerosp. Sci.* **2003**, *39*, 467–510. [CrossRef]
12. Parente, A.; Gorle, C.; Beeck, J.v.; Benocci, C. Improved k-e model and wall function formulation for the RANS simulation of ABL flows. *Wind Eng. Ind. Aerodyn.* **2011**, *99*, 267–278. [CrossRef]
13. Tran, T.T.; Kim, D.-H. Fully coupled aero-hydrodynamic analysis of a semi-submersible FOWT using a dynamic fluid body interaction approach. *Renew. Energy* **2016**, *92*, 244–261. [CrossRef]
14. Politecnico di Milano. Qualification of innovative floating substructures for 10MW wind turbines and water depths greater than 50m. 2016. Available online: https://lifes50plus.eu/wp-content/uploads/2015/11/GA-640741_LIFES50_D3.2-mn-fix-1.pdf (accessed on 15 April 2018).
15. Miao, W.; Li, C.; Pavesi, G.; Yang, J.; Xie, X. Investigation of wake characteristics of a yawed HAWT and its impacts on the inline downstream wind turbine using unsteady CFD. *J. Wind. Eng. Ind. Aerodyn.* **2017**, *168*, 60–71. [CrossRef]
16. Jonkman, J.; Butterfield, S.; Musial, W.; Scott, G. *Definition of a 5-MW Reference Wind Turbine for Offshore System Development*; Report No. NREL/TP-500-38060; National Renewable Energy Laboratory: Golden, CO, USA, 2009.
17. Zhong, W.; Shen, W.; Wang, T.; Li, Y. A tip loss correction model for wind turbine aerodynamic performance prediction. *Renew. Energy* **2020**, *147*, 223–238. [CrossRef]
18. Yu, Z.; Zheng, X.; Ma, Q. Study on Actuator Line Modeling of Two NREL 5-MW Wind Turbine Wakes. *Appl. Sci.* **2018**, *8*, 434. [CrossRef]

19. Quallen, S.; Xing, T. An Investigation of the Blade Tower Interaction of a Floating Offshore Wind Turbine. In Proceedings of the 25th International Ocean and Polar Engineering Conference, Kona, HI, USA, 21–26 June 2015.
20. Rahimi, H.; Schepers, J.; Shen, W.; García, N.R.; Schneider, M.; Micallef, D.; Ferreira, C.S.; Jost, E.; Klein, L.; Herráez, I. Evaluation of different methods for determining the angle of attack on wind turbine blades with CFD results under axial inflow conditions. *Renew. Energy* **2018**, *125*, 866–876. [CrossRef]
21. IEC. *IEC61400-1, Wind Turbines-Part1: Design Requirements*; International Electrotechnical Commission: Geneva, Switzerland, 2005.

9

Optimize Rotating Wind Energy Rotor Blades using the Adjoint Approach

Lena Vorspel [1,*], Bernhard Stoevesandt [2] and Joachim Peinke [1,2]

[1] ForWind, University of Oldenburg, Ammerländer Heerstr. 114-118, 26129 Oldenburg, Germany; peinke@uni-oldenburg.de
[2] Fraunhofer IWES, Küpkersweg 70, 26129 Oldenburg, Germany; bernhard.stoevesandt@iwes.fraunhofer.de
* Correspondence: lena.vorspel@forwind.de

Abstract: Wind energy rotor blades are highly complex structures, both combining a large aerodynamic efficiency and a robust structure for lifetimes up to 25 years and more. Current research deals with smart rotor blades, improved for turbulent wind fields, less maintenance and low wind sites. In this work, an optimization tool for rotor blades using bend-twist-coupling is developed and tested. The adjoint approach allows computation of gradients based on the flow field at comparably low cost. A suitable projection method from the large design space of one gradient per numerical grid cell to a suitable design space for rotor blades is derived. The adjoint solver in OpenFOAM is extended for external flow. As novelty, we included rotation via the multiple reference frame method, both for the flow and the adjoint field. This optimization tool is tested for the NREL Phase VI turbine, optimizing the thrust by twisting of various outer parts between 20–50% of the blade length.

Keywords: rotor blade optimization; blade parametrization; computational fluid dynamics; OpenFOAM; gradient-based; adjoint approach

---

1. Introduction

The generated electric power from wind turbines increases linearly with the swept rotor area. Therefore, wind turbines grow larger and larger. Otherwise, limits in maximal mountable rotor weight and constraints coming from material science restrict the growth of rotor blades. Further limiting factors are alternating wind loads, gusts, turbulences and rotational forces. As wind turbines are normally designed for a life time of 20–25 years, the estimated total amount of rotations is about 150×10^6, which leads to 5×10^8–10×10^8 load changes. Periodic loads due to the vertical wind velocity gradient and the tower shadow, each one of them proportional to the blade rotation frequency, add up to this [1]. Added to those load changes due to the rotational movement are the wind loads, which were shown to be highly intermittent [2]. At the same time, most favorable sites for wind turbines with little turbulence intensity and large average wind speeds are already exploited, such that either repowering or worse sites are chosen in order to increase the generated wind power [3]. For slow average wind speeds, slender blades are favorable, which react even on small wind velocities due to a large aerodynamic efficiency. This enforces the trend towards long, slender blades. Such blades are prone to elastic deformations and even more influenced by wind load changes.

In return, this might lead to critical fatigue of the material, both from blades, but also from bearings and support structures. However, flexibility and light structures offer the possibility of passive control. The pitch systems of wind turbine blades are too slow to react on highly intermittent load fluctuations within the wind field, the use of structural bend twist coupling, which reacts instantly on aerodynamic loads, is a promising concept to reduce the amount of load changes and therefore of fatigue by a pre-designed deformation of the blade [4]. Especially adaptive changes in the twist of the blade can reduce the angle of attack and thus lead to a significant load reduction, which results in a longer life time or less cumbersome maintenance of the wind turbine blades and therefore in the end in lower specific costs of energy [5,6].

Based on the actual development in rotor design, we present in this work an optimization tool for the aerodynamics of rotating rotor blades within the open-source toolbox for numerical simulations OpenFOAM. This tool is tested for the optimization of the twist distribution in order to reduce the overall thrust force. Gradient-based optimization offers on the one hand the large advantage of a high-quality search direction for optima. On the other hand, only local optima are found. Thus, this method is suitable if there is not more than one optimum or if the initial point is not too far away from the global optimum. In our case, the aerodynamics of wind turbine rotor blades can be considered as sophisticated, so that the mentioned disadvantage of the gradient-based method can be neglected, while the advantage of a high-quality search direction is utilized. The continuous adjoint approach is used for the computation of gradients. In contrast to other methods, the adjoint approach is marked by the independence of the computational cost on the amount of design parameters, thus no full simulation per each single design parameter is required [7]. This qualifies the adjoint approach for the combination with computationally expensive Computational Fluid Dynamics (CFD). The optimization is based on the solution of steady-state flow equations of the current blade geometry. Obtained gradients are then evaluated to find the optimum. The implementation of the continuous adjoint equations in OpenFOAM was firstly done by Ohtmer [8] for ducted flows. Several publications show the applicability of the adjoint approach in combination with CFD, both for external and ducted flows [9–12]. However, these works showed the application either for low Reynolds numbers, for non-rotating fields or using porosity of numerical cells to impact the shape of ducts. Thus, the usage of the adjoint approach in wind energy needed further development and investigation. The adjoint solver for ducted flows could be used in the field of diffuser-augmented wind turbines. The generalized actuator disc theory, proposed by Jamieson [13] and extended by Lui and Yoshida [14], could be used within OpenFOAM to compute optimal difuser geometries with maximal effective diffuser efficiency. In this work, we focus on the optimization of rotor blade geometries of wind turbines in open flow. Therefore, the adjoint equations in OpenFOAM were adopted for external flow and the resulting gradients were verified against finite differences by Schramm [15,16]. It was shown in Schramm [16] that the adjoint approach can be used in wind energy relevant cases, in that work shown for the optimization of an airfoil slat. In a previous work, the adjoint optimization in OpenFOAM has been tested against other optimization strategies, like finite differences and Nelder-Mead. The independence from the amount of design parameters was shown there, as well as the applicability for wind energy relevant profiles [17].

In this work, the adjoint approach is used for rotating geometries and large Reynolds numbers. To the authors' knowledge, this combination poses a novelty and offers new areas of application. It should be noted that the rotational flow around rotor blades is complex and it is generally already challenging to achieve low residuals for such flows using stationary flow solvers. In addition, the NREL Phase VI turbine is stall regulated and the chosen velocity already corresponds to the rated velocity. Thus, the flow is separated at the inner blade region, leading to poor convergence behavior of the flow field. The adjoint field is driven by the flow field so that the errors in the flow field transfer directly to the adjoint field and decrease the convergence quality of it. This challenging set-up is used purposely to thoroughly test the optimization tool, its convergence and applicability.

The paper is structured as follows. First, the theory of the underlying optimization is given, including the gradient estimation via the adjoint approach and their processing is explained in Section 2. In Section 3, the simulation set-up in OpenFOAM is shown, as well as the optimization set-up and cases. Finally, results of the optimization are given in Section 4 followed by a conclusion in Section 5.

2. Optimization Method

To design a blade using bend-twist-coupling as a passive control mechanism, a favorable deformed blade shape that ensures the optimal load reductions has to be determined. The deformed blade shape can then be integrated into the structural blade design where layer lay-up or geometrical sweep are options to achieve the favorable deformations. To find those favorable deformations, an optimization tool for the aerodynamic shape of the rotor blades based on the continuous adjoint approach is implemented in OpenFOAM. The gradients are then projected to central nodes, adopting geometry updates like in Fluid–Structure Interaction (FSI) based on a beam. In this work, the optimization is directly coupled with the CFD toolbox, allowing direct communication and data usage. This tool is used to find deformations that reduce loads under certain load conditions. Our approach is that the general geometrical shape of the 2D profiles is constant, and only elastic deformations are allowed, such as changing the twist angle along the blade. In order to test the optimization tool and its performance, the optimizations are conducted for the same blade under the same inflow conditions but with a variable movable part of the blade and different aimed values. Therefore, between 20% and 50% of the outer blade are allowed to be changed with respect to its twist distribution. This investigation allows an assessment of the effect of complex lay-up within large to short parts of the blades, respectively. The optimization in this work is based on the reduction of the thrust that effects the blade root bending, tower bending and other forces. Although no constraints were used, it is possible to include limiting ranges for other forces in further applications of this tool.

In addition, the developed optimization tool allows other definitions of design parameters and cost functions.

Details regarding the adjoint approach are discussed next.

2.1. Gradient Estimation by the Adjoint Approach

The adjoint approach allows for computing the gradient of a cost function with respect to the design parameters independently from the amount of design parameters. The main idea is to build a Lagrangian function, which includes both the cost function to be optimized and the constraint and thus can be used for constrained optimization problems. For optimization based on CFD, the flow equations have to be satisfied in each optimization point. Thus, the constraining functions are the Navier–Stokes equations (NSE)

$$\mathbf{R_u} = (\mathbf{u} \cdot \nabla)\,\mathbf{u} + \nabla p - \nabla \cdot (2\nu D(\mathbf{u})), \quad (1)$$

$$\mathbf{R}_p = -\nabla \cdot \mathbf{u}, \quad (2)$$

with $\mathbf{u}$ and p being the flow velocity and pressure, respectively. The rate of strain tensor for any given vector x is defined as $D(\mathbf{x}) = \frac{1}{2}\left(\nabla x + (\nabla x)^T\right)$. This notation is used for the velocity $\mathbf{u}$ and adjoint velocity $\mathbf{u_a}$ in the flow and adjoint field, respectively. The viscosity ν is a sum of the molecular and turbulent viscosity.

The Lagrangian function

$$L := I + \int_\Omega (\mathbf{u_a}, p_a) \cdot \mathbf{R}\, d\Omega \quad (3)$$

combines the cost function I with the NSE $\mathbf{R}$ in the flow domain Ω. The Lagrangian multipliers, which allow the combination of the cost function with the NSE, are the adjoint velocity $\mathbf{u_a}$ and adjoint pressure p_a.

The adjoint field equations

$$-2D(\mathbf{u_a})\,\mathbf{u} = -\nabla p_a + \nabla \cdot (2\nu D(\mathbf{u_a})) - \frac{\partial I_\Omega}{\partial \mathbf{u}}, \quad (4)$$

$$\nabla \cdot \mathbf{u} = \frac{\partial I_\Omega}{\partial \mathbf{p}}, \quad (5)$$

result from the variation of the Lagrangian function with respect to the flow and design variables. A step-by-step derivation of the adjoint field is shown in the work of Othmer [18].

In this work, the flow equations are solved following the SIMPLE algorithm [19]. The continuous adjoint equations are similar to the ones from the flow field, such that similar numerical solving strategies can be applied. The frozen turbulence approach is applied, meaning that the viscosity from the flow field is also applied for the adjoint field, without additional adjoint turbulence equations [7,20].

In the following, flow gradients refer to changes in the flow field over distance, whereas gradients are the gradients of the cost function w.r.t. the design variables.

The two discussed fields have to be solved once, respectively, to derive all gradients at each numerical grid node on the blade geometry surface. Thus, the additional cost for the gradient information at any amount of design parameters equals the cost of solving a second set of partial differential equations that are derived from the flow field equations. One consequence from this is that the adjoint field is driven by the flow field on which it is based. This can also be seen in Equations (4) and (5) and leads to a transfer of the error made in the flow field computations to the adjoint field. Thus, the adjoint field is less stable. The large advantage of the adjoint method compared to other gradient estimation methods is that the amount of necessary solutions of the flow field doesn't scale with the amount of design parameters. This is especially desirable when working with gradient based optimization based on fully resolved simulations.

2.2. Gradient Projection and Evaluation

For an optimization procedure, it is an essential point to work out suitable parameters. The definition of the rotor blade's shape can be done in various ways. It was shown in [17] that the free parametrization, i.e., all numerical grid points are design parameters, is not suitable for optimizing rotor blades. This results from large flow field gradients in the chordwise direction and in the spanwise direction, which transfer to the computed gradient. In areas of large gradients, the deformation is much larger, which leads to bumps and uneven surfaces. Instead of the use of all grid points, one can select parameters that define the blade geometry, like the twist distribution, thickness of airfoils along the blade and so forth, which define the blade geometry. For the developed optimization tool, an automatized routine is implemented that allows the projection of the gradients at each numerical grid cell to a reduced design space.

The flow around airfoils and rotor blades is highly sheared with large flow gradients both in velocity and pressure. Therefore, the numerical grid has to be appropriately fine at those areas in order to catch the flow phenomena. Typical values for the amount of cells in the chordwise direction vary between 250 and 350. The adjoint approach leads to gradient information on each of these cells. To use a parametrization with less design parameters, a projection from these 250–350 gradients on less design parameters is implemented. The main idea for the parametrization used in the optimization tool comes from FSI using a finite beam. The gradients are treated like forces acting on the blade and therefore projected on central nodes. The principle of the parametrization is shown in Figure 1.

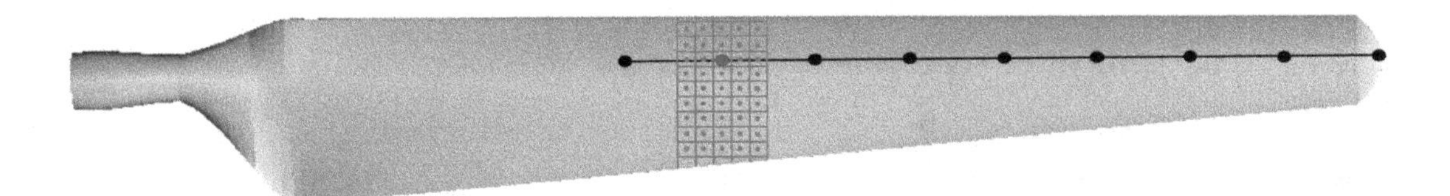

Figure 1. Principle of gradient projection for a wind turbine rotor blade.

As a pre-processing step, the blade is split into a selected amount of elements. Each of these elements has a central node, which is represented by a big red dot in Figure 1. All big dots represent one final design parameter each. The amount of design parameters is chosen beforehand by the user. The number can be controlled by adapting a dictionary that is defined for the usage of the developed tool.

In this exemplary case, there are nine design parameters. For one exemplary element, the structured grid is shown with thin blue lines. In the figure, the represented grid has been coarsened for the sake of clarity. Within each cell of this grid, the small blue dot represents the local gradient of each grid cell, computed via the adjoint approach. The adjoint gradients are then transferred onto the nodes depending on the distance of the local cell to the central node of the element. Evaluating the projected gradients then leads to a movement of the nodes, which act like beam nodes. The difference to FSI is that the structure is not solved, and it is assumed that any movement given by the gradients can be realized. This assumption is necessary for the presented optimization. The final, optimized geometries can be used within the blade design process as an input to the design of the structure, for example when passive control shall be applied. Therefore, the structure is not final when the optimization takes place. In addition, the movement of the blade due to a structural response might interfere with the movement given by the gradients.

The pre-condition for this projection is that an elastic deformation shall be guaranteed, where the blade cross sections are constant. However, even this parametrization on a reduced design space takes into account the gradient information of each numerical grid cell, as each local gradient is used within the projection. This makes the usage of the adjoint approach convenient.

The gradients of each numerical cell that belongs to one central node of an element are transferred to two vectors. One is acting like a force, that can be compared to the bending force in FSI and one is acting like a torsional moment that can be compared to a moment of force responsible for twisting in FSI. After projection, each element of the blade has these two gradient vectors. In a second step, for each of these two vector types, the maximal value among all elements is used for normalization, respectively. This leads to a similar range of values, with a maximal magnitude of one, for both. With this, weighting factors can be introduced that allow for augmenting one of the movements, when needed. Optional smoothing by neighboring elements is implemented and can be switched on and off via a dictionary. In between the elements, an interpolation is used to avoid kinks on the blade surface. This interpolation as well as the mesh update is taken from an in-house FSI solver by Dose [21].

The projected gradients can then be used within an optimization algorithm for gradient based optimization. The most direct way to use the gradients is via the steepest descent algorithm, which is also used in this work. According to the direction and dimension of the gradient, changes in the geometry are performed for all selected design parameters. Depending on the slope of the design space, the optimal search direction (pointing to the minimum) can differ largely from the gradient direction. In these cases, the steepest descent search results in a highly inefficient, but typical zigzag pattern. Consequently, the method of steepest descent has found little applicability in practice, since its performance is in most cases largely inferior to second-order methods, like the Quasi-Newton method. For high-dimensional design spaces, though, the method of steepest descent can be advantageous, since it does not require the estimation of the inverse Hessian, which scales with the number of design variables [22].

3. Simulation and Optimization Set-Ups

This is a conceptual work in which we want to show a new optimization approach for wind energy applications. Therefore, the optimizations are conducted for the NREL Phase VI turbine. This turbine is a stall-regulated, two-bladed turbine with a rated power of 19.8 kW. It was used for experiments in the NASA/Ames 24.4 m $\times$ 36.6 m wind tunnel [23] and is often used for validation of CFD. The S809 airfoil with 21% thickness was used for the construction of the blade [24]. In this work, the cone angle of the rotor is $0°$ and the pitch angle of the blade is $3°$.

To include rotational effects in steady-state simulations, the Multiple Reference Frame (MRF) is used. The rotation is modelled via a source force, that adds a rotational component to each cell. This way, the rotor does not have to move within the mesh reducing the computational effort, as no mesh update is necessary to include the rotation. The MRF method is also adopted to the adjoint field, so that the used schemes and solution methods for both fields stay similar.

The mesh used in this work was created by the bladeBlockMesher (BBM) developed by Rahimi et al. [25] for the automatic mesh generation for wind turbine rotor blades. The grid quality and validation based on the NREL Phase VI turbine is also shown in their publication. In this work, a grid with 5.2 million cells is used with a y^+-value below 1. A symmetry plane is used to exploit the symmetry of the two bladed rotor. In Figure 2 the domain and the mesh around a cut along the chord of the blade is shown.

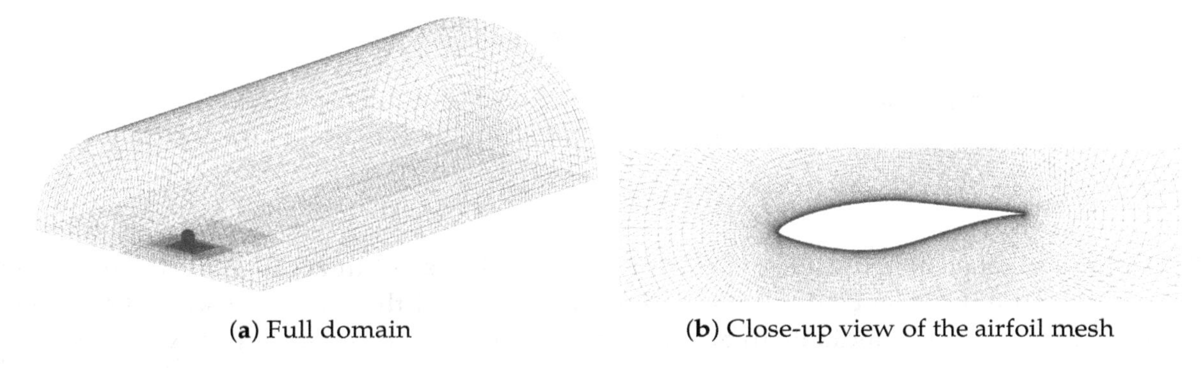

(**a**) Full domain (**b**) Close-up view of the airfoil mesh

Figure 2. Numerical grid. The flow is coming from the left in (**a**). In (**b**), the o-type mesh around a cut through the blade is shown.

A rounded tip is created and meshed via the BBM. The k-ω-SST turbulence model is chosen for the flow field, whereas for the adjoint field the frozen turbulence approach is applied. This approach was used in various publications and it was shown that the gradient quality is adequate [7,18]. Standard boundary conditions are applied, similar to those in Rahimi [25]. The inflow velocity is 10 m/s and the rotational speed is 7.54 rad/s, i.e., 72 rpm and the density equals 1.225 kg/m^3. This velocity leads the stall regulated turbineto already partially separate flow at large parts of the blade. Steady state CFD computations of this turbine at separated flow are known to differ from the experimental data at those separated areas [25]. As it was shown above, the adjoint field is driven by the flow field and therefore it is expected that the adjoint field converges even less as good as the flow field. Still, this velocity was chosen in this work, as even small changes in AoA have a large impact on the forces, when separation is reduced.

As the adjoint field moves upstream, the boundary conditions are switched compared to the flow field. The adjoint velocity has a Neumann boundary condition at the inlet, and a Dirichlet boundary condition at the outlet and at the blade. The adjoint pressure has a Dirichlet boundary condition at the inlet and Neumann boundary condition at the outlet and blade. For a higher stability, the adjoint velocity at the hub is put to zero. This was necessary, as the transition between hub and blade root led to large instabilities in the adjoint field.

It was shown in Section 2.1 that the adjoint field is driven by the primal field. Therefore, preconverged solutions of the flow field are used in order to stabilize the adjoint equations. A convergence of the velocity up to a residual in the range of 10^{-7} is aimed before the adjoint field equations are switched on. An initial study using the same numerical grid based on the initial geometry was conducted to find adequate amounts of CFD iterations to reach good convergence of both fields. As a result, 25,000 iterations are used solving the flow field based on the SIMPLE algorithm [19]. An additional 15,000 iterations are used for solving the adjoint and flow field before the optimization tool is started. These pre-converged fields are used for all optimizations, as the initial blade geometry and flow boundary conditions are identical. Starting from the total amount of 40,000 iterations used for pre-convergence of the fields, the optimizer uses an additional 500 iterations after each geometry update for a pre-convergence of the flow field starting from the last flow field data before the geometry update. An additional 2500 iterations are then used solving both fields. After a total of an additional 3000 iterations, the gradients are evaluated. As the allowed movement of the design parameters, leading to an update of the geometry, is not too large, the flow and adjoint field before and after a geometry update are also rather similar. Thus, using the variables of the last step before a geometry update leads to a good pre-converged solution and a reduced amount of needed iterations between two optimization steps compared to the iterations needed for the first step. The convergence of the optimization, in this case the evaluation of the thrust force, is tested after the 500 iterations solving the flow field.

The general definition of the minimization problem

$$\min f(s) \quad s \in R^N, \tag{6}$$

with s being the N design parameters of the cost function f in R^N, is specified in this work for the thrust force F_T. The cost function $f(s)$ is defined as a least square function of the thrust

$$f(s) = \frac{1}{2}\left(F_T(s) - F_{T^*}\right)^2. \tag{7}$$

This allows to set an aimed value of the thrust F_{T^*} which the optimizer then tries to reach. As the thrust force has an impact on blade root bending moments and the loads transferred on the wind turbine, this force is a good candidate to be one key element when it comes to passive control.

The initial value $F_T(s_0)$ of the original blade with initial design parameters s_0 and the above given inflow and rotational speed equals 958.9 N. Two different aim values F_{T^*} are chosen, namely 925 N and 900 N. This equals a reduction by 3.54% and 6.14%, respectively. Two different convergence limits are applied for the simulations, i.e., an interval of $\pm 1.5\%$ and $\pm 2.5\%$ around the aimed thrust force. Whenever the new geometry leads to a thrust within the given range around the aimed thrust, the optimization is ended. This can lead, for example, to a final value only a bit below of $1.025 \cdot F_{T^*}$ in the case of the $\pm 2.5\%$ range. This convergence criterion is chosen due to the instable flow phenomena transferring to the convergence of the gradients. Three different radial parts r of the blade are chosen to be optimized. The tip region is left out because the tip vortex leads to more complex flows and therefore leads to inferior gradient quality. From the blade length of 5 m, the following radial parts r are chosen, with 0 m being the blade root and 5 m the blade tip: r_1 = 2 m–4.5 m, r_2 = 3 m–4.5 m and r_3 = 3.75 m–4.75 m. The short range r_3 is split into five elements, while, for the middle and long range, nine elements are used. An overview of the optimization cases is given in Table 1.

Table 1. Optimization set-ups for different parts of the blade, aimed values and convergence interval around the aimed value.

Case	F_{T^*}	Convergence Interval		Radial Part r	Blade Part	Length
	(N)	%	(N)	(m)	%	(m)
1	900	1.5	886.5–913.5	2–4.5	50	2.5
2				3–4.5	30	1.5
3				3.75–4.75	20	1
4		2.5	877.5–922.55	2–4.5	50	2.5
5				3–4.5	30	1.5
6				3.75–4.75	20	1
7	925	1.5	911.125–938.875	2–4.5	50	2.5
8				3–4.5	30	1.5
9				3.75–4.75	20	1
10		2.5	901.875–948.125	2–4.5	50	2.5
11				3–4.5	30	1.5
12				3.75–4.75	20	1

4. Results

In this section, the performance and results of the developed optimization tool are presented. The implementation of the rotating adjoint field is one major step towards an optimization tool for wind energy application. First, the flow and adjoint fields around the blade are shown and analysed with focus on the adjoint field and the output of the MRF method applied to it. Then, the optimization tool is used for the test cases mentioned above. The additional twist deformations for various cases are compared, as well as the convergence behavior and stability of the optimization tool.

4.1. Flow and Adjoint Field

In some engineering applications, the adjoint field is used as a sensitivity map within the design process. For the design of complex geometries, like cars, the solution of the adjoint field offers the information where the surface should be changed to achieve a more aerodynamical efficient behavior. For this type of application, the adjoint field is only solved once and the surface is not adopted automatically within an optimization. However, it is also possible to compute gradients from the adjoint field, which then are used for optimization. This is also done in this work. To get an impression of the adjoint field, it is shown for the initial geometry at a cut through the blade, showing the airfoil geometry, at an exemplary radial position of 3 m of the span. In Figure 3, cuts through the blade are colored by the velocity (left) and adjoint velocity (right).

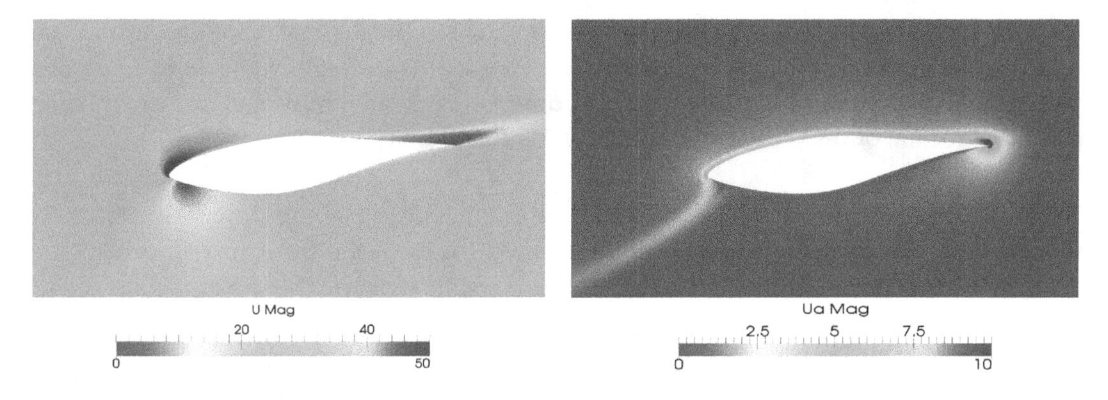

Figure 3. Magnitude (Mag) of the velocity U and adjoint velocity Ua around the blade at a cut at 3 m of the span.

In the left plot, typical flow around rotor blades can be seen. The stagnation point is below the leading edge, which shows a positive angle of attack. From the stagnation point onwards along the first part of the suction side, the flow is accelerating and finally the flow is slightly separated at the last third of the suction side. The adjoint velocity in the right figure shows an upstream movement, which is also expected from the theory. It is high along the suction side, where the flow gradients are also large. Then, the adjoint velocity moves upstream from the stagnation point on and follows a path around the angle of attack.

The trailing edge of an airfoil is a discontinuity that impairs the convergence of the adjoint field. The flow is separated in the rear part of the blade, thus turbulent effects are dominant in this region. At this part of the blade, the frozen turbulence approach decreases the convergence behavior of the adjoint field. This can also be seen at the adjoint velocity plot, showing a peak at the trailing edge. A used workaround for this problem in literature is to exclude the cells close to discontinuities from the evaluated gradients [16]. This would imply an additional processing of the gradients. For this work, such an exclusion of certain blade areas is not done, as this could overlap with the effects of the chosen projection of gradients. Testing this projection was one of the main goals in this work, which is why overlapping effects are avoided. However, it could easily be added for further applications of the optimization tool using the existing utilities available in OpenFOAM.

In order to show the flow around the blades in its rotational component, the Q-criterion in terms of the vorticity was used in Figure 4.

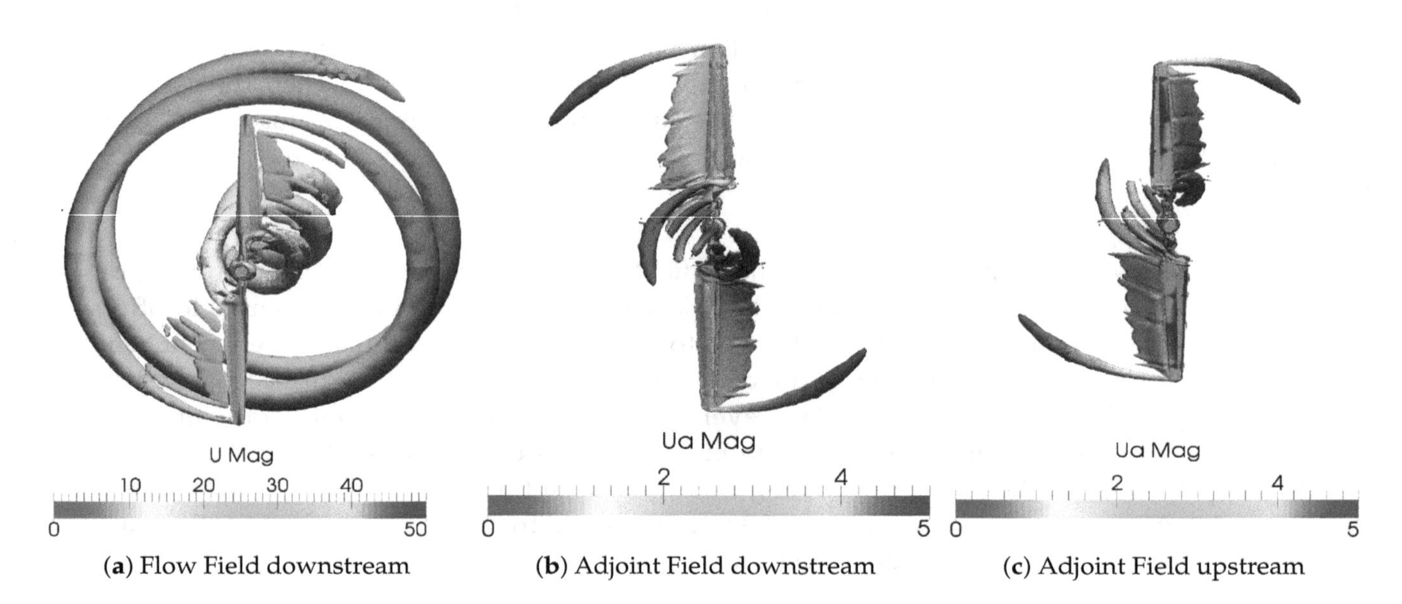

(**a**) Flow Field downstream (**b**) Adjoint Field downstream (**c**) Adjoint Field upstream

Figure 4. Field visualisation based on the Q-criterion in terms of vorticity. The full rotor for the flow field in (**a**) and for the adjoint field in (**b**,**c**) is shown. For good comparability, the adjoint field is shown in (**b**) in the same direction, as the flow field, i.e., downstream. As it moves upstream, it is secondly shown in the direction of the propagation of the adjoint field in (**c**).

The shown flow structures then are colored by the magnitude of the flow velocity and the adjoint velocity, respectively. The computations were performed for both the flow and adjoint velocity. As the adjoint field moves upstream, both views of the rotor are shown for the adjoint field. One view is in the downstream direction, according to the view for the flow field and one is in the upstream direction according to the propagation of the adjoint field. In Figure 4a, the tip and root vortex of the rotor can be seen, as well as low velocities close to the stagnation point and increasing velocities towards the tip of the blade. The tip and root vortex impair the flow field convergence at the root and tip area of the blade, respectively. The computed flow variables are less stable and reach inferior convergence in stationary simulations due to the instationary flow phenomena. As the gradients are evaluated along the movable part of the blade, which are according to Table 1 starting at 2 m spanwise and reach up to 4.75 m of the span, the influence of the vortices is supposed to be small for the gradient computation. In Figure 4b, it can be seen that similar vortex structures are built for the adjoint field, whereas they are moving upstream and rotating in the other direction compared to the flow field. Figure 4c shows the adjoint field looking at the suction side of the rotor. The large adjoint velocities at the suction side can be seen. In addition, the influence of the tip and root vortex, leading to a reduced adjoint velocity at the tip and root of the blade, respectively, is evident. Large and small adjoint velocities transfer to large and small gradients, respectively. Thus, large adjoint velocities indicate areas of the shape, where changes of the geometry are the most immediate. Overall, the implemented MRF approach for the adjoint field works in a stable manner and allows the computation of adjoint approach based gradients for rotating geometries, and the implemented version shows results expected from theory.

4.2. Convergence Behavior of the Optimizations

In this work, different types of convergence are used. Numerical convergence of the computed fields is investigated as well as convergence of the optimization tool. Convergence criteria for the flow field are mentioned in Section 3. The latter is defined depending on the ratio between current cost function value and aimed cost function value. The convergence behavior of the various optimizations for different aimed values and varying lengths of the twistable part of the blade is compared in Figure 5a–d.

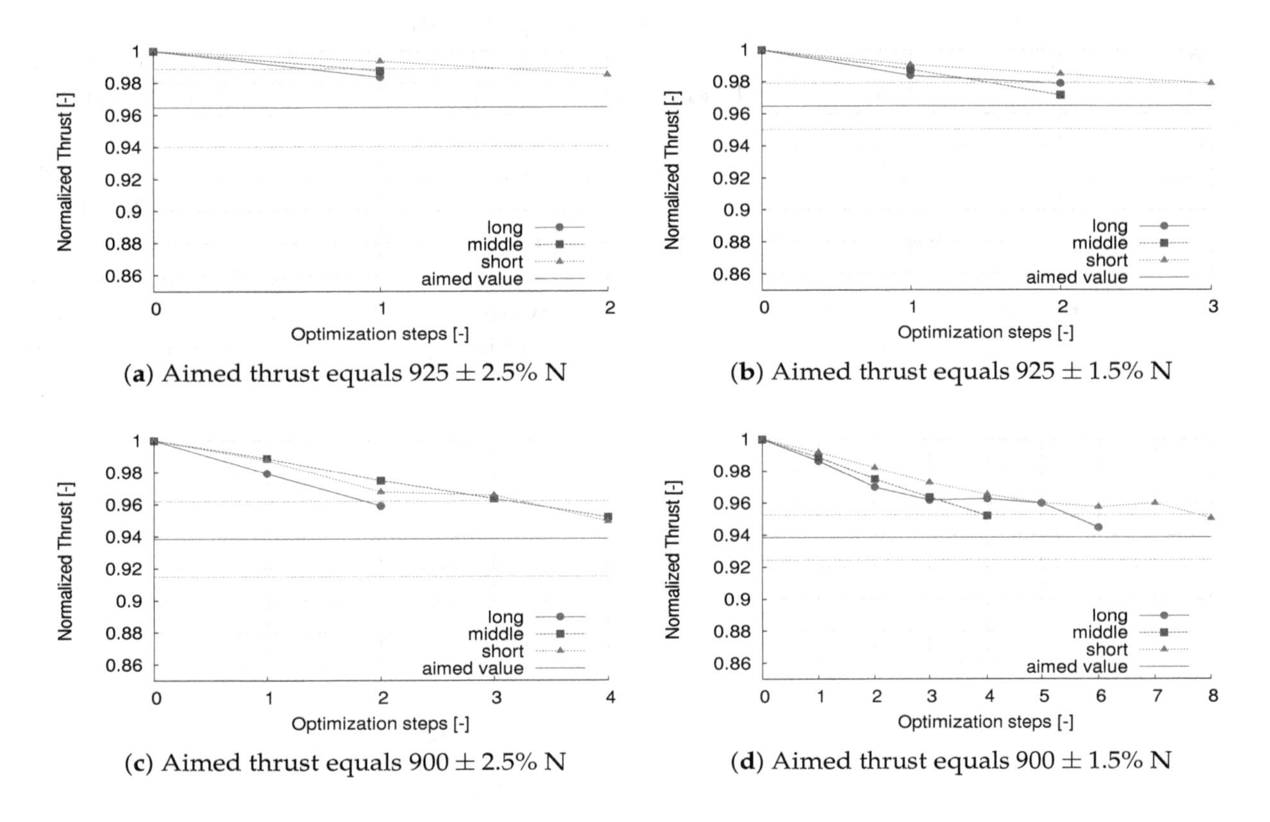

(**a**) Aimed thrust equals 925 ± 2.5% N
(**b**) Aimed thrust equals 925 ± 1.5% N
(**c**) Aimed thrust equals 900 ± 2.5% N
(**d**) Aimed thrust equals 900 ± 1.5% N

Figure 5. Development of the normalised thrust force over the optimization steps for design parameters within varying parts of the blade. Long: 2 m–4.5 m radial position, middle: 3 m–4.5 m radial position and short: 3.75 m–4.75 m radial position. The aimed values and the limits of the allowed range are shown with black and grey lines, respectively.

For all plots, the linear lines between the points are added to ease the following of the trend, whereas only the dots represent computed values of the actual geometries of the blade during the optimization. The thrust is normalized by the initial value of 958.9 N to ensure comparability of the different aimed values and convergence ranges. The aimed values equal 925 N for the upper plots (a) and (b) and 900 N for the lower plots (c) and (d). The left plots are for the wider allowed range of 2.5% around the aimed value, whereas the right plots are for a range of 1.5% around the aimed value.

All optimizations, besides the short and long blade parts for an aimed value of 900 ± 1.5% N, show a similar behavior. With each optimization step, the thrust is reduced until the upper limit of the allowed aim range is met. The two named exceptions show a zig-zagging behavior close to the bound, which is a known problem of steepest descent.

Comparing all optimizations, it can be seen that more optimization steps are needed when the aim is further away from the initial value and when the bounds are closer. Furthermore, longer movable blade parts lead to faster convergence, except for the case shown in Figure 5d. In this case, the middle length converges faster than the long length case. It was expected that longer blade parts lead to faster convergence of the optimization as larger parts of the blade are moved relatively to the inflow angle, which then has larger influence on the forces, respectively. For the case shown in Figure 5d, the sensitivity of the optimization process with respect to the user's input, like initial step-size and pre-convergence of the flow and adjoint field, was found to be the largest. This explains the outlier of the long radial section r_1 compared to r_2.

It has to be remembered that each optimization step represents a larger amount of CFD iterations, which are necessary to compute reliable gradients via the adjoint approach. For a convergence after four optimization steps, e.g., shown for the short and middle length in Figure 5c and the given pre-convergence iterations and iterations between two optimization steps specified in Section 3, this sums up to 54,500 CFD iterations per optimization. Roughly 35,000 of these iterations include solving two fields. These numbers are still very convincing for gradient based optimization based on CFD.

When the optimizations for these cases were conducted, a dependence of the convergence, appearance of zig-zagging and initial step-size were observed. As mentioned in Section 2.2, the projected gradients are normalized by the current maximal value. These normalized gradients are then multiplied by the step size, such that the maximal gradient leads to a movement according this step size and the further design points move less. Once converged, optimizations could be sped up by changing the step-size. This is shown in Figure 6 for the long blade part when the aimed thrust equals $900 \pm 1.5\%$ N.

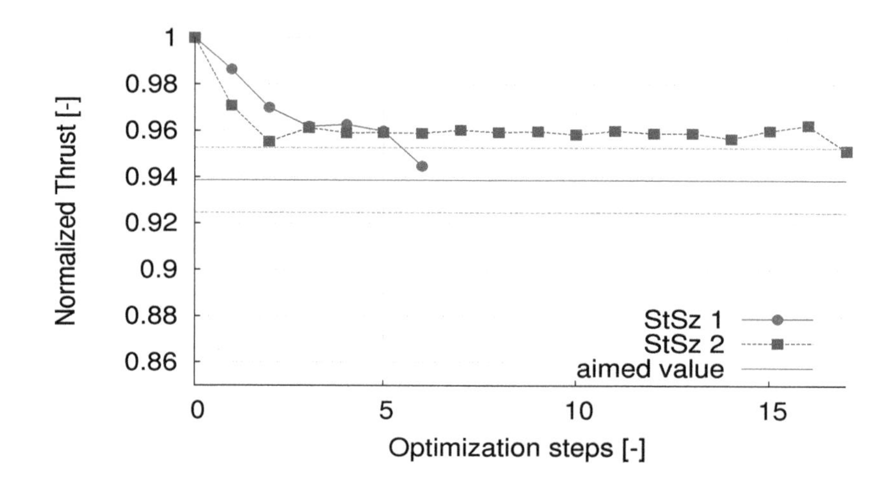

Figure 6. Development of the normalised thrust force over the optimization steps for the long blade part. The aimed force is $900 \pm 1.5\%$ N and the initial step sizes are 10^{-2} (red dots, StSz1) and 2×10^{-2} (blue squares, StSz2).

The two step sizes 10^{-2} (red) and 2×10^{-2} (blue) are compared. The larger step size leads to a faster decrease of the thrust within the first steps, but then zig-zagging happens for 14 optimization steps before convergence is reached. The case with the lower step size shows zig-zagging for two steps, before convergence is reached.

This underlines one major drawback of the continuous adjoint approach, also reported in other publications [26,27]. The user input, e.g., initial step size and set-up of the underlying flow simulation, is highly important for these optimizations, whereas other methods are more robust. Still, the shown amount of needed CFD iterations also underlines the advantage of this approach, which is its independence from the amount of design parameters. This still holds for the large Reynolds numbers and rotating geometry investigated in this work.

4.3. *Additional Twist Computed by the Optimization*

The final additional twist of the blade for the various optimization cases is shown in Figure 7a–c. The dots represent the moving element nodes where the gradients of each element are projected on (see Figure 1). The lines represent the smoothed movement between the nodes, as the geometry movement is also smoothed by the optimization tool. This plotted twist is then added to the initial twist of the according blade. The shown twist are the final ones, summed up over all optimization steps.

All twist values increase from the first element node towards a maximal value close to a spanwise position of 4 m. Afterwards, the additional twist decreases and even reaches negative values for the last element. The positive values in this set-up lead to a reduced angle of attack.

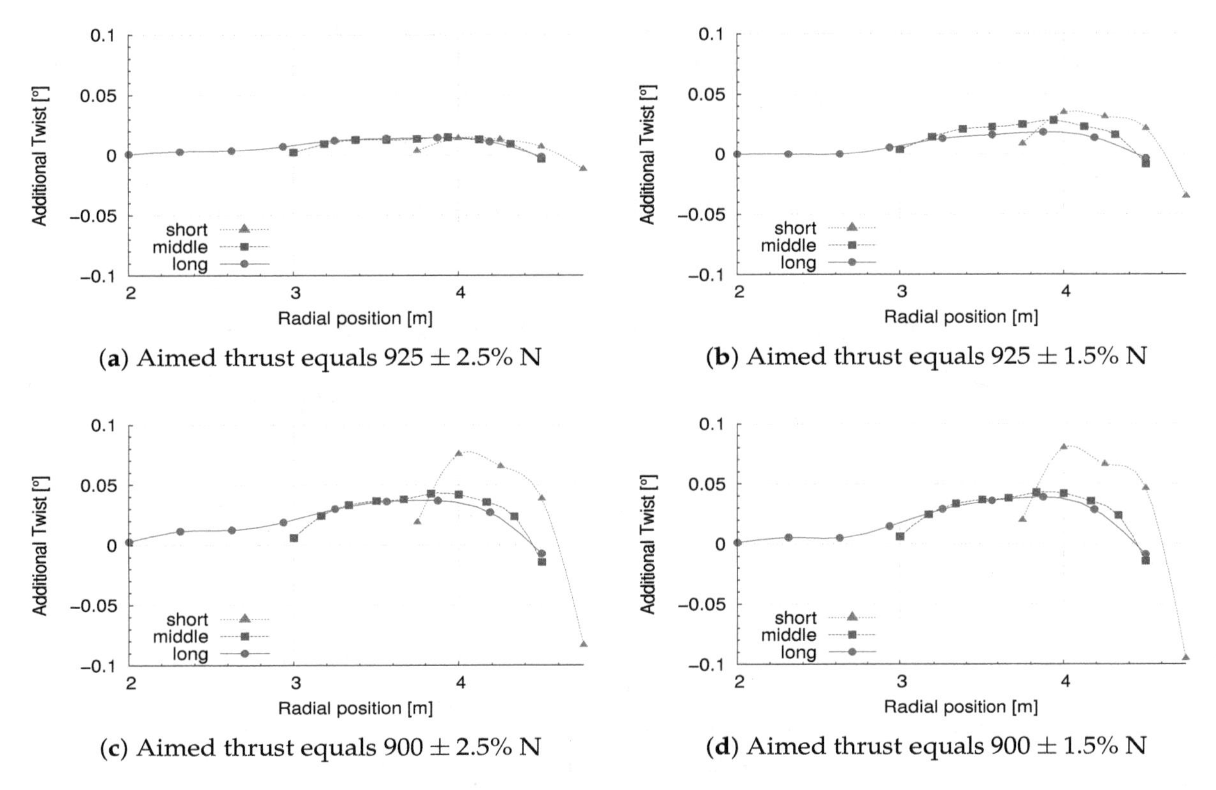

(**a**) Aimed thrust equals 925 $\pm$ 2.5% N

(**b**) Aimed thrust equals 925 $\pm$ 1.5% N

(**c**) Aimed thrust equals 900 $\pm$ 2.5% N

(**d**) Aimed thrust equals 900 $\pm$ 1.5% N

Figure 7. Development of the normalized thrust force over the optimization steps for design parameters within varying deformable parts of the blade. Long: r_1 = 2 m–4.5 m radial position, middle: r_2 = 3 m–4.5 m radial position and short: r_3 = 3.75 m–4.75 m radial position.

In general, those plots all show two trends: the shorter the movable section, the larger the additional twist. Furthermore, the more restrictive the aimed value, the more deformation is necessary to reach this value. This verifies the developed tool for optimizing rotating rotor blades.

To investigate this behavior further, the spanwise production of thrust of the initial NREL Phase VI blade is plotted in Figure 8.

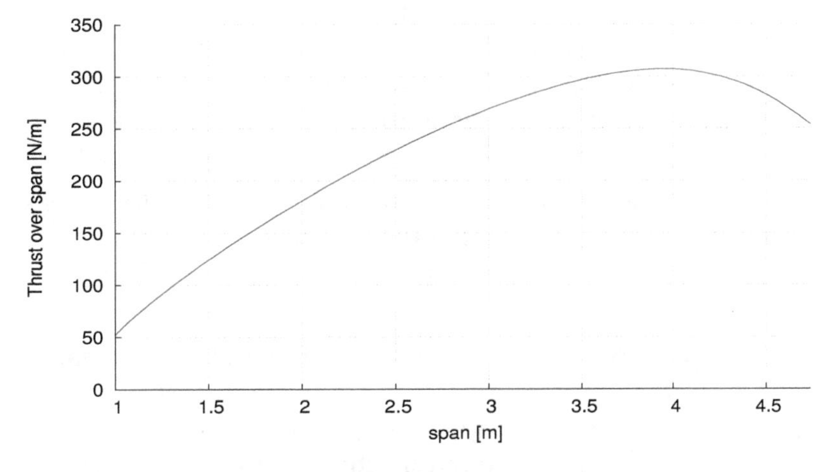

Figure 8. Spanwise production of thrust of the original NREL Phase VI blade from 1 to 4.75 m in radial direction.

It can be seen that the thrust production increases in the spanwise direction until a peak at a radial position of $r = 4$ m. Afterwards, the production of thrust decreases. Thus, the gradients follow the trend of production of thrust in size and direction, leading to a larger deformation around $r = 4$ m and decreased deformation towards the tip.

Larger movements in the outer blade part, which are aerodynamically more efficient, also have a larger impact on the thrust. This is apparently also represented by the gradients leading to a larger deformation in the outer regions. Close to the tip of the blade, the vortices seem to destabilize the adjoint field and thus also the accuracy of the computed gradients. This is also underlined by the larger negative twist values, which where not expected beforehand, of the short deformable blade part r_3 in comparison to the two longer ones. The deformable blade part r_3 ends at 4.75 m, whereas the other two deformable blade parts r_1 and r_2 end at 4.5 m. Thus, the influence from the tip vortex seems to be stronger for r_3, which reaches 0.25 m further towards the tip.

The conducted optimizations are chosen to test the developed optimization tool. It is therefore important to reduce the occurring effects as far as possible and avoid converse effects that could arise when using cost functions based on various force components and constraints. For future applications of the developed tool, it is a challenge to extend this method to more important design constraints and more elaborate definitions of the cost function.

5. Conclusions

An optimization tool for rotating wind energy rotor blades at realistic Reynolds numbers was developed and implemented based on the adjoint approach in OpenFOAM. In this work, it was tested for the NREL Phase VI blade. The rotation is represented using the MRF method, both for the flow and adjoint field. A projection for gradients to nodes similar to beam nodes in FSI tools was developed and implemented. This optimization tool is very flexible with respect to the cost function and other optimization settings, using the same dictionary-based input as OpenFOAM.

In this work, the cost function is a least square of the thrust and various aimed values and allowed ranges for the cost function were defined. The different test cases were used to test the stability and convergence behavior of the developed optimization tool. Defining a rather easy cost function allows for interpreting the optimization results based on rotor blade theory.

The 12 test cases all converged within a reasonable amount of CFD iterations. It was found that pre-convergence of both the flow and adjoint field is absolutely necessary in order to achieve usable gradient information. Still, a large influence of the chosen step size for the evaluation of the gradients was observed. This effect got more evident with aimed values further away from the initial value and tighter ranges around this aimed value. An initial study for a range of step sizes, which leads to good convergence behavior, might help to avoid zig-zagging. In the future, gradients can be used in more efficient optimization algorithms like SQP or Quasi-Newton. These algorithms also avoid zig-zagging that can appear when purely following the gradient direction. In addition, known implementations of these algorithms already have an automatized adjustment of the step sizes in run time.

In addition, an adjoint turbulence model is a promising approach to increase the quality of the gradients in future work and thereby the optimization convergence and stability. Using this tool for a blade of a turbine that is not stall regulated could perform better with respect to convergence of the flow and adjoint field.

The applicability of the developed tool, even for a challenging test case based on a rotating, stall regulated turbine at high Reynolds numbers, was shown in this work. When the suggested improvements of the gradient estimation are applied, the optimization tool is expected to perform in a more stable manner. This will allow the usage within the design process—for example, using more elaborate cost functions and being under consideration of constraints.

Author Contributions: Conceptualization, L.V.; Methodology, L.V. and B.S.; Software, L.V.; Writing—Original Draft Preparation, L.V.; Writing—Review and Editing, J.P.; Supervision, J.P.; Project Administration, B.S.

Acknowledgments: For the simulations, we use the high performance computing cluster Eddy of the University of Oldenburg [28].

References

1. Gasch, R.; Twele, J. *Wind Power Plants*, 2nd ed.; Springer: Berlin/Heidelberg, Germany, 2012; ISBN 3-642-22938-1.
2. Mücke, T.; Kleinhans, D.; Peinke, J. Atmospheric turbulence and its influence on the alternating loads on wind turbines. *Wind Energy* **2011**, *14*, 301–316. [CrossRef]
3. Kumar, Y.; Ringenberg, J.; Depuru, S.; Devabhaktuni, V.K.; Lee, J.W.; Nikolaidis, E.; Andersen, B.; Afjeh, A. Wind energy: Trends and enabling technologies. *Renew. Sustain. Energy Rev.* **2016**, *53*, 209–224, doi:10.1016/j.rser.2015.07.200. [CrossRef]
4. Fedorov, V.; Berggreen, C. Bend-twist coupling potential of wind turbine blades. *J. Phys. Conf. Ser.* **2014**, *524*, 012035. [CrossRef]
5. Berry, D.T.A. *Design of 9-Meter Carbon-Fiberglass Prototype Blades: CX-100 and TX-100*; Sandia National Laboratories: Livermor, CA, USA, 2007.
6. Ashwill, T. Passive Load Control for Large Wind Turbines. In Proceedings of the 51st AIAA/ASME/ASCE/AHS/ASC Structures, Structural Dynamics, and Materials Conference, Orlando, FL, USA, 12–15 April 2010.
7. Soto, O.; Löhner, R. On the computation of flow sensitivities from boundary integrals. In Proceedings of the 42nd AIAA Aerospace Sciences Meeting and Exhibit, Reno, NV, USA, 5–8 January 2004.
8. Othmer, C.; de Villiers, E.; Weller, H. Implementation of a continuous adjoint for topology optimization of ducted flows. In Proceedings of the 18th AIAA Computational Fluid Dynamics Conference, Miami, FL, USA, 25–28 June 2007.
9. Othmer, C. Adjoint methods for car aerodynamics. *J. Math. Ind.* **2014**, *4*, 6, doi:10.1186/2190-5983-4-6. [CrossRef]
10. Soto, O.; Löhner, R.; Yang, C. A stabilized pseudo-shell approach for surface parametrization in CFD design problems. *Commun. Numer. Methods Eng.* **2002**, *18*, 251–258. [CrossRef]
11. Anderson, W.; Venkatakrishnan, V. Aerodynamic design optimization on unstructured grids with a continuous adjoint formulation. In Proceedings of the 35th Aerospace Sciences Meeting and Exhibit, Reno, NV, USA, 6–9 January 1997.
12. Grasso, F. Usage of Numerical Optimization in Wind Turbine Airfoil Design. *J. Aircr.* **2011**, *48*, 248–255. [CrossRef]
13. Liu, Y.; Yoshida, S. An extension of the Generalized Actuator Disc Theory for aerodynamic analysis of the diffuser-augmented wind turbines. *Energy* **2015**, *93*, 1852–1859. [CrossRef]
14. Jamieson, P. Generalized Limits for Energy Extraction in a Linear Constant Velocity Flow Field. *Wind Energy* **2008**, *11*, 445–457. [CrossRef]
15. Schramm, M.; Stoevesandt, B.; Peinke, J. Lift Optimization of Airfoils using the Adjoint Approach. In Proceedings of the EWEA 2015, Europe's Premier Wind Energy Event, Paris, France, 17–20 November 2015.
16. Schramm, M.; Stoevesandt, B.; Peinke, J. Simulation and Optimization of an Airfoil with Leading Edge Slat. *J. Phys. Conf. Ser.* **2016**, *753*. [CrossRef]
17. Vorspel, L.; Schramm, M.; Stoevesandt, B.; Brunold, L.; Bünner, M. A benchmark study on the efficiency of unconstrained optimization algorithms in 2D-aerodynamic shape design. *Cogent Eng.* **2017**, *4*, doi:10.1080/23311916.2017.1354509. [CrossRef]
18. Othmer, C. A continuous adjoint formulation for the computation of topological and surface sensitivities of ducted flows. *Int. J. Numer. Methods Fluids* **2008**, *58*, 861–877, doi:10.1002/fld.1770. [CrossRef]
19. Ferziger, J.; Peric, M. *Computational Methods for Fluid Dynamics*; Springer:Berlin/Heidelberg, Germany, 2001.
20. Papadimitriou, D.; Giannakoglou, K. A continuous adjoint method with objective function derivatives based on boundary integrals, for inviscid and viscous flows. *Comput. Fluids* **2007**, *36*, 325–341. [CrossRef]
21. Dose, B.; Rahimi, H.; Herráez, I.; Stoevesandt, B.; Peinke, J. Fluid-structure coupled computations of the NREL 5MW wind turbine blade during standstill. *J. Phys. Conf. Ser.* **2016**, *753*, 022034. [CrossRef]
22. Nocedal, J.; Wright, S. *Numerical Optimization*; Springer: Berlin/Heidelberg, Germany, 2006.

23. Hand, M.M.; Simms, D.; Fingersh, L.; Jager, D.; Cotrell, J.; Schreck, S.; Larwood, S. *Unsteady Aerodynamics Experiment Phase VI: Wind Tunnel Test Configurations and Available Data Campaigns*; National Renewable Energy Laboratory: Golden, CO, USA, 2001.
24. Tangler, J.L.; Somers, D.M. *NREL Airfoil Families for HAWTs*; Citeseer: Pennsylvania, PA, USA, 1995.
25. Rahimi, H.; Daniele, E.; Stoevesandt, B.; Peinke, J. Development and application of a grid generation tool for aerodynamic simulations of wind turbines. *Wind Eng.* **2016**, *40*, 148–172, doi:10.1177/0309524X16636318. [CrossRef]
26. Giles, M.B.; Duta, M.C.; Müller, J.D.; Pierce, N.A. Algorithm Developments for Discrete Adjoint Methods. *AIAA J.* **2003**, *41*, 198–205. [CrossRef]
27. Müller, J.D.; Cusdin, P. On the performance of discrete adjoint CFD codes using automatic differentiation. *Int. J. Numer. Methods Fluids* **2005**, *47*, 939–945, doi:10.1002/fld.885. [CrossRef]
28. HPC Cluster EDDY of the University of Oldenburg. Available online: https://www.uni-oldenburg.de/en/school5/sc/high-perfomance-computing/hpc-facilities/eddy/ (accessed on 9 July 2018).

10

Evaluation of the Power-Law Wind-Speed Extrapolation Method with Atmospheric Stability Classification Methods for Flows over Different Terrain Types

Chang Xu [1], Chenyan Hao [1], Linmin Li [1,*], Xingxing Han [1], Feifei Xue [1], Mingwei Sun [2] and Wenzhong Shen [3]

1. College of Energy and Electrical Engineering, Hohai University, Nanjing 211100, China; zhuifengxu@hhu.edu.cn (C.X.); HAOcy@hhu.edu.cn (C.H.); hantone@hhu.edu.cn (X.H.); xuefeifeihhu@163.com (F.X.)
2. College of Naval Coast Defence Arm, Naval Aeronautical University, Yantai 264001, China; sunmingwei1993@163.com
3. Department of Wind Energy, Fluid Mechanics Section, Technical University of Denmark, Nils Koppels Allé, Building 403, 2800 Kgs. Lyngby, Denmark; wzsh@dtu.dk

* Correspondence: lilinmin@hhu.edu.cn

Abstract: The atmospheric stability and ground topography play an important role in shaping wind-speed profiles. However, the commonly used power-law wind-speed extrapolation method is usually applied, ignoring atmospheric stability effects. In the present work, a new power-law wind-speed extrapolation method based on atmospheric stability classification is proposed and evaluated for flows over different types of terrain. The method uses the wind shear exponent estimated in different stability conditions rather than its average value in all stability conditions. Four stability classification methods, namely the Richardson Gradient (RG) method, the Wind Direction Standard Deviation (WDSD) method, the Wind Speed Ratio (WSR) method and the Monin–Obukhov (MO) method are applied in the wind speed extrapolation method for three different types of terrain. Tapplicability is analyzed by comparing the errors between the measured data and the calculated results at the hub height. It is indicated that the WSR classification method is effective for all the terrains investigated while the WDSD method is more suitable in plain areas. Moreover, the RG and MO methods perform better in complex terrains than the other methods, if two-level temperature data are available.

Keywords: wind speed extrapolation; atmospheric stability; wind shear; wind resource assessment

---

1. Introduction

In the feasibility study and microsite selection stage of wind farms, using the wind measurement data is a key step to evaluate the wind resources at the hub height. At present, wind farms are being developed towards low wind speed and high hub height. Since the height of existing wind mast towers is often lower than the hub height which results in a lack of wind data at high hub heights, it is important to develop a reliable method to evaluate the accurate wind resource for wind farm development. The motivation of wind speed extrapolation is to characterize the wind shear and the importance of wind shear characterization for wind speed calculation was mentioned in a number of studies [1–3].

Wind turbines typically operate at altitudes below 200 m, while the height of the convective boundary layer is on the order of 1000 m [4]. The wind shear within 100–200 m is a function of wind speed, atmospheric stability, surface roughness, and height spacing [5,6]. At present, the power law and logarithmic law are commonly used to extrapolate the wind speed. The logarithmic law can represent the vertical wind-speed profiles fairly well under neutral conditions [7] and was extended by using the Monin–Obukhov similarity theory [8] under stable or unstable conditions to include thermal stratification effects [9]. According to Optis et al. [10], the logarithmic wind-speed profile based on the Monin–Obukhov similarity theory was found inaccurate under strong stable conditions due to the shallow surface layer. For unstable conditions, Lackner et al. [11] found that the power law is robust to give a realistic wind profile than the logarithmic law.

In the near-surface layer, the wind speed affected by surface friction and atmospheric stability, varies significantly with the height [12]. Holtslag [13] analyzed the 213-m-high Cabauw meteorological tower data and found that the measured wind profile was consistent with the Monin–Obukhov similarity theory (surface layer scaling) even for the height above 100 m. Gualtieri [14] investigated the relationship between the wind shear coefficient and atmospheric stability based on the dataset recorded from the met mast of Cabauw (Netherlands), and found that the Panofsky and Dutton [15] model appeared particularly skillful during the diurnal unstable hours and during the nighttime under moderately stable conditions. Mohan and Siddiqui [16] summarized various methods for atmospheric stability classification. In the radiation method, the intensity of radiation is determined based on the amount of observed cloud which is difficult to obtain, and in the M-O length method, the heat flux data are required. In addition to these two methods, there are four methods based on the gradient Richardson number, horizontal wind standard deviation, vertical temperature gradient and wind speed ratio to classify the atmospheric stability. The parameters used in these methods can be calculated using measured wind speed, wind direction, and temperature data. Wharton and Lundquist [17] analyzed the wind farm operating data and found that, when the atmosphere was stable and the wind speed was in the range of 5~8.5 m/s, the output power was obviously greater than that in unstable conditions with strong convection, and the average difference was close to 15%. Gryning et al. [18] applied the similarity theory to establish a new model for wind-speed profile by considering the effects of atmospheric stability, and the model was applied to the height of the entire atmospheric boundary layer in a flat terrain. Đurišić and Mikulović [19] proposed a model of vertical wind-speed extrapolation for improving the wind resource assessment using the WAsP (Wind Atlas Analysis and Application Program) software [20], and found that if the estimation is carried out with the fixed wind shear exponent, the overestimated wind speed in the day time (unstable atmosphere) and underestimated wind speed in the night time (stable atmosphere) are usually obtained. Gualtieri and Secci [21] tested the power law extrapolation model over a flat rough terrain in the Apulia region (Southern Italy), and investigated the effect of atmospheric stability and surface roughness on wind speed. They found that the empirical JM Weibull distribution extrapolating model was proved to be preferable. Different extrapolating models were compared and their advantages and disadvantages were investigated. However, there are few studies on the effect of atmospheric stability on wind shear and wind speed extrapolation in different terrain types. At the same time, there are few discussions about the results of wind speed extrapolation above different ground topographies.

The present work aims to evaluate the wind speed extrapolation method using the power law and the atmospheric stability classification. For different terrain types, the characteristics of wind shear exponent are discussed, and four atmospheric stability classification methods are employed and investigated. The model is validated against the measured data, and the suitability of different atmospheric stability classification methods for flows over different types of terrain is indicated.

2. Model Description

2.1. Power Law

It is usually necessary to calculate the wind shear exponent by the wind speeds at two different levels. In the neutral condition, the wind-speed profile can be calculated using the empirical formula:

$$u(z) = \frac{u^*}{\kappa} \ln \frac{z}{z_0} \quad (1)$$

where κ is the Von Karman constant equal to 0.4, u^* is the friction velocity calculated as $u^* = (\tau/\rho)^{1/2}$, τ is the wall friction, ρ is the density, z_0 is the roughness length, and z is the height.

The power law (PL) method is widely used for estimating the wind speed at a wind generator hub height [22], which is defined as

$$u_2 = u_1(z_2/z_1)^{\alpha} \quad (2)$$

where u_1 is the wind speed at the height z_1, u_2 is the wind speed at the height z_2, and α is the wind shear exponent.

From Equation (2), α can be calculated by u_1 and u_2:

$$\alpha = \ln(u_2/u_1)/\ln(z_2/z_1) \quad (3)$$

2.2. Atmospheric Stability Classification

The atmospheric thermal stability is suppressed or enhanced by the vertical temperature difference. The two atmospheric stability classification methods commonly used are those from Pasquill [23] and IAEA (International Atomic Energy Agency) [24]. The Pasquill method (abbreviated to P·S) proposed in the Chinese standards includes six categories: highly unstable, moderately unstable, slightly unstable, neutral, moderately stable and extremely stable. They are denoted as A, B, C, D, E, and F.

Taking the turbulence and thermal factors into account, there are three methods for classifying the atmospheric stability, namely the Monin-Obukhov method, Bulk Richardson number method, and Richardson Gradient method. Among them, theBulk Richardson number method does not have a uniform atmospheric stability classification standard, so it is not used in the present work. When the measured wind data only contains the wind direction and speed data, the wind direction standard deviation method and wind speed ratio method can be used to classify the atmospheric stability.

2.2.1. Richardson Gradient (RG) Method

The Richardson number, *Ri*, synthesizes the effects of thermodynamic and kinetic factors caused by the turbulence, reflecting more turbulent information [25]. Therefore, the RG method can distinguish the atmospheric stability more accurately. The *Ri* in the surface layer can be expressed as [26]

$$Ri = \frac{gz}{T}\frac{\Delta T}{\Delta u^2} \ln \frac{z_2}{z_1} \quad (4)$$

where ΔT is the difference of temperatures at z_2 and z_1. Δu is the wind speed difference. T is the average atmospheric temperature, g is the acceleration of gravity, and z is the average geometric height calculated as $z = \sqrt{z_1 z_2}$. Table 1 shows the classification standard of *Ri* in different terrains [27].

Atmospheric stability depends on the net heat flux to the ground, which is equal to the sum of incident radiation, radiation emitted, latent heat, and sensible heat exchange between the atmosphere and underlying surface. When the radiation incident on the ground is dominant, the air parcels at the lower part rise, leading to a vertical air motion, and making the atmosphere unstable. Therefore, the thermal effect aggravates the air movement and prevents the wind speed from changing dramatically in the vertical direction. In this case, the Richardson number is negative. When the ground cools down, the temperature increases with increasing the height, which weakens the vertical air movement. The situation is recognized as a stable state and a positive Richardson number is obtained. The neutral stability corresponds to the case where the thermal effect is not significant. This situation occurs when the cloud is dense, and the Richardson number is zero.

Table 1. Classification of stability based on Ri in different terrain conditions.

Stability Conditions	Mountain	Plain
A	$Ri < -100$	$Ri < -2.51$
B	$-100 \leq Ri < -1$	$-2.51 \leq Ri < -1.07$
C	$-1 \leq Ri < -0.01$	$-1.07 \leq Ri < -0.275$
D	$-0.01 \leq Ri < 0.01$	$-0.275 \leq Ri < 0.089$
E	$0.01 \leq Ri < 10$	$0.089 \leq Ri < 0.128$
F	$10 \leq Ri$	$0.128 \leq Ri$

2.2.2. Wind Direction Standard Deviation (WDSD) Method

The wind direction pulsation is a direct indicator of atmospheric turbulence. The magnitude of the wind direction pulsation angle is directly related to the diffusion parameter, so it can be used as an indicator to classify the atmospheric stability. However, the measurement of pulsation angle is easily influenced by the local influence of sampling location and instrument performance, which makes the method unrepresentative. So this method is more suitable for flat terrain. Table 2 shows the classification criteria recommended by Sedefian [28] based on the horizontal wind direction standard deviation, σ_θ. When σ_θ is used, the classification standard should be corrected according to the actual surface roughness. The classification standard is recommended by the United States Environmental Protection Agency (1980), and the actual area is corrected with roughness as $(z_0/0.15)^{0.2}$ (the corrected value is obtained by multiplying the values in Table 2).

Table 2. Relationship between σ_θ and the atmospheric stability.

Parameter	Atmospheric Stability					
	A	B	C	D	E	F
$\sigma_\theta/°$	$\sigma_\theta \geq 22.5$	$22.5 > \sigma_\theta \geq 17.5$	$17.5 > \sigma_\theta \geq 12.5$	$12.5 > \sigma_\theta \geq 9.5$	$9.5 > \sigma_\theta \geq 3.8$	$3.8 > \sigma_\theta$

2.2.3. Wind Speed Ratio (WSR) Method

The wind speed ratio U_R is defined as the ratio of the wind speeds at two different heights:

$$U_R = \frac{u_2}{u_1} \quad (5)$$

According to the rule of Pasquill stability classification, Chen [29] divided it into six categories according to the ratio of the wind speeds at two different heights (Table 3).

Table 3. Relationship between U_R and the atmospheric stability.

Atmospheric Stability	A	B	C
Range	$U_R < 1.0032$	$1.0032 \leq U_R < 1.0052$	$1.0052 \leq U_R < 1.0101$
Atmospheric Stability	**D**	**E**	**F**
Range	$1.0101 \leq U_R < 1.5717$	$1.5717 \leq U_R < 2.1963$	$U_R \geq 2.1963$

2.2.4. Monin–Obukhov (MO) Method

In the Monin–Obukhov theory, the atmospheric stability is described by the Obukhov length L, which is determined directly from sonic anemometer measurements of friction velocity and heat flux:

$$L = -\frac{(u^*)^3}{\kappa \frac{g}{T} \overline{w'T'_S}} \tag{6}$$

where $\overline{w'T'_S}$ is the covariance of temperature and vertical wind speed fluctuations and g is the gravitational acceleration. When only the temperature gradient is available, L can be obtained by [30].

$$\begin{cases} L = \frac{z}{Ri} & (Ri < 0) \\ L = \frac{z(1-5Ri)}{Ri} & (Ri > 0) \end{cases} \tag{7}$$

According to the relationship between L and Ri, Table 4 shows the classification of stability based on L for flows above different terrains [27].

Table 4. Classification of stability based on L in different terrains.

Stability Conditions	Mountain	Plain
A	$L > -0.032$	$L > -0.316$
B	$-3.162 < L \leq -0.032$	$-3.162 < L \leq -0.316$
C	$-316.228 < L \leq -31.623$	$-63.246 < L \leq -3.162$
D	$L \leq -316.228$, $L > 158.114$	$L \leq -63.246$, $L > 158.114$
E	$-154.952 < L \leq 158.114$	$-154.952 < L \leq 158.114$
F	$L \leq -154.952$	$L \leq -154.952$

2.3. Wind Speed Extrapolation Method Based on Atmospheric Stability

The power law does not fully consider the effect of atmospheric stability on wind shear. According to the characteristics of atmospheric stability, the wind speed extrapolation (WSE) method based on the atmospheric stability is proposed. The measured dataset by wind towers is named as dataset Q, including time, wind speed, wind direction and wind direction standard deviation. The low wind speed data less than the cut-in wind speed in the dataset Q is first removed to obtain the filtered dataset W. Then the dataset W is classified into different stability conditions using the RG, MO, WSR, and WDSD methods. After that, Equation (3) is used to calculate the average wind shear exponent using the dataset W under each atmospheric stability condition. Finally, using the wind speeds at the lower heights in dataset Q, the higher level wind speed is predicted by Equation (2).

Four kinds of atmospheric stability classification methods, RG, WDSD, WSR and MO are adopted to develop four new methods, namely WSE-RG, WSE-WDSD, WSE-WSR and WSE-MO, for the wind speed extrapolation.

3. Cases and Measurements

3.1. Case Definition

Three cases are chosen to test all the extrapolation methods as shown in Table 5. The terrains include mountain, plateau and plain, and the surface types include wasteland, shrubbery and farmland.

Table 5. Wind tower site information.

Number	Site Name	Terrain	Surface	Elevation (m)
1	SJB	Plateau	Wasteland	1678
2	GFC	Mountain	Shrubbery	713
3	HNH	Plain	Farmland	66

Figure 1 shows the topographic maps of the three cases. The wind tower site SJB is located at an edge of a plateau (Figure 1a,d). The elevation of SJB in the plateau is 1678 m, and the surface type is wasteland. There is a hillside in the southwest of SJB, and flat grassland in the northeast. The GFC wind tower sits on the mountaintop, which has an altitude of 713 m (Figure 1b,e). The surface type of GFC is shrubbery. Due to the special weather conditions in the area, underlying shrub is evergreen throughout the year. The HNH wind tower is located at the elevation of 66 m, where the terrain is flat (Figure 1c,f). The land surface type is farmland and there are a large number of villages in the area.

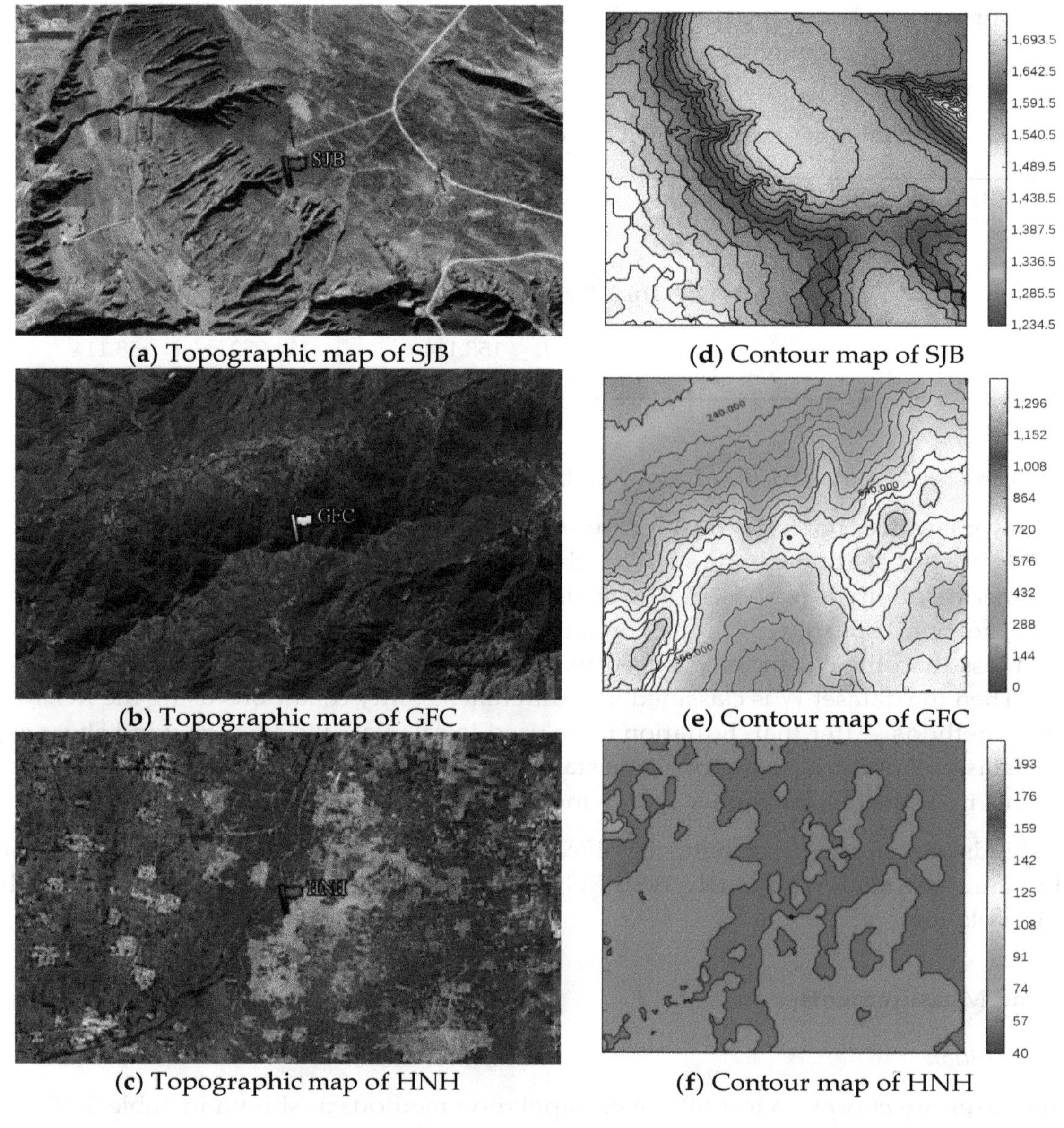

(**a**) Topographic map of SJB (**d**) Contour map of SJB

(**b**) Topographic map of GFC (**e**) Contour map of GFC

(**c**) Topographic map of HNH (**f**) Contour map of HNH

Figure 1. Topographic maps of three cases: (**a**) SJB; (**b**) GFC and (**c**) HNH; Contour maps of the three cases: (**d**) SJB; (**e**) GFC and (**f**) HNH.

3.2. Measurements

The wind resource in the three cases is usually measured at a site using a wind tower, equipped with wind speed and wind direction sensors. The locations of the wind towers are shown in Figure 1. The datasets of the three cases are obtained as follows:

(1) SJB: The measured data includes wind speed, wind direction and temperature data. The cup anemometers are placed at 30, 50 and 70 m, and the wind vines are placed at 30 and 70 m. The temperature observations are also used to obtain the temperature data at 30 and 70 m.
(2) GFC: The cup anemometers are at 30, 50 and 80 m, and the wind direction is obtained with wind vines placed at 30 and 80 m. Unfortunately, there is no temperature sensor installed on the tower.
(3) HNH: Wind speeds are measured by the cup anemometers placed at 10, 60 and 100 m. Wind vanes are placed at 10 and 100 m. The temperature data is also unavailable in the HNH area.

The datasets at upper heights are used to verify the reliability of the wind speed extrapolation method. The measurement parameters of each case including height, time interval and data collection period are presented in Table 6.

Table 6. Summary of measurement parameters.

Number	Site Name	$h1$ (m)	$h2$ (m)	$h3$ (m)	Time Interval (min)	Data Collection Period
1	SJB	30	50	70	10	27 August 2015~26 August 2016
2	GFC	30	50	80	10	1 August 2012~31 July 2013
3	HNH	10	60	100	10	15 September 2014~16 September 2015

4. Results and Discussion

4.1. Overall Meteorological Characteristics

In Table 7, the annual mean meteorological parameters observed for the three cases are provided where $U1$ and $U2$ are the average wind speeds at the heights of $h1$ and $h2$, $U3$ is the average wind speed at the hub height of $h3$, $T1$ and $T2$ are the mean temperatures at the heights of $h1$ and $h2$. Besides, α_{h1-h2} and α_{h2-h3} are the mean wind shear exponents calculated with velocities at $h1$ and $h2$, and $h2$ and $h3$ respectively.

Table 7. Wind data information.

Site Name	$U1$ (m/s)	$U2$ (m/s)	$U3$ (m/s)	$T1$ (°C)	$T2$ (°C)	α_{h1-h2}
SJB	5.34	5.46	5.71	8.27	8.06	0.0435
GFC	5.07	5.10	5.17	-	-	0.0115
HNH	2.73	4.61	5.44	14.30	-	0.2924

The results indicate the difference of the wind speed distribution between the plain and mountain areas. In the HNH case, where the terrain is flat, the inconspicuous mixing of atmospheric turbulence and the weak exchange of momentum between the vertical layers result in a vertical gradient of wind speed and a large wind shear exponent. Because of the acceleration effect on the wind speed on the mountaintop, the wind shear exponents in GFC and SJB are smaller than the one in HNH.

For the SJB plateau complex terrain, at 30 m (Figure 2a) and 70 m (Figure 2b), the south direction is the most frequent wind direction with the occurrence percentages of 15.81% and 14.50%, respectively. The changes of wind speed and direction in this area are limited, because at 30 m, the wind direction is less affected by the terrain, and the height difference between the two measurement levels is only 40 m. At the same time, under the complex terrain condition, the mixing of atmospheric turbulence, the momentum exchange in the vertical direction are strong, resulting in a small wind shear exponent.

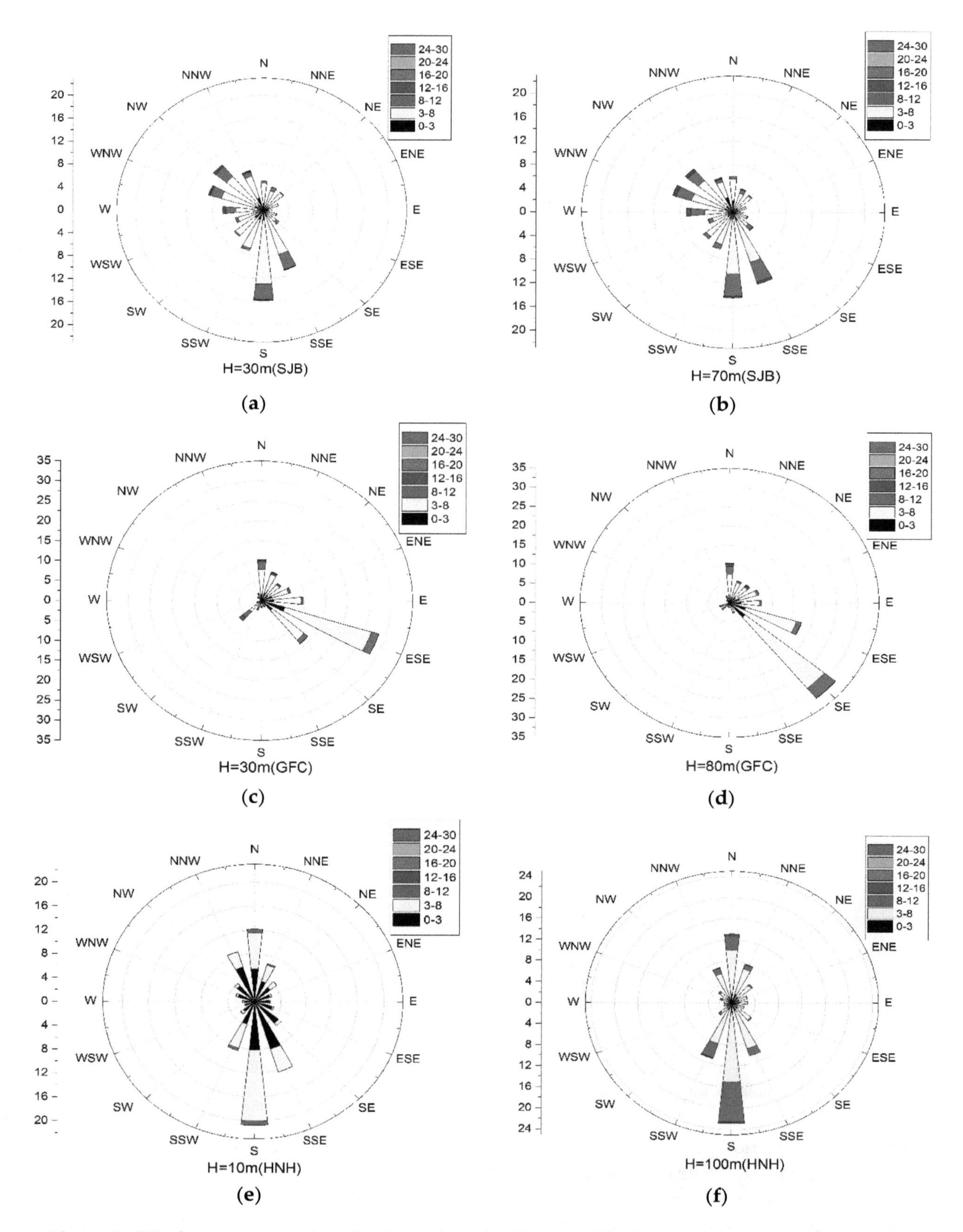

Figure 2. Wind rose measured at the three sites: (**a**) 30 m at SJB; (**b**) 70 m at SJB; (**c**) 30 m at GFC; (**d**) 80 m at GFC; (**e**) 10 m at HNH; (**f**) 100 m at HNH.

The complex topography in the GFC site strongly affects the wind behaviors at both 30 and 80 m (Figure 2c,d). At 30 m (Figure 2c), the most frequent direction is ESE (28.37%) while at 80 m

(Figure 2d), the SE direction is the most frequent direction (32.17%), and ESE (17.63%) is the secondary predominant direction. At the same time, the overall wind speed interval is dominated by a low wind speed interval of 3~8 m/s. This is because the complex mountain terrain has a great impact on the wind direction. At the same time, the topographic factors also cause strong mixing in vertical, which results in smaller values of wind shear exponent, so there is no significant change in the wind speed.

For the plain area HNH, at 10 m (Figure 2e), the S direction occurs most frequently (20.75%), and at 80 m (Figure 2f), it is also the S direction (22.90%). The wind direction at 10 m and 100 m are similar, but the wind speed at 100 m is significantly higher. This is related to the flat terrain condition of this area. The atmosphere is inadequately mixed in the vertical direction in this terrain condition, which leads to a larger wind shear exponent and causes a significant change in the vertical wind speed.

4.2. Wind Shear Characteristics of Different Terrains

Figure 3 shows the wind shear characteristics as the diurnal and monthly changes. As expected, the monthly variations of wind shear exponent at SJB and HNH are relatively high in Winter and low in Summer. The main reason is that, in Summer, the temperature is high, the mixing of near-surface air is sufficient, and the wind shear exponent is small. In Winter, the temperature is low and the air mixing is weak, usually resulting in a large wind shear exponent. However, the wind shear exponent at GFC is low in Winter and high in Summer, which deviates from the trend at SJB and HNH. This is because the area has more rains in Spring and Summer, making the air more humid and the frequency of low-level jets higher. The terrain has a dynamic lifting effect on the mountainous airflow and is prone to a strong wind shear. Furthermore, the daily change of wind shear exponent is closely related to the ambient temperature as seen in Figure 3b. The main reason for the diurnal variation of wind shear exponent is the cyclic variation of temperature. The wind shear exponent in the daytime is lower than that of the nighttime. This result is coincident with the results in the previous studies [31–33].

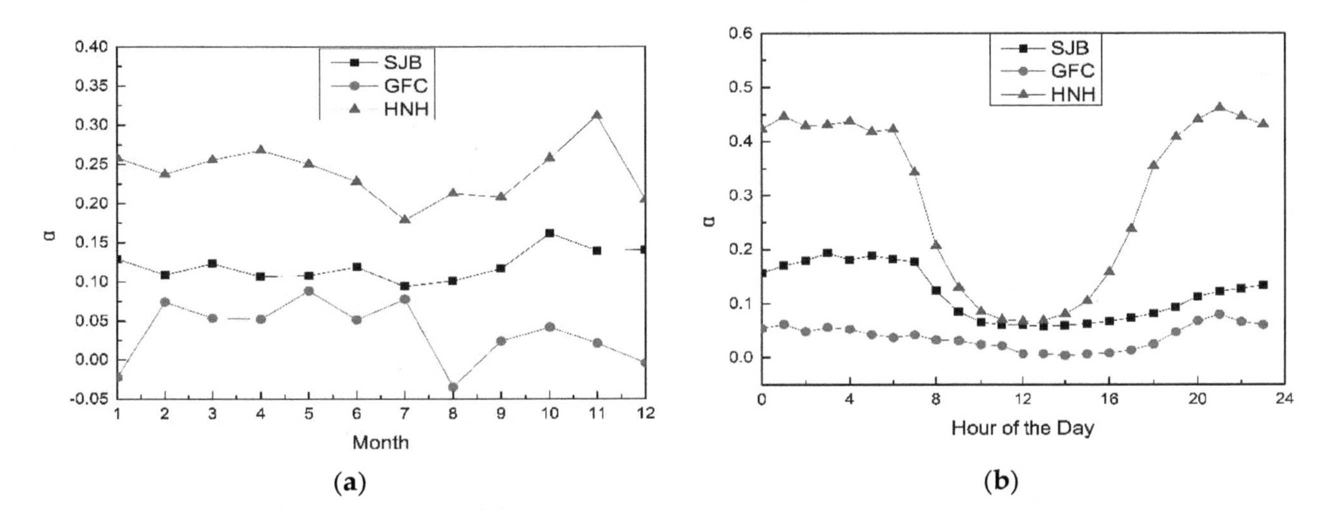

Figure 3. Monthly and diurnal variations of $h1$–$h2$ wind shear exponents in the three cases: (**a**) monthly variation; (**b**) diurnal variation.

Figure 4 shows the monthly and daily variations of atmospheric stability at SJB. Comparing the atmospheric stability variation with the variation of wind shear exponent, it is found that the wind shear exponent is lower when the atmosphere is more unstable.

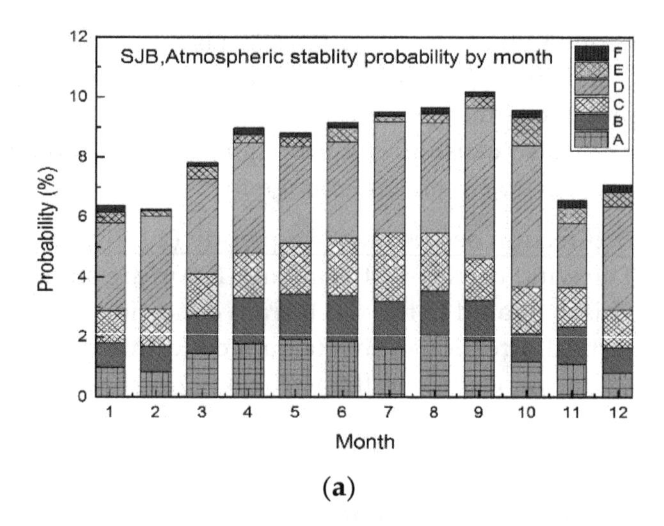

(a)

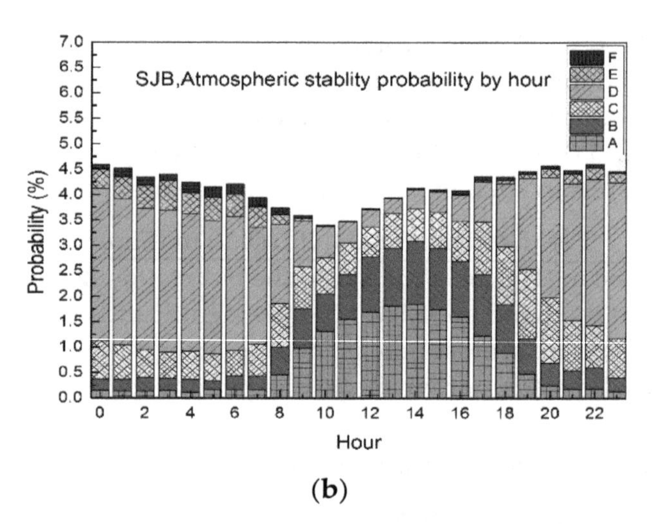

(b)

Figure 4. Probability of atmospheric stability classification at SJB: (**a**) monthly variation; (**b**) daily variation.

4.3. High Level Wind Speed Extrapolation and Validation

In the SJB area, starting from the 30 m and 50 m 10-min observations, the α_{30-50} is calculated and then used to estimate the wind resource at a higher level. In particular, the 10-min α_{30-50} is calculated using the filtered dataset W by implementing the methods of PL, WSE-RG, WSE-WDSD, WSE-WSR and WSE-MO. All the methods are adopted to calculate the wind shear exponents for the three areas and the results are listed in Table 8. At the same time, because the two-level temperature data at GFC and HNH is unavailable and the WSE-RG and WSE-MO methods cannot be used, the wind shear exponents of the two areas are calculated using the PL, WSE-WDSD, and WSE-WSR methods. It is shown that the wind shear exponent is larger under stable conditions and smaller under unstable conditions, especially in HNH. It is found that, using the WSE-WSR method, the variation of wind shear exponent with the atmospheric stability is more obvious than that using the WSE-WDSD method.

Using the results of wind shear exponent in Table 8, the wind speeds at the hub height for the three cases are also calculated. The wind speed mean relative error (MRE), root-mean-square error (RMSE) and mean wind speed at the hub height are shown in Table 8. For SJB, the MREs between the measurements and predicted wind speeds obtained by the PL, WSE-WSR, WSE-WDSD, WSE-RG and WSE-MO methods have the following sequence: MRE(PL) > MRE(WSE-WSR) > MRE(WSE-WDSD) > MRE(WSE-RG) > MRE(WSE-MO). For GFC, MRE(PL) > MRE(WSE-WSR) > MRE(WSE-WDSD). Besides, for HNH, it is MRE(PL) > MRE(WSE-WSR) > MRE(WSE-WDSD). The improvements of MRE range from 1.58 to 0.22%, 0.33 to 0.02%, 7.38 to 3.17% for the three cases using the new wind speed extrapolation methods.

It is proposed that the new WSE method based on the atmospheric stability better reflects the true changes of wind speed in two dimensions: height and time. It takes the effect of atmospheric stability on the wind profile into account and makes separate calculations for the wind resources at different conditions of atmospheric stability, which can reflect the mixing of atmosphere in the vertical direction. On this basis, the accuracy of atmospheric stability classification will directly affect the accuracy of wind speed estimation. When the two-level temperature data sets are available, it can be seen from Table 9 that the WSE-RG and WSE-MO methods can better estimate the wind speed for the SJB plateau area. For the GFC mountainous area without temperature measurements, both the WSE-WDSD and WSE-WSR methods can better estimate the wind speeds and the WSE-WDSD method is more accurate than the WSE-WSR method. For the HNH plain area, where the underlying surface is farmland, both the WSE-WDSD method and WSE-WSR method can better estimate the wind speeds than the PL method.

Table 8. Wind shear exponents in different areas.

Area	Method	Wind Shear Exponent under Different Stability Conditions					
		A	B	C	D	E	F
SJB	PL	0.0679					
	WSE-RG	0.1861	0.0558	0.0676	0.3825	0.0809	0.0981
	WSE-MO	0.2875	0.0474	0.0789	0.2753	0.1156	0.0963
	WSE-WDSD	0.0648	0.0774	0.0931	0.0608	0.0779	0.0523
	WSE-WSR	0.0863	-	0.0637	0.1309	0.4704	-
GFC	PL	−0.0023					
	WSE-WDSD	0.0066	0.0224	0.0307	0.0030	0.0239	-0.0175
	WSE-WSR	−0.1111	0.0081	0.0147	0.1500	0.9619	-
HNH	PL	0.1790					
	WSE-WDSD	0.1466	0.1455	0.1373	0.1740	0.2587	0.3703
	WSE-WSR	−0.0079	-	-	0.1227	0.3278	0.4807

Area	Method	Wind Shear Exponent under Different Stability Conditions					
		A	B	C	D	E	F
SJB	PL	0.0679					
	WSE-RG	0.1861	0.0558	0.0676	0.3825	0.0809	0.0981
	WSE-MO	0.2875	0.0474	0.0789	0.2753	0.1156	0.0963
	WSE-WDSD	0.0648	0.0774	0.0931	0.0608	0.0779	0.0523
	WSE-WSR	0.0863	-	0.0637	0.1309	0.4704	-
GFC	PL	−0.0023					
	WSE-WDSD	0.0066	0.0224	0.0307	0.0030	0.0239	-0.0175
	WSE-WSR	−0.1111	0.0081	0.0147	0.1500	0.9619	-
HNH	PL	0.1790					
	WSE-WDSD	0.1466	0.1455	0.1373	0.1740	0.2587	0.3703
	WSE-WSR	−0.0079	-	-	0.1227	0.3278	0.4807

Table 9. Mean wind speed relative error (MRE), root-mean-square error (RMSE) of wind speed and mean speed at the hub height in the three cases.

Area	Methods	MRE	RMSE (m/s)	Calculated Mean Speed (m/s)	Measured Mean Speed (m/s)
SJB	PL	−1.58%	0.378	6.6909	6.8210
	WSE-WDSD	−1.58%	0.3781	6.6911	
	WSE-MO	0.22%	0.3337	6.8229	
	WSE-RG	0.63%	0.3231	6.8081	
	WSE-WSR	−1.52%	0.3311	6.7242	
GFC	PL	−0.33%	0.543	6.2069	6.2904
	WSE-WDSD	−0.02%	0.5276	6.2386	
	WSE-WSR	−0.03%	0.536	6.2346	
HNH	PL	−7.38%	0.8732	7.2567	7.5972
	WSE-WDSD	−3.17%	0.7931	7.2681	
	WSE-WSR	−3.26%	0.7035	7.2815	

Figure 5 shows MRE and RMSE between the measured and calculated wind speeds in different atmospheric stability conditions. For SJB, MREs and RMSEs of WSE-WDSD under different atmospheric stability conditions are close to each other and are similar to the results of PL. It can be concluded that the WDSD method cannot be applied to the complex terrain of plateau to classify the atmospheric stability. In contrast, combining the results of MREs and RMSEs, it is found that the accuracy of the WSE-RG method is improved in the cases of A, B and C, and the WSE-WSR method is suitable for the cases of C, D and E, while the WSE-MO method improves the calculation accuracy in all the conditions. For GFC, both MREs and RMSEs of the WSE-WSR method are large when the atmospheric stability condition is E. Through the comparison of RMSEs, it is seen that the WSE-WDSD method performs well in almost all the conditions. For HNH, the absolute values of MREs of the WSE-WDSD and WSE-WSR methods are smaller than those obtained by the PL method. The accuracy of both the WSE-WDSD and WSE-WSR methods has been greatly improved. It is shown that these two atmospheric stability classification methods are suitable for flat terrain. And the WSE-WSR method is more effective than the WSE-WDSD method especially in unstable conditions.

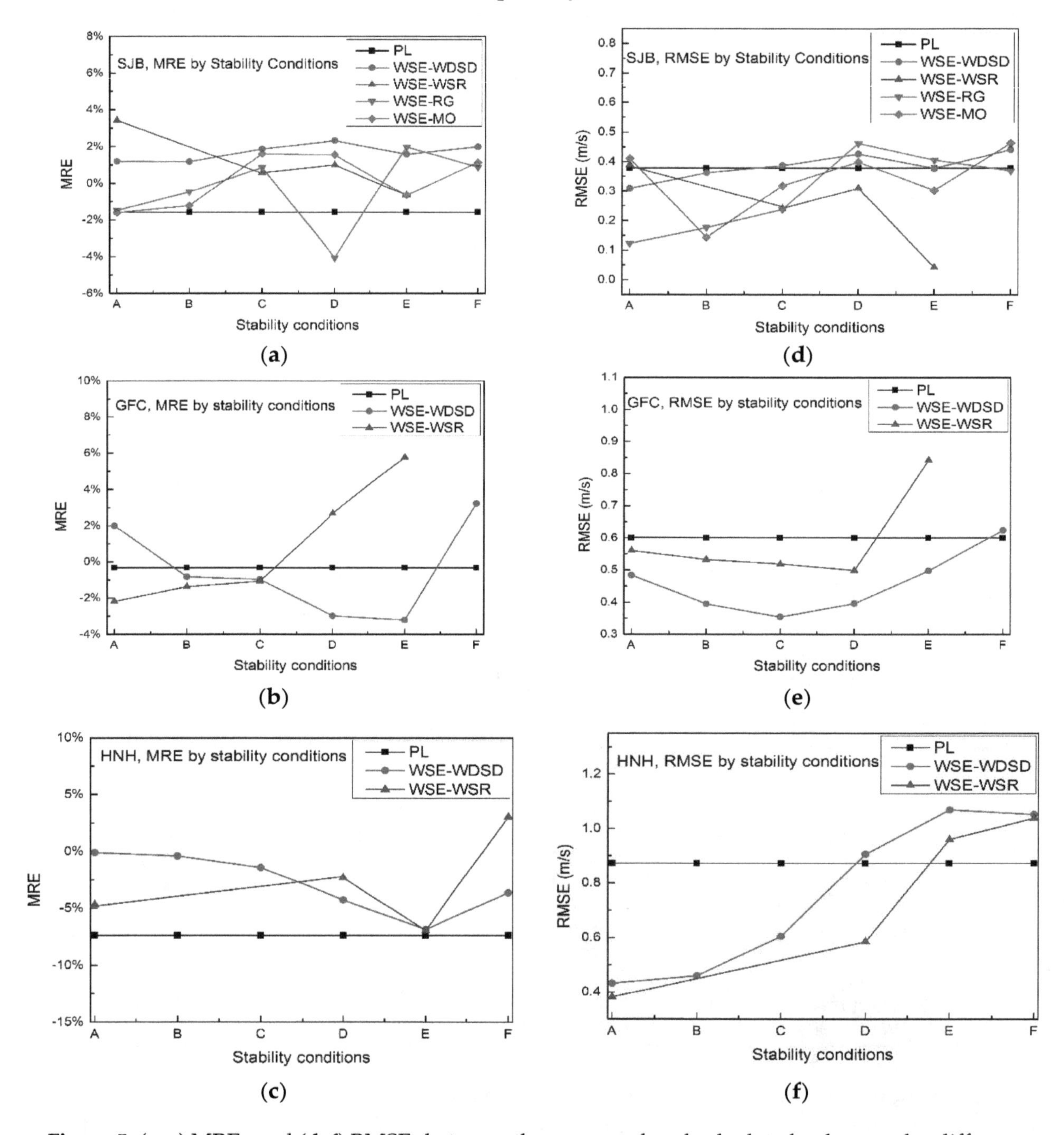

Figure 5. (**a**–**c**) MREs and (**d**–**f**) RMSEs between the measured and calculated values under different atmospheric stabilities conditions in the three different areas.

The monthly and daily variations of RMSE of the three cases using different wind speed extrapolation methods are shown in Figure 6. For SJB, the results of WSE-WDSD are very close to the traditional PL method, indicating that the WSE-WDSD method is inaccurate in the complex terrain plateau. In contrast, the results of the WSE-RG method are obviously better than those obtained by the other methods. At the same time, the results of the WSE-WSR and WSE-MO methods are similar especially for daily variations. From the daily variation, it is found that the WSE-WSR and WSE-MO methods are superior to the WSE-RG method in the night, but worse in the daytime. For the GFC and HNH, both the WSE-WDSD and WSE-WSR methods can effectively reduce RMSE, among which the WSE-WSR method is more effective. By analyzing RMSEs of the WSE-WDSD method in the HNH area where the surface is farmland, the WSE-WDSD method performs better in Winter and Spring. This is because that there are no crops in these periods, the roughness length is small, and the WDSD method is more suitable for that case.

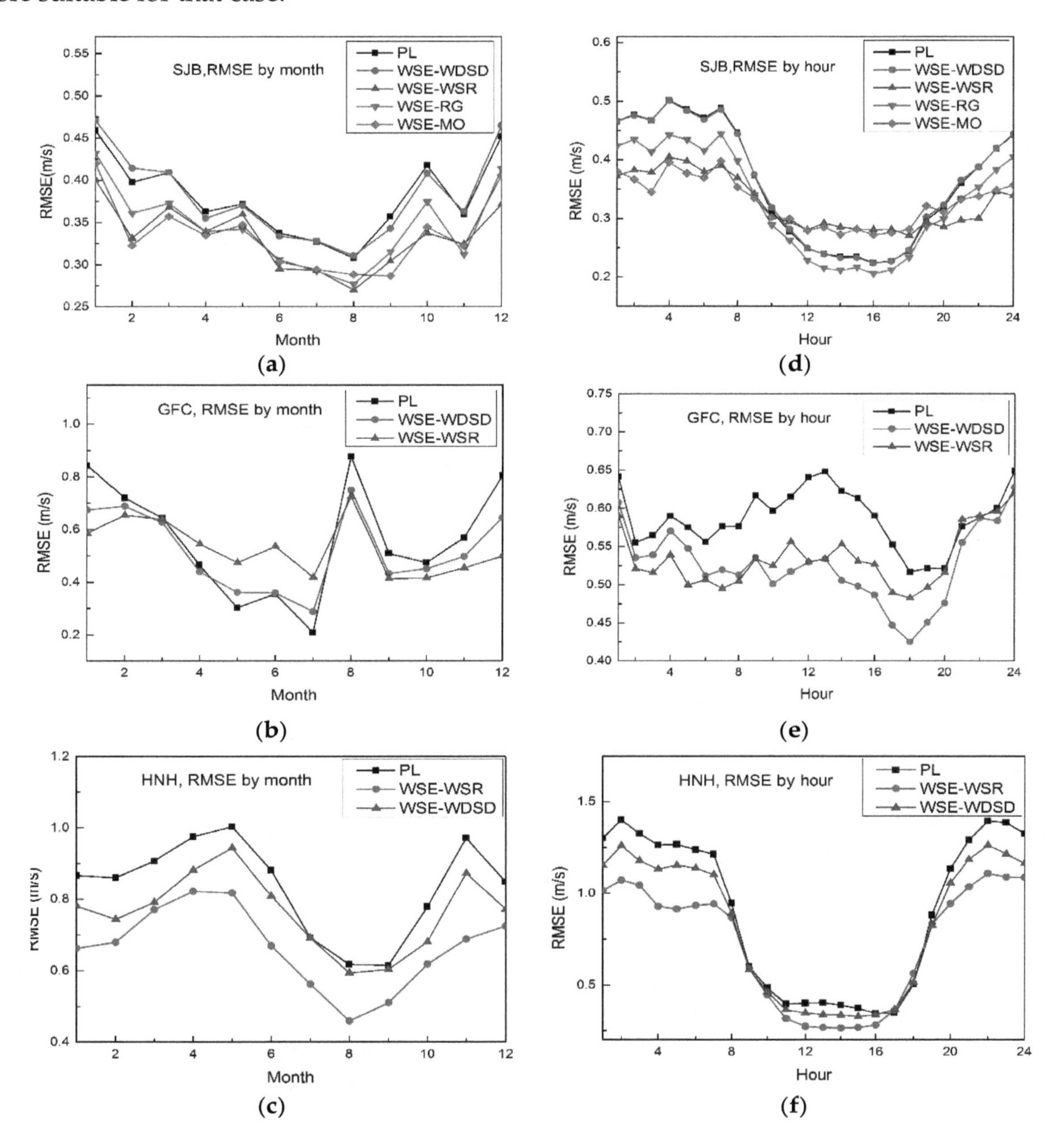

Figure 6. (**a–c**) Monthly and (**d–f**) daily variations of RMSE using different methods in the three different cases.

As a result, the WSE-WSR method is confirmed to be suitable for all the above mentioned terrain types. The WSE-WDSD method is more prominent under flat terrain and mountaintop. Meanwhile, both the WSE-RG and WSE-MO methods are good choices when the measurement data has the two-level temperatures.

5. Conclusions

A new wind shear extrapolation method based on theatmospheric stability was proposed in order to calculate the wind speed at the hub height and compared with the traditional PL method. Particularly, four methods for the classification of atmospheric stability were incorporated into the WSE method. Calculations were performed for flows in three different areas, namely SJB, GFC and HNH, to verify the suitability of the proposed methods. Conclusions can be drawn as follows:

1. For SJB, the plateau where the surface is wasteland, the WSE-RG, WSE-WSR and WSE-MO methods can well calculate the wind speed at the hub height. When two-level temperature data is available, the WSE-RG and WSE-MO methods are more effective, of which MRE of WSE-MO is 0.22% (MRE of PL is −1.58%) and RMSE of WSE-RG is 0.3231 m/s (RMSE of PL is 0.3780 m/s). When there are not enough temperature data, the WSE-WSR method is most effective, of which MRE is −1.52% and RMSE is 0.3311 m/s.
2. For GFC, the mountain where the surface is shrubbery, the WSE-WDSD and WSE-WSR methods perform well and the WSE-WDSD method is most effective, of which MRE is −0.02% (MRE of PL is 0.33%) and RMSE is 0.5276 m/s (RMSE of PL is 0.5430 m/s).
3. For HNH, the plain where the surface is farmland, the WSE-WDSD and WSE-WSR methods are also suitable. MREs of the WSE-WDSD and WSE-WSR methods are −3.17% and −3.26%, respectively (MRE of PL is −7.38%) and RMSEs of the WSE-WDSD and WSE-WSR methods are 0.7931 m/s and 0.7035 m/s, respectively (RMSE of PL is 0.6005 m/s). The WSE-WSR method is recommended when the atmosphere is unstable in most of the time.
4. The new WSE model proposed in the present work has advantages over the traditional PL method. Besides, the WSE-WDSD method for extrapolating the wind speed at the hub height is more effective in plain terrain. WSE-WSR is suitable in complex terrain. Besides, the WSE-RG and WSE-MO methods have more advantages when Ri and L can be calculated.

Author Contributions: Methodology, C.X.; Writing-Original Draft Preparation, C.H. and L.L.; Writing-Review and Editing, X.H., F.X., M.S. and W.S.

Acknowledgments: Authors are grateful to the support by the Jiangsu Science Foundation for Youths (Grant No. BK20180505), the China Postdoctoral Science Foundation (No. 2018M630502), the Fundamental Research Funds for the Central Universities (No. 2018B01614) and the International Cooperation of Science and Technology Special Project (No. 2014DFG62530).

List of Symbols

PL	Power law
WSE	Wind speed extrapolation method
WSE-RG	Wind speed extrapolation method based on the Richardson gradient method
WSE-WSR	Wind speed extrapolation method based on the wind speed ratio method
WSE-WDSD	Wind speed extrapolation method based on the wind direction standard deviation method
WSE-MO	Wind speed extrapolation method based on the Monin–Obukhov method
MRE	Mean relative error
RMSE	Root-mean-square error, m/s
κ	Von Karman's constant
u^*	Friction velocity, m/s
z	Height, m
z_0	Roughness length, m
u	Wind speed, m/s
α	Wind shear exponent

A~F	Classification of atmospheric stability: highly unstable, moderately unstable, slightly unstable, neutral, moderately stable and extremely stable
Ri	Gradient Richard number
ΔT	Temperature difference between two levels of height of z_1 and z_2, °C
Δu	Wind speed difference between two levels of height of z_1 and z_2, m/s
T	Atmospheric average absolute temperature, °C
σ_θ	Horizontal wind direction standard deviation, °
U_R	Wind speed ratio
L	Obukhov length
$\overline{w'T_S'}$	Covariance of temperature and vertical wind speed fluctuations at the surface
g	Gravitational acceleration
Q	Measured wind tower data set
W	Filtered dataset
$h1$	Height of a low level, m
$h2$	Height of a medium level, m
$h3$	Height of a high level, m
$U1$	Mean wind speed at the height of $h1$, m/s
$U2$	Mean wind speed at the height of $h2$, m/s
$U3$	Mean wind speed at the height of $h3$, m/s
$T1$	Mean temperature at the height of the low level, °C
$T2$	Mean temperature at the height of the high level, °C

References

1. Motta, M.; Barthelmie, R.J.; Vølund, P. The influence of non-logarithmic wind speed profiles on potential power output at Danish offshore sites. *Wind Energy* **2005**, *8*, 219–236. [CrossRef]
2. En, Z.; Altunkaynak, A.; Erdik, T. Wind Velocity Vertical Extrapolation by Extended Power Law. *Adv. Meteorol.* **2012**, *2012*, 885–901.
3. Rehman, S.; Al-Abbadi, N.M. Wind shear coefficients and their effect on energy production. *Energy Convers. Manag.* **2005**, *46*, 2578–2591. [CrossRef]
4. Huang, W.Y.; Shen, X.Y.; Wang, W.G.; Huang, W. Comparison of the Thermal and Dynamic Structural Characteristics in Boundary Layer with Different Boundary Layer Parameterizations. *Chin. J. Geophys.* **2015**, *57*, 543–562.
5. Justus, C.G. *Winds and Wind System Performance*; Franklin Institute Press: Philadelphia, PA, USA, 1978.
6. Irwin, J.S. A theoretical variation of the wind profile power-law exponent as a function of surface roughness and stability. *Atmos. Environ.* **1979**, *13*, 191–194. [CrossRef]
7. Troen, I.; Petersen, E.L. European Wind Atlas. Available online: http://orbit.dtu.dk/files/112135732/European_Wind_Atlas.pdf (accessed on 20 August 2018).
8. Monin, A.S.; Obukhov, A.M. Basic regularity in turbulent mixing in the surface layer of the atmosphere. *Trans. Geophys. Inst. Acad. Sci. USSR* **1954**, *24*, 163–187.
9. Jensen, N.O.; Petersen, E.L.; Troen, I. *Extrapolation of Mean Wind Statistics with Special Regard to Wind Energy Applications*; World Meteorological Organization: Geneva, Switzerland, 1984.
10. Optis, M.; Monahan, A.; Bosveld, F.C. Moving Beyond Monin-Obukhov Similarity Theory in Modelling Wind Speed Pro les in the Stable Lower Atmospheric Boundary Layer. *Bound. Layer Meteorol.* **2014**, *153*, 497–514. [CrossRef]
11. Lackner, M.A.; Rogers, A.L.; Manwell, J.F.; Mcgowan, J.G. A new method for improved hub height mean wind speed estimates using short-term hub height data. *Renew. Energy* **2010**, *35*, 2340–2347. [CrossRef]
12. Li, P.; Feng, C.; Han, X. Effect Analysis on the Wind Shear Exponent for Wind Speed Calculation of Wind Farms. *Electr. Power Sci. Eng.* **2012**, *28*, 7–12.
13. Holtslag, A.A.M. Estimates of diabatic wind speed profiles from near-surface weather observations. *Bound. Layer Meteorol.* **1984**, *29*, 225–250. [CrossRef]
14. Gualtieri, G. Atmospheric stability varying wind shear coefficients to improve wind resource extrapolation: A temporal analysis. *Renew. Energy* **2016**, *87*, 376–390. [CrossRef]
15. Panofsky, H.A.; Dutton, J.A. *Atmospheric Turbulence: Models and Methods for Engineering Applications*; Prentice-Hall: Upper Saddle River, NJ, USA, 1983.

16. Mohan, M.; Siddiqui, T.A. Analysis of various schemes for the estimation of atmospheric stability classification. *Atmos. Environ.* **1998**, *32*, 3775–3781. [CrossRef]
17. Wharton, S.; Lundquist, J.K. Atmospheric stability affects wind turbine power collection. *Environ. Res. Lett.* **2012**, *7*, 17–35. [CrossRef]
18. Gryning, S.E.; Batchvarova, E.; Brümmer, B.; Jørgensen, H.; Larsen, S. On the extension of the wind profile over homogeneous terrain beyond the surface layer. *Bound. Layer Meteorol.* **2007**, *124*, 251–268. [CrossRef]
19. Đurišić, Ž.; Mikulović, J. A model for vertical wind speed data extrapolation for improving wind resource assessment using WAsP. *Renew. Energy* **2012**, *41*, 407–411. [CrossRef]
20. Wind Resource Assessment, Siting & Energy Yield Calculations. Available online: http://www.wasp.dk/wasp (accessed on 30 July 2018).
21. Gualtieri, G.; Secci, S. Extrapolating wind speed time series vs. Weibull distribution to assess wind resource to the turbine hub height: A case study on coastal location in Southern Italy. *Renew. Energy* **2014**, *62*, 164–176. [CrossRef]
22. Hellmann, G. *Über die Bewegung der Luft in den untersten Schichten der Atmosphäre*; Kgl. Akademie der Wissenschaften: Copenhagen, Denmark, 1919. (In Germany)
23. Pasquill, F. *Atmospheric Diffusion*, 2nd ed.; Ellis Horwood: Chichester, UK, 1974.
24. Agency, I.A.E. *Atmospheric Dispersion in Nuclear Power Plant Siting: A Safety Guide*; International Atomic Energy Agency: Vienna, Austria, 1980.
25. Balsley, B.B.; Svensson, G.; Tjernström, M. On the Scale-dependence of the Gradient Richardson Number in the Residual Layer. *Bound. Layer Meteorol.* **2008**, *127*, 57–72. [CrossRef]
26. Ma, Y. The basic physical characteristics of the atmosphere near the ground over the Qinghai-Xizang Plateau. *Acta Meteorol. Sin.* **1987**, *2*, 188–200.
27. Deng, Y.; Fan, S.J. Research on the surface layer's stability classifying schemes over coastal region by Monin-Obukhov length. In Proceedings of the National Conference on Atmospheric Environment, Nanning, China, 25 October 2003; pp. 136–141.
28. Sedefian, L.; Bennett, E. A comparison of turbulence classification schemes. *Atmos. Environ.* **1980**, *14*, 741–750. [CrossRef]
29. Chen, P. A comparative study on several methods of stability classification. *Acta Sci. Circumstantiae* **1983**, *3*, 77–84.
30. Businger, J.A. Flux profile relationships in the atmospheric surface layer. *J. Atmos. Sci.* **1971**, *28*, 181–189. [CrossRef]
31. Rehman, S.; Al-Abbadi, N.M. Wind shear coefficient, turbulence intensity and wind power potential assessment for Dhulom, Saudi Arabia. *Renew. Energy* **2008**, *33*, 2653–2660. [CrossRef]
32. Greene, S. Analysis of vertical wind shear in the Southern Great Plains and potential impacts on estimation of wind energy production. *Int. J. Energy Issues* **2009**, *32*, 191–211. [CrossRef]
33. Fox, N.I. A tall tower study of Missouri winds. *Renew. Energy* **2011**, *36*, 330–337. [CrossRef]

11

A Fully Coupled Computational Fluid Dynamics Method for Analysis of Semi-Submersible Floating Offshore Wind Turbines under Wind-Wave Excitation Conditions based on OC5 Data

Yin Zhang and Bumsuk Kim *

Faculty of Wind Energy Engineering Graduate School, Jeju National University, Jeju City 63243, Korea; scarletyuki@jejunu.ac.kr
* Correspondence: bkim@jejunu.ac.kr

Abstract: Accurate prediction of the time-dependent system dynamic responses of floating offshore wind turbines (FOWTs) under aero-hydro-coupled conditions is a challenge. This paper presents a numerical modeling tool using commercial computational fluid dynamics software, STAR-CCM+(V12.02.010), to perform a fully coupled dynamic analysis of the DeepCwind semi-submersible floating platform with the National Renewable Engineering Lab (NREL) 5-MW baseline wind turbine model under combined wind–wave excitation environment conditions. Free-decay tests for rigid-body degrees of freedom (DOF) in still water and hydrodynamic tests for a regular wave are performed to validate the numerical model by inputting gross system parameters supported in the Offshore Code Comparison, Collaboration, Continued, with Correlations (OC5) project. A full-configuration FOWT simulation, with the simultaneous motion of the rotating blade due to 6-DOF platform dynamics, was performed. A relatively heavy load on the hub and blade was observed for the FOWT compared with the onshore wind turbine, leading to a 7.8% increase in the thrust curve; a 10% decrease in the power curve was also observed for the floating-type turbines, which could be attributed to the smaller project area and relative wind speed required for the rotor to receive wind power when the platform pitches. Finally, the tower-blade interference effects, blade-tip vortices, turbulent wakes, and shedding vortices in the fluid domain with relatively complex unsteady flow conditions were observed and investigated in detail.

Keywords: computational fluid dynamics; floating offshore wind turbine; dynamic fluid body interaction; semi-submersible platform; OC5 DeepCWind

---

1. Introduction

Energy generation from offshore wind farms has been garnering the attention of researchers, owing to the abundance of resources and low environmental impact. Compared to offshore wind turbines in shallow water, floating offshore wind turbines (FOWTs) have more advantages [1]; i.e., there are several deep-water sites suitable for installing turbines, wind is more abundant in offshore areas, and public concerns on the visual and environmental impacts are minimized with this technology. Some floating wind farms have been established; for instance, the first full-scale 2.3-MW FOWT was installed in Hywind near the coast of Norway, and last year, five 6-MW FOWTs were installed in the North Sea off the coast of Peterhead, Scotland.

However, it is difficult and expensive to operate a real-scale test model and accurately calculate critical loads because the complex multi-physical phenomena are not easy to simulate in reality. In addition, this technology is dependent on extreme weather situations (such as 25 m/s cut-out speeds). Thus, the use of computational methods, involving virtual full-scale modeling, may increase the development of the controllers' reliability (such as structure and loads) of FOWTs, reduce the risks involved, and build confidence in the design stage. Among the codes used, one of the most famous ones is the Fatigue, Aerodynamics, Structures, and Turbulence code (FAST), which was developed by the National Renewable Engineering Lab (NREL) based on the blade element momentum (BEM) theory [2]. However, the BEM theory is seldom applied in FOWT situations owing to its theoretical limitation. In contrast, the fluid structure interaction (FSI) simulations, as a modern computational analysis method, has proven to be an accurate and convincing method for considering aero-hydro-servo-elastic problems; however, complex fluid conditions and blade deformation presents significant computational challenges.

Further, correctly simulating the movement of floaters on free surfaces is also a major challenge; many researchers from different institutions have developed various codes and solvers to simulate the hydrodynamic performance of floaters. Nearly all solvers are based on the following theories: the potential-based panel approach and Morrison equation. The former cannot determine viscous flow details and is usually used together with the damping coefficient obtained from experimental test data. The FAST code HydroDyn module has applied this method. The Morison equation is a semi-empirical equation; this equation mainly describes the inline force in oscillatory flow conditions; this also has theoretical limitations and it cannot adequately describe the time-dependent force. Examples include the wave analysis MIT (WAMIT), TimeFloat, and CHARM3D. However, there are still some physical phenomena that cannot be fully described. Conversely, the unsteady computational fluid dynamics (CFD) approach can simulate with consideration of all physical effects, including flow viscosity, hydrostatic forces, wave diffraction, radiation, wave run-up, and slamming and provide reliable and accurate results regarding the platform movement.

Owing to the reasons mentioned above, the CFD method is widely considered an effective and reliable method to simulate the FOWT problem; to date, several CFD-related investigations have been performed. However, previous studies have used the following methods, ignoring some effects, leading to inaccurate results. First, to investigate the hydrodynamic load and motion response of a platform on an FOWT, some studies just simplified the problem into wind turbine aerodynamic loading or ignored the tower and rotor-nacelle-assembly. Second, some studies focused on aerodynamic loading but restricted the motion of the floating platforms to a prescribed position or did not allow the platform to move with 6 DOF.

Unai Fernandez-Gamiz et al. [3] developed an improved BEM-based solver to verify the NREL 5-MW wind turbine and determined the bending moment and thrust force in the blade root; they also investigated rectangular sub-boundary layer vortex generators using the CFD method [4], which showed the highest vortex generator suitable for separation control. Nematbakhsh et al. [5] developed a CFD spar model and successfully captured strong nonlinear effects, which cannot be captured using the FAST code. Furthermore, their study also observed that when the wave amplitude was large, a discrepancy could exist between CFD and FAST. Vaal et al. [6] used the BEM method to investigate the surge motion of FOWTs. This showed that the BEM method could only provide a reasonable solution under slow surge motion conditions; this is because, in this condition, the wake dynamics can be ignored. Zhao and Wan [7] used a Naoe-FOAM-SJTU simulated OC4 platform to study the effects of the presence of wind turbines. They carried out platform pitch motions at high wind speeds and investigated the wind turbine effect on the floating platform. Tran et al. [8] set the platform to execute a prescribed sinusoidal pitching motion and changed the motion amplitudes and frequencies, instead of modeling a floating platform with 6 DOF using the unsteady BEM theory, generalized dynamic wake (GDW), and CFD; large discrepancies were observed when the pitch amplitude increased to 4°. Tran et al. [9] analyzed an FOWT system under a prescribed sinusoidal surge motion, and found that thrust and power varied significantly, which is related to the oscillation frequency; the surge motion amplitude also varied significantly. Liu et al. [10] superimposed three DOF platform motions (surge, heave, and pitch) and concluded that the platform motion significantly impacted the thrust and torque of the wind turbine. Ren et al. [11] used FLUENT analysis for a 5-MW tension-leg-platform-type turbine under coupled wave-wind conditions and validated the simulation results against experimental data. They only considered the surge motion and concluded that during the variation in the average/mean surge response of the system, aerodynamic forces played the main role. Quallen et al. [12] performed a CFD simulation involving an OC3 spar-type FOWT model under wind–wave excitation conditions. The mean surge motion predicted using the CFD model was 25% less than that predicted using FAST. Tran and Kim [13] modeled an OC4 semi-submersible FOWT using the dynamic fluid body interaction (DFBI) method and an overset mesh technique under wind–wave excitation conditions. A good overall agreement was found between the CFD results and FAST data. Both codes used the quasistatic method for modeling the mooring lines. S. Gueydon et al. [14] modeled a semi-submersible platform using the aNyPHATAS code to investigate operating rotor effects on drift motions and additional damping. Chen et al. [15] modeled a semi-submersible FOWT with two different blade configurations in a wave basin to further optimize the blade design for FOWTs. A.J. Dunbar et al. [16] developed an open-source CFD/6-DOF solver based on OpenFOAM and compared rotational and translational motions with FAST, demonstrating the accuracy of this tightly coupled solver.

The main purpose of this study was to conduct a virtual test of a real-scale 5-WM semi-submersible FOWT using the advanced CFD method. The hydrodynamic responses were validated using the latest physical test data of the Offshore Code Comparison, Collaboration, Continued, with Correlations (OC5) projects. Full-configuration FOWT simulations, simultaneously considering the rotating blade motion with 6-DOF platform dynamics were performed; a relatively large discrepancy in the predicted power was observed owing to the different properties of the mooring line and rotating inertia moment between the OC4 and OC5 projects. This proves the high infinity result of OC5 project. Further, this study may provide some reference for the validation of the CFD method for use in the OC5 Phase II system and high-fidelity simulation investigations of FOWTs in coupled aero-hydro conditions.

The OC5 DeepCWind semi-submersible floating wind turbine model was used for the investigation, which is briefly described in Section 2. The numerical methods used in the study are introduced in Section 3. The aerodynamic validation studies performed using different modeling tools are briefly presented in Section 4. Section 5 presents the results of the dynamic responses of the floating system under regular wave conditions. Section 6 presents the simulation results of the fully coupled configuration. Section 7 presents the conclusions of the study.

2. Floating Offshore Wind Turbine Model

2.1. Model Description

A semi-submersible FOWT tested in Phase II of the OC5 project was investigated. The design parameters of the full-scale OC5 DeepCWind semi-submersible platform are summarized in Table 1. The NREL 5-MW baseline wind turbine model was set above the tower, 87.6 m from the water surface; we used airfoil data from the DOWEC project, which is also mentioned in Jonkman's work [1] from NREL. The major properties of the NREL 5-MW baseline wind turbine are given in Table 2; the cross-sections of the rotor blade were composed of a series of Delft university of technology (DU) and national advisory committee for aeronautics (NACA) 64 airfoils from the hub to the tip of the outboard section. The computer-aided design (CAD) model of the blade was first developed using Solidworks software (Dassault systems, Velizy-Villacoublay, France), as shown in Figure 1. Except for the cylinder in the blade root and transition section, the control airfoil spread along the blade includes the DU series airfoil from 40% to 21% thickness and a NACA airfoil of 18% thickness. Details concerning the wind blade aerodynamics including the blade twist, chord length, and airfoil designation are presented in Table 3.

Table 1. Full system structural properties.

Parameters	Value
Mass	1.3958×10^7 kg
Draft	20 m
Displacement	1.3917×10^4 m^3
Center of mass (CM) location below seawater level (SWL)	8.07 m
Roll inertia about system CM	1.3947×10^{10} kg/m^2
Pitch inertia about system CM	1.5552×10^{10} kg/m^2
Yaw inertia about system CM	1.3692×10^{10} kg/m^2

Table 2. Blade structural properties.

Parameters	Value
Length (w.r.t. root along axis)	61.5 m
Overall (integrated) mass	2.2333×10^4 kg
Second mass moment of inertia	1.48248×10^7 kg/m^2
First mass moment of inertia	4.5727×10^5 kg/m
CM location	20.475 m

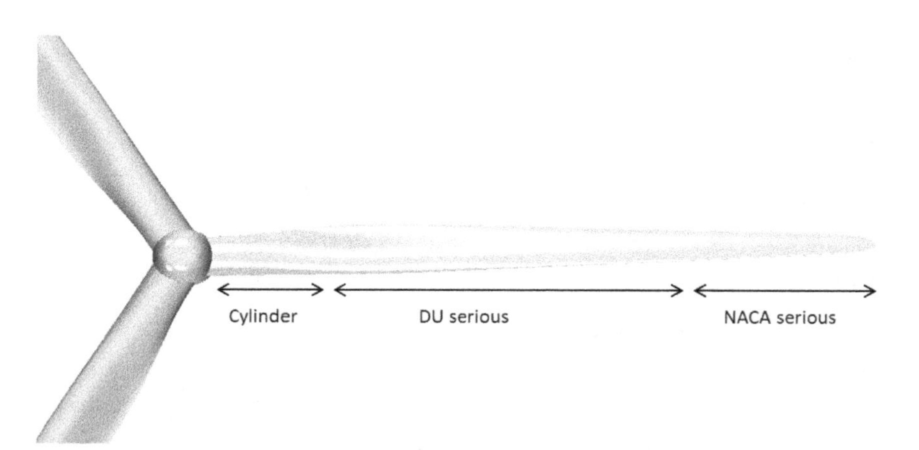

Figure 1. Airfoil construction of National Renewable Engineering Lab (NREL) 5 MW blade.

Table 3. Blade airfoil distribution of National Renewable Engineering Lab (NREL) 5 MW wind turbine.

Node Radius (m)	Twist Angle (deg)	Chord Length (m)	Airfoil Designation
2.867	13.308	3.542	Cylinder
5.600	13.308	3.854	Cylinder
8.333	13.308	4.167	Cylinder
11.750	13.308	4.557	DU 40
15.850	11.480	4.652	DU 35
19.950	10.162	4.458	DU 35
24.050	9.011	4.249	DU 30
28.150	7.795	4.007	DU 25
32.250	6.544	3.748	DU 25
36.350	5.361	3.502	DU 21
40.450	4.188	3.256	DU 21
44.550	3.125	3.010	NACA 64-618
48.650	2.319	2.764	NACA 64-618
52.750	1.526	2.518	NACA 64-618
56.167	0.863	2.313	NACA 64-618
58.900	0.370	2.086	NACA 64-618
61.633	0.106	1.419	NACA 64-618

2.2. OC4 and OC5 Projects

Previous studies mostly used test data from the study by Coulling et al. [17], which was led by the University of Maine at the maritime research institute Netherlands (MARIN) offshore wave basin in 2011; however, the geometrically scaled model did not perform as expected under the low-Reynolds number wind conditions. In addition, only semi-submersible properties (center of mass, inertia force, etc.) were considered in the OC4 project, while the OC5 project considered the properties of the full system, and the mooring line properties could be adjusted. Hence, a new model was built that resulted in better-scaled thrust and torque loads. However, this turbine was retested in 2013, and this turbine was examined in Phase II of the OC5 project [18]. The different physical properties of the semi-submersible platforms in the two projects are summarized in Table 4.

Table 4. Comparison of OC4 and OC5 project.

Semisubmersible Platform	OC5 Phase II	OC4 Phase II
Mass	12,919,000 kg	13,444,000 kg
Draft	20 m	20 m
CM below SWL	14.09 m	14.4 m
Roll inertia	7.5534×10^9 kg/m^2	8.011×10^9 kg/m^2
Pitch inertia	8.2236×10^9 kg/m^2	8.011×10^9 kg/m^2
Yaw inertia	1.3612×10^{10} kg/m^2	1.391×10^{10} kg/m^2
Buoyancy center below SWL	13.15 m	-
Mooringline anchors from center	837.6 m	837.6 m
Mooringline fairlead from center	40.868 m	40.868 m
Unstretched mooringline length	835.5 m	835.5 m
Mooringline mass density	Line 1: 125.6 kg/m	113.35 kg/m
	Line 2: 125.8 kg/m	
	Line 3: 125.4 kg/m	
Mooringline extensional stiffness	Line 1: 7.520 E8 N	7.536 E8 N
	Line 2: 7.461 E8 N	
	Line 3: 7.478 E8 N	
6 degrees of freedom (DOF) Natural Periods	Surge: 107 s	Surge: 107 s
	Sway: 112 s	Sway: 113 s
	Heave: 17.5 s	Heave: 17.5 s
	Roll: 32.8 s	Roll: 26.9 s
	Pitch: 32.5 s	Pitch: 26.8 s
	Yaw: 80.8 s	Yaw: 82.3 s

The turbine is a 1/50th scale horizontal-axis model of the NREL 5-MW reference wind turbine with a flexible tower affixed atop a semi-submersible platform. The DeepCwind semisubmersible platform is composed of the main column and three offset columns linked to the main column via several pontoons and braces, as mentioned in the OC5 report [18]; the 1/50th scale and full-scale models are shown in Figure 2. A 5-MW baseline wind turbine is vertically mounted on the main column so that the hub height from the sea surface is 90 m. In addition, the platform is moored with three catenary mooring lines, with fairleads located at the base columns. The anchors are located 200 m below the sea surface, on the seabed. One mooring line is aligned in the wave direction, which is also the platform surge direction; the other two mooring lines are distributed around the platform uniformly and the attachment angle between each mooring line is 120°.

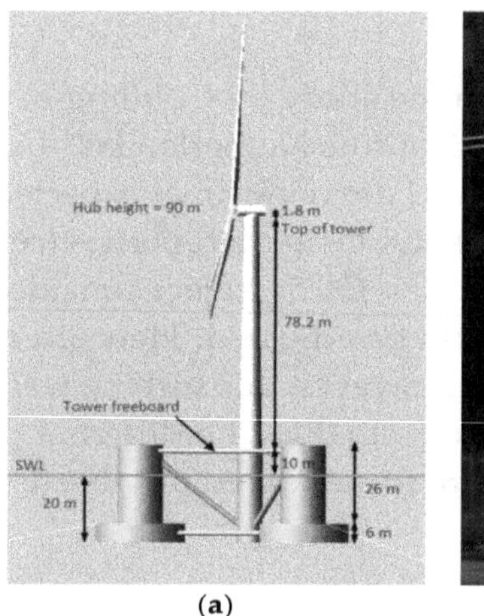

(a)

(b)

Figure 2. Semi-submersible model: (**a**) Full-scale model; (**b**) The 1/50th scale model in maritime research institute Netherlands (MARIN) wave basin.

3. Simulation Method

3.1. Numerical Setting and Governing Equations

This paper presents a numerical modeling tool using commercial CFD software, STAR-CCM+(V12.02.010) (Siemens, Munich, Germany), to perform a fully coupled dynamic analysis of the DeepCwind semi-submersible floating platform with the NREL 5-MW baseline wind turbine model under combined wind-wave excitation conditions.

This investigation used the unsteady incompressible Navier–Stokes equations, according to the first principles of the conservation of mass and momentum. To solve the pressure–velocity coupling, a semi-implicit method was used, which involved a predictor–corrector approach. Second-order upwind and central difference schemes were used for the convection terms and temporal time discretization, respectively. Additionally, the shear stress transport (SST) k-ω turbulence model (Menter's Shear Stress Transport) is a robust two-equation, eddy-viscosity turbulence model used for many aerodynamic applications to resolve turbulent behavior in the fluid domain and was first introduced in 1995 by F.R. Menter [19]. The model combines the k-ω and k-e turbulence models; therefore, the k-ω turbulence model can be used in the inner region of the boundary, and the k-e turbulence model can be used in free shear flow. Menter's SST turbulence model can be expressed as follows:

$$\frac{\partial(\rho k)}{\partial t} + \frac{\partial(\rho u_j k)}{\partial x_j} = P - \beta^* \rho \omega k + \frac{\partial}{\partial x_j}\left[(\mu + \sigma_k \mu_t)\frac{\partial k}{\partial x_j}\right] \tag{1}$$

$$\frac{\partial(\rho \omega)}{\partial t} + \frac{\partial(\rho u_j \omega)}{\partial x_j} = \frac{\gamma}{v_t} P - \beta \rho \omega^2 + \frac{\partial}{\partial x_j}\left[(\mu + \sigma_\omega \mu_t)\frac{\partial \omega}{\partial x_j}\right] + 2(1 - F_1)\frac{\rho \sigma_{\omega 2}}{\omega}\frac{\partial k}{\partial x_j}\frac{\partial \omega}{\partial x_j} \tag{2}$$

To obtain details of the free surface between air and water, the unsteady CFD method with the volume of fraction (VOF) approach coupled with the 6 DOF solver was used for the hydrodynamic analysis considering the surge, sway, heave, roll, pitch, and yaw motions of the platform.

3.2. Dynamic Fluid Body Interaction (DFBI) Method

The DFBI method was applied to simulate the motion of the rigid FOWT body in response to pressure and shear forces in the fluid domain and consider the recovery force from the mooring lines. STAR-CCM+ was used to calculate the resultant force and moment acting on the body due to various influences and also to solve the governing equations of rigid body motion to determine the new position of the rigid body. A flow chart illustrating the DFBI method is shown in Figure 3.

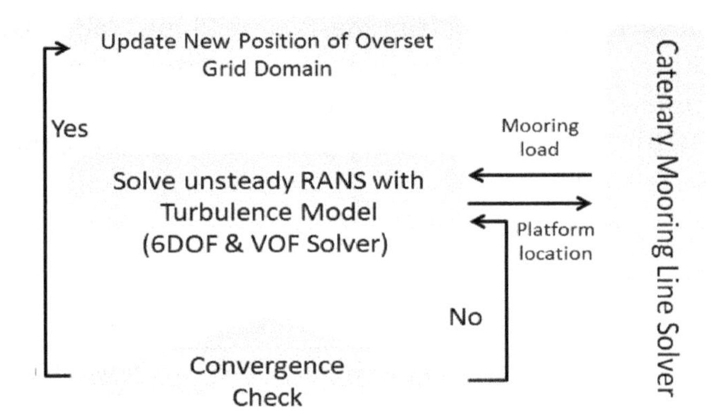

Figure 3. Flowchart of Dynamic Fluid Body Interaction (DFBI) method.

3.3. Overset Mesh Technology

The overset mesh technique, also called overlapping or chimera grids, was applied. A new internal interface node was created within the overset mesh region. This volume-type interface enables the coupling of solutions on the domains using automatically generated sets of acceptor cells in one mesh and donor cells in another mesh. Varying values of the donor cells affect the values of the acceptor cells based on interpolation. This method can handle complex geometries and body motions in dynamic simulations.

3.4. Mooring Line Modeling and Damping

The catenary coupling model was used to model an elastic, quasi-stationary catenary, hanging between two endpoints and subject to its own weight in the field of gravity. In a local Cartesian coordinate system, the shape of the catenary is given by the following set of parametric equations:

$$x = au + bsin(u) + \alpha \tag{3}$$

$$y = acosh(u) + \frac{b}{2}sinh^2(u) + \beta \tag{4}$$

$$for\ u_1 \leq u \leq u_2 \tag{5}$$

In addition, a wave-damping area was applied, considering the wave reflection near the outlet boundary; this treatment includes a wave-damping zone. The wave-damping area was designed to minimize the effects of wave reflections on the far downstream outlet boundary. As a result, the VOF wave could be damped in the pressure outlet boundary to reduce wave oscillations. This damping introduces vertical resistance to the vertical motion of the wave.

4. Aerodynamic Validation of Wind Turbine

4.1. Numerical Setting and Mesh Convergence Test

An aerodynamic simulation of the rotor part was performed to validate the accuracy of the 3D model and the numerical model using the CFD method. The hexahedral computation domain size and boundary type was 8D(x) × 5D(y) × 3D(z) and extended up to 2.5D and 5.5D in the upstream and downstream x-directions from the wind turbine, respectively.

Both poly grids and trim grids were tried in the mesh test procedure, and the trim mesh was finally selected owing to its robust features and relatively low computational cost. Nineteen layers of prism grids were generated to determine the blade-attached flow, and the near-wall first boundary layer thickness was 3.82×10^{-6} m. A mesh convergence test was also performed, and the results are given in Table 5; five sets of grids were generated with different grid densities while all the other parameters remained unchanged. A total of 22 million grids were selected with both a short calculation time and acceptable power loss. Details of the trim mesh around the blade tip are presented in Figure 4; at the same time, denser grids around the leading and trailing edges of the airfoils were used to detect fluid details. Y plus is a dimensionless value used to measure the mesh quality; a value below 1 is produced when the k-ω SST turbulence model is applied. In this study, the Y plus value was much lower than 1 in all five cases.

Table 5. Mesh convergence test.

Mesh	Thrust (kN)	Power (kW)	Power Variation
14 millions	726	4980	−5.0%
20 millions	728	4990	−4.8%
22 millions	731	5080	−3.0%
30 millions	730	5120	−2.3%
38 millions	734	5240	0%

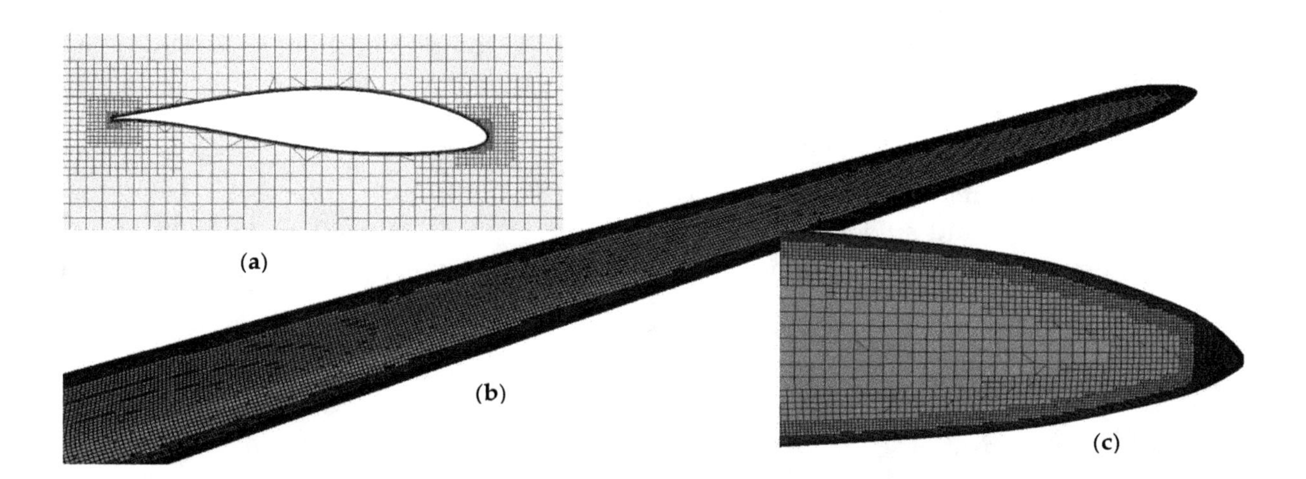

Figure 4. Trim cell detail: (**a**) Mesh section near 45 m spinwise airfoil; (**b**) Trim mesh around blade tip; (**c**) Blade surface mesh.

4.2. Validation of Rotor Aerodynamic Performance

Currently, there are three main methods used for simulating aerodynamic performance: the BEM method, generalized dynamic wake (GDW) model, and CFD model. The results of each model are shown in Figure 5. The CFD results are in good agreement with those of the other codes with regard to both power and thrust prediction, which were collected by Gyeongsang national university (GNU) using FAST code. The BEM method was observed to overestimate results at a relatively high wind speed, which is also noted in other studies [20]. Sivalingam et al. also compared results between the CFD and BEM methods; there was a good agreement in terms of the thrust and torque results below the rated speed. However, because of tip loss factors at relatively high wind speeds, a deviation of axial induction factors was shown by the BEM method, while the CFD method captured wake rotation accurately [21].

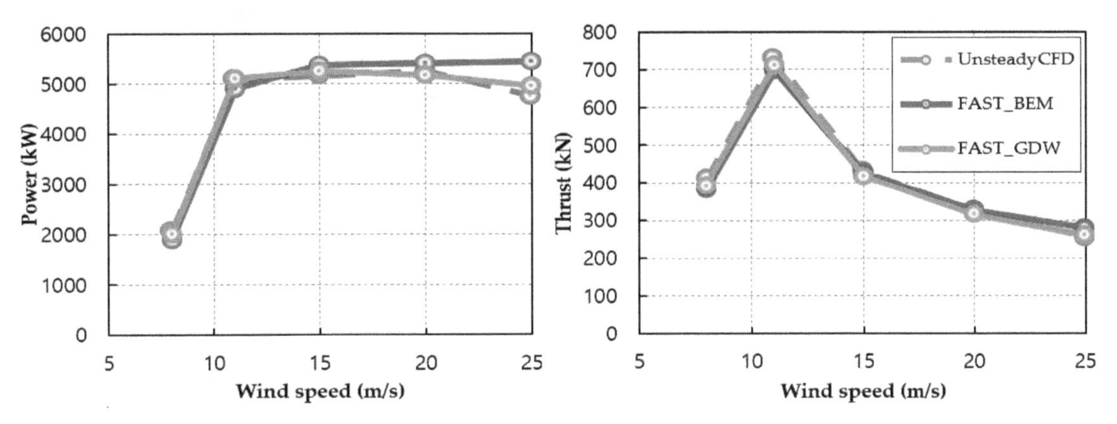

Figure 5. Power and thrust in 8, 11, 15, 20, and 25 m/s uniform wind speeds.

4.3. Study of Wind Profile and Tower Dam Effect Under Onshore Wind Turbine Generator Conditions

The aerodynamics of the NREL 5-MW fixed wind turbine was studied on a full-scale without the floating platform. The results will later be compared with the data for a floating wind turbine. All the numerical settings used in the CFD-rigid body motion (RBM) approach in the previous simulations were applied to the rotor part, except the inlet uniform wind was replaced with the wind profile.

The wind profile shows variations in the horizontal wind speed with height, which may result in increased fatigue loading and reduced power output, usually characterized by the power law, as follows:

$$v = v = v_{hub}(H/H_{hub})^{\alpha} \tag{6}$$

The wind shear exponent (alpha) was 0.12 for flat onshore conditions. Figure 6 illustrates the computational mesh domain; the upstream boundary of the inlet is defined as velocity inlet and the pressure outlet is defined as the downstream boundary. Symmetric boundary conditions were applied in the far field region and a no-slip wall condition was imposed on the surface. Tran et al. [20] had carried out a convergence test to determine the fluid domain size; herein, we used a hexahedral computational domain size of 1000 m × 600 m × 275 m (length, width, height), the same as that used by Tran et al, in the x-direction; the fluid domain extended 313 m upstream and 687 m downstream to help analyze the fluid domain and consider the impact of the vortex after the tower, considering the time-dependent motion of the rotating blades; we used the RBM method in this study.

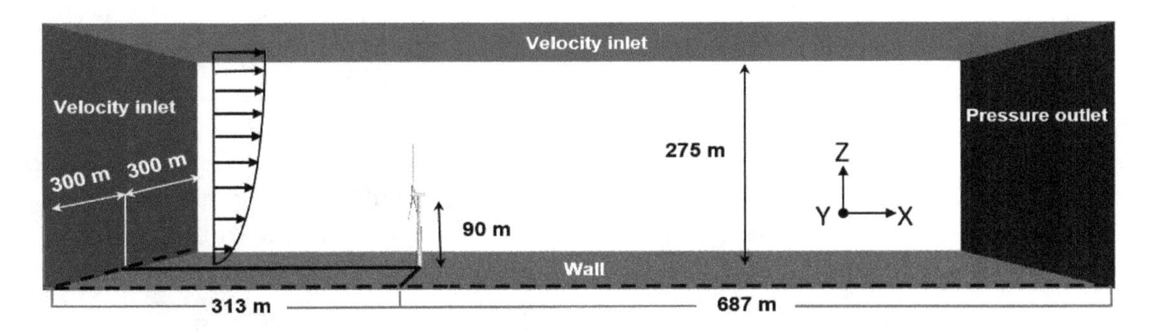

Figure 6. Onshore 5 MW wind turbine fluid domain and boundary conditions.

Aerodynamic simulations of onshore wind turbines were conducted using the RBM method under unsteady conditions; the obtained torque output under 11 m/s wind conditions was compared with the results of the FSI method obtained at the University of California. As shown in Figure 7, there was good agreement between both, but the predictions from our simulation were slightly higher, which may be because blade deformation was ignored in the RBM method [22,23].

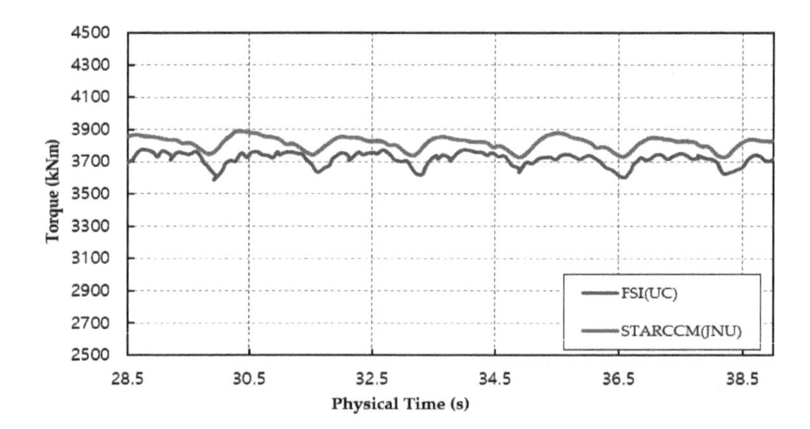

Figure 7. Torque curve between fluid structure interaction (FSI) method and unsteady method in 11 m/s wind speed.

5. Hydrodynamic Response of Floating Platform

5.1. Free-Decay Test

Free-decay tests are methods generally used in a wave tank to determine the natural period of a floating system. The OC4 project was based on a 1/50th-scale semi-submersible platform in the MARIN wave basin in 2011. This model was revised two years later, to provide more precise test data for the OC5 project [18]. Owing to the different properties of the mooring line (line stiffness) and rotating inertia forces in the two tests, the OC4 and OC5 data reveal good agreement in terms of translation motions (surge, sway, and heave) but a relatively large discrepancy in terms of rotating motions (roll, pitch, and yaw), as observed in Table 6. Six-DOF free decay tests were conducted to determine the hydrodynamic damping characteristics of the OC5 semi-submersible platform.

Table 6. Natural period in OC4 and OC5 project.

DOF	OC5 Natural Period	OC4 Natural Period
Surge	107 s	107 s
Sway	112 s	112 s
Heave	17.5 s	17.5 s
Roll	32.8 s	26.9 s
Pitch	32.5 s	26.8 s
Yaw	80.8 s	82.3 s

The wave mode was set as still water, and the air density was zero. Only the platform was considered to simplify the simulation; however, the gross mass should also be considered. The platform was given a prescribed displacement and released to move freely from the initial position. This test considered only three rigid-body DOFs, that is, the surge, pitch, and heave motions.

The results are presented in Figure 8, along with the simulation results from GNU and the wave basin test results from phase II of the OC5 project. The heave and surge time-domain responses for the platforms are in good agreement, as similar results were obtained in the heave and surge periods in the OC4 and OC5 projects. However, in the case of pitch, a relatively large discrepancy was observed in the time-domain response for the GNU simulation. As mentioned above, this effect may be owing to the different properties of the mooring line and rotating inertia forces of the platforms used in the two projects. Based on the natural period of the pitch, the pitch results were in good agreement with the OC5 test data.

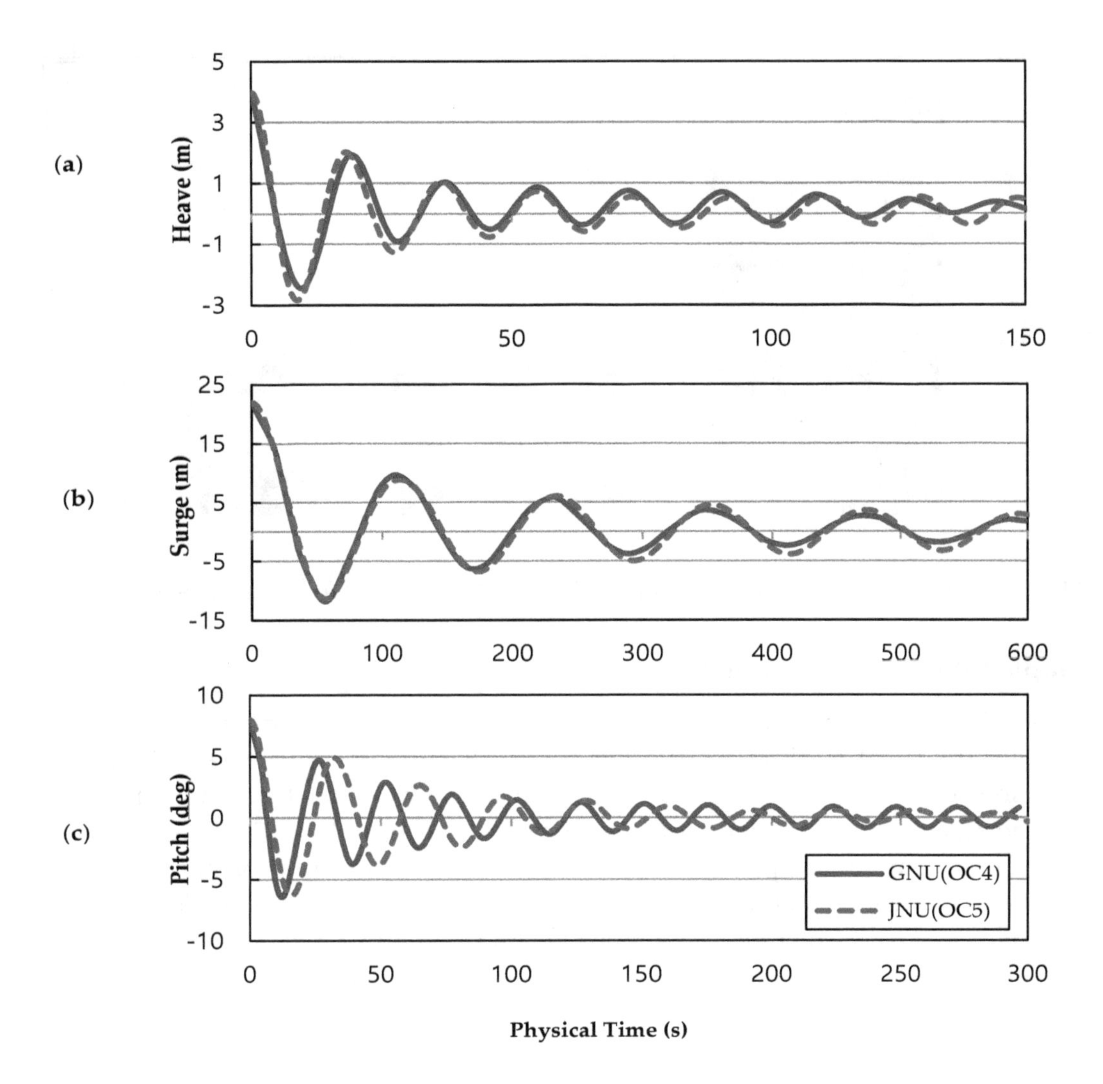

Figure 8. 3-DOF movement of platform by time domain: (**a**) Heave movement; (**b**) Surge movement; (**c**) Pitch movement.

5.2. Hydrodynamic Response Under Regular Waves

The characteristics of the DeepCWind platform under regular wave conditions were investigated by calculating the response amplitude operators (RAOs). An RAO is the normalized value of the amplitude of the periodic response of a field variable divided by the amplitude of a regular wave [24]. The platform was initialized at a static position, and a regular wave was introduced. The regular wave had an amplitude of 3.79 m and a period of 14.3 s. A fifth-order wave was applied in the regular wave test; here, the fifth-order wave was modeled with a fifth order approximation to the Stokes theory of waves [25]. This wave more closely resembled a real wave than one that was generated by the first-order method. The transient start-up period should not be considered in the results. After simulation runs for 400 s, the platform movement achieved a periodic quasi-steady state. The surge, heave, and pitch motion amplitudes were calculated by averaging the amplitudes over the last eight wave periods [24]. These values were then normalized using the amplitude of the regular wave to obtain the RAO. The RAOs of Phase II of the OC5 project were higher for the surge, heave, and pitch DOFs, similar to our simulation results obtained using the unsteady CFD method, as shown in Figure 9.

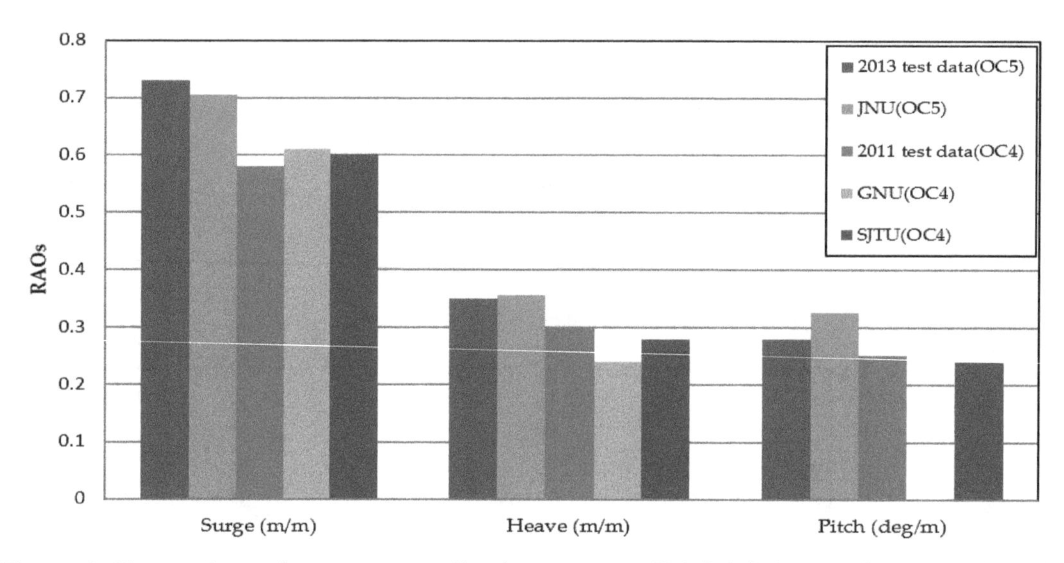

Figure 9. Comparison of response amplitude operators (RAOs) for surge, heave, and pitch.

6. Fully Coupled Wind–Wave Simulation

The full-scale DeepCwind OC5 model in a coupled wind–wave excitation condition was finally conducted using the DFBI method mentioned above. Figure 10 shows the fluid domain, together with an xz-plane section of the mesh distribution in the whole fluid domain. To obtain the fluid details near the free surface, as well as those near the turbine blade tip and vortex regions after the tower, mesh refinement was performed around the blades and platform, as shown in Figure 11. Nearly 27 million cells were generated using the built-in trim mesh feature in STAR-CCM+. The wind speed, V, was assumed to be the rated wind speed (11.4 m/s); the wave height and wave period were assumed to be 7.58 m and 12.1 s, respectively, similar to the MARIN wave basin.

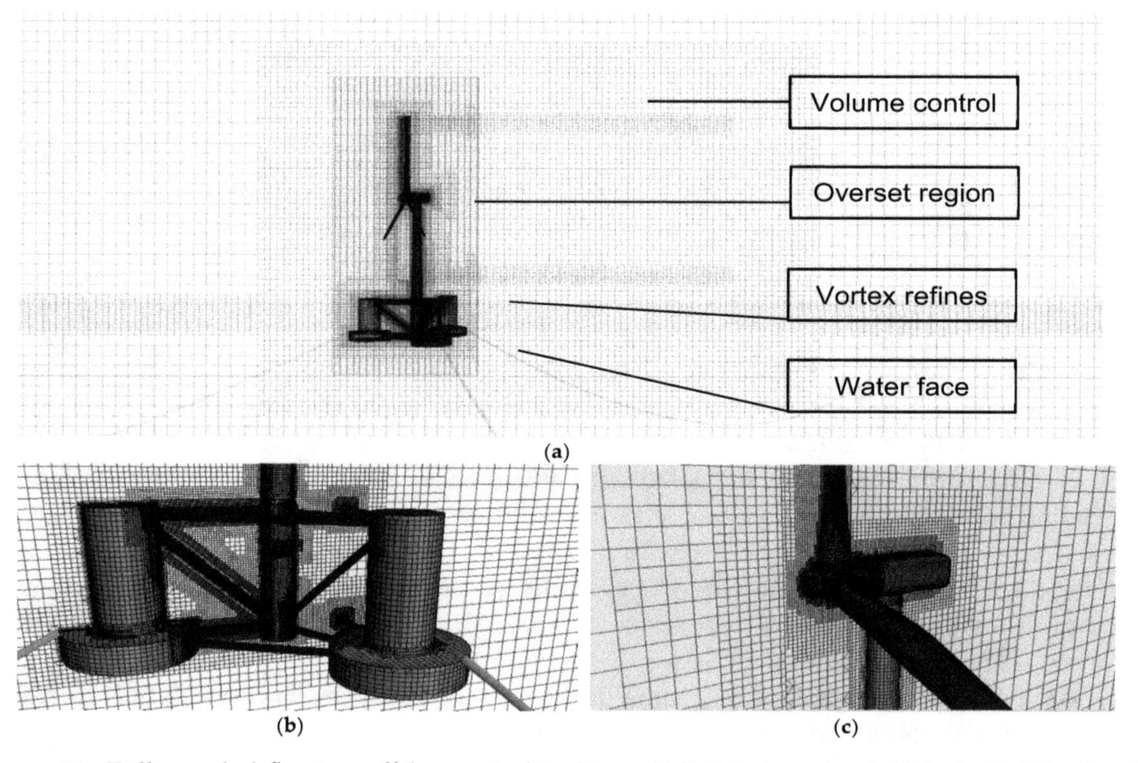

Figure 10. Full-coupled floating offshore wind turbines (FOWT) domain: (**a**) Mesh distribution in xz-plane; (**b**) Close-up view of mesh around platform; (**c**) Close-up view of mesh around wind turbine.

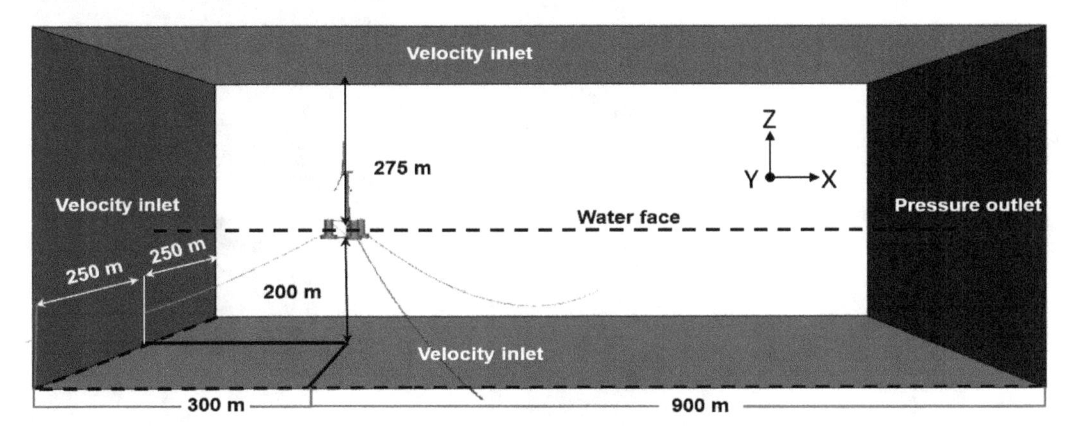

Figure 11. Fully coupled floating offshore wind turbines (FOWT) domain fluid, domain size, and boundary conditions.

After a 10 s start-up time, the aerodynamic output for the wind turbine was stabilized and the platform was released to move. A computational flow chart for the FOWT is presented in Figure 12.

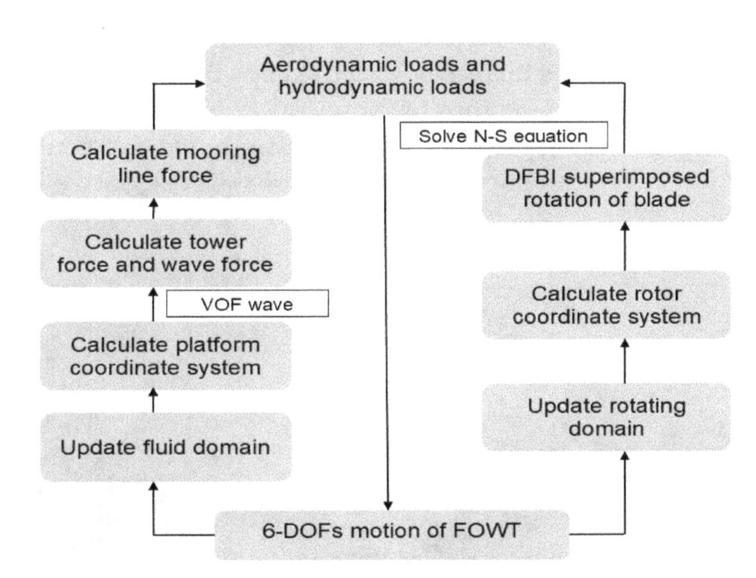

Figure 12. Flow chart of fully coupled simulation in the wind–wave condition.

The time for one revolution of the blades with a rotation speed of 12.1 rpm is 4.96 s. The time-step size (dt) of 0.07009 s utilized here corresponds to a 5° increment in the azimuth angle of the blade for each time-step. The wave heading angle is 0° and the wave is parallel to the direction of mooring line 2, which is also parallel to the platform surge direction.

All computations of the FOWT considering the wind–wave coupling were performed using a 4U multi D500 server. The elapsed real central processing unit (CPU) time for parallel processing per time-step with 15 sub-iterations was 6 min when using 66 CPUs. The total number of iterations for a simulation runtime of 300 s was approximately 30,000. The total simulation time taken to obtain the results using 66 CPUs was 20 days.

Figure 13 demonstrates the vortex contours with 0.5 Q-criteria, colored according to the velocity magnitude component, where the free surface is colored according to the surface elevation. After post-processing, as easily observed in the figure, strong vortices appear near the blade tips and roots. The presence of the tower caused a complex flow wake because of the interaction between the tower and flow. Such a detailed flow map is useful to identify the means for improving the wind turbine power output in the design stage.

Figure 13. Velocity scene and water elevation.

This is an advantage of the CFD method, which is not present in other codes such as FAST. The size of the vortex tubes gradually decreases with time, and the patterns can be described by an iso-vorticity value. Herein, one wave period is separated into eight steps, i.e., T1 to T8; the duration from T1 to T8 represents one period of the platform surge motion. When the wind turbine moves backward, the number of vortex tubes increases, and the gaps between the blade tip vortices tend to continuously decrease at the same time. Figure 14 shows the fluid field and turbulence wakes between the fluid and tower and nacelle during the platform surge motion at different times. It shows that the generated vortexes near the tower and nacelle configurations diffuse outward as the platform moves backward, and vice versa.

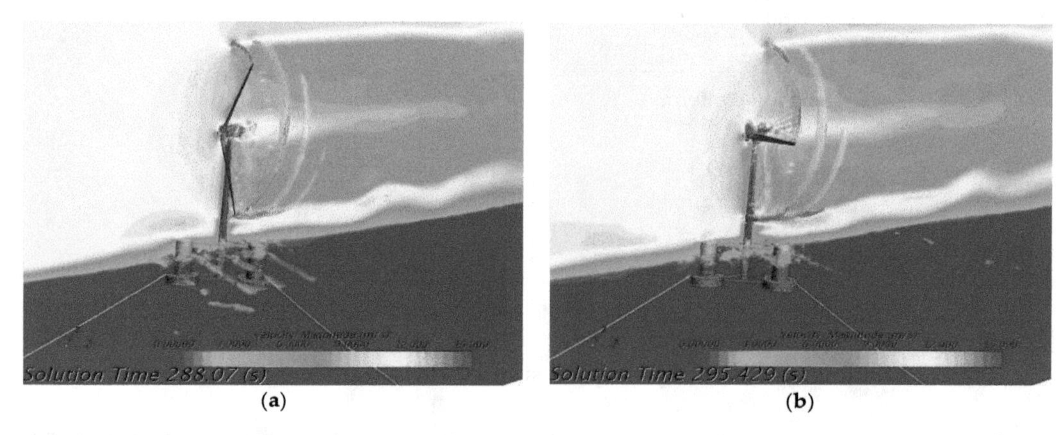

Figure 14. Instantaneous iso-velocity contours within one period: (**a**) upwind direction movement; (**b**) downwind direction movement.

The RAOs of surge, heave, and pitch in the fully coupled configuration under wind–wave excitation conditions are given in Table 7. Compared to the result of the regular wave test, where only regular waves exist, under a no-wind condition some discrepancies can be observed. The motion RAOs and the time-average values over the last four wave periods for 3-DOF were compared. Due to the unavailability of MARIN test data for the wind and wave conditions simulated herein, the comparison was only performed with the result from a previous study. All the 3-DOFs, i.e., heave, surge, and pitch showed small amplitudes compared to the results of the regular wave test without wind conditions. The incoming wind from the x-direction obviously has a significant effect on the restoring force of the mooring line; hence, the FOWT system cannot be restored to the equilibrium position as in the regular wave test. The incoming wind increased the aerodynamic thrust towards the floating system and pushed the platform further away in the backward direction, also leading to an increase in the mean surge. Nevertheless, the close agreement between the results for the 3-DOFs demonstrates the capability of this method.

Table 7. Motion response amplitude operators (RAOs) in different environment conditions.

Motion	Wave Only	Wind-Wave Coupled
Surge (m/m)	0.70	0.12
Heave (m/m)	0.36	0.13
Pitch (deg/m)	0.33	0.07

Thrust and power are two critical aerodynamic performance factors for evaluating a wind turbine. Thrust is defined as the integrated force component normal to the rotor plane. The power output and thrust force time-histories for the coupled simulation are presented in Figure 15 along with the dynamic responses of the pitch motion rate of the platform.

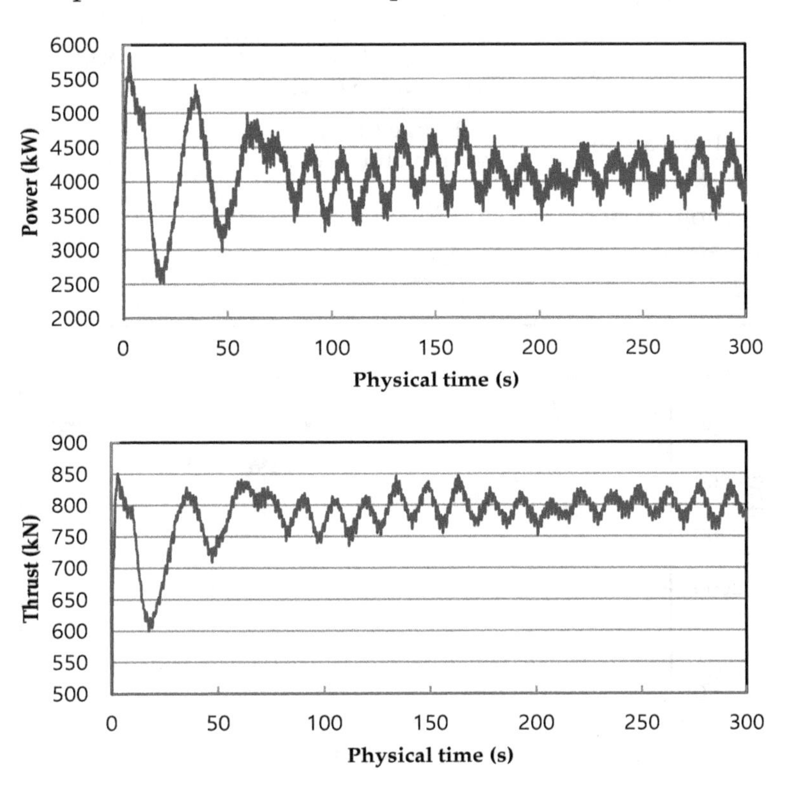

Figure 15. Fully coupled FOWT aerodynamic performance.

The response curves of power and thrust act at the same frequency as the incident wave. Due to the tower shade effect, the curves of power and thrust force exhibit periodical fluctuations with a period of 120° for the blade rotation. However, the effect of tower dam effects on the power output of the wind turbine is less than 5%. The variation of pitch motion also acts at the same frequency as the inlet wave. When the platform moves in the upwind direction, the power output and thrust force both increase, while the aerodynamic load decreases as the sign of the pitch motion changes. This is because the upwind pitch motion of the FOWT increases the relative velocity between the wind turbine and the inlet wind, and the angle of attack for each blade section increases correspondingly.

The dynamic responses of the wind turbine performance and typical platform motions after 300 s of the simulation are presented in Table 8. The power output varies from 3446 kW to 4698 kW at a rated wind speed. The variation in power is larger than that in the thrust force, that is, the power output is more sensitive than the thrust force to platform motion. Then, aerodynamic performances of the onshore fixed wind turbine and offshore floating wind turbine were compared and the average thrust value was calculated over the last four periods. In the case of thrust, a 7.8% increase was observed in a floating offshore turbine, the floating offshore wind turbine had an average thrust around 796 kN and the onshore wind turbine had an average thrust of 738 kN, which indicated a relatively small load on the hub and blades. This is because of the thrust force acting on the top of the tower, meaning the platform always moves in the upwind direction to offset the capsizing moment induced by the thrust force. In the case of the power curve, the average power value was calculated over the last four periods. A 10% decrease was observed in the floating offshore wind turbine, which is likely due to the smaller project area and relative income wind speed when the platform pitches, as shown in Figure 16. Accordingly, the platform surges from 7.9 to 9.8 m, which is also due to the thrust force that must be offset by the mooring line tension.

Table 8. Dynamic response of floating offshore wind turbines (FOWT) in the wind–wave condition.

	Parameters	Value
Power output	Range (kW)	3446–4689
	Mean value (kW)	4181
Thrust force	Range (kN)	759–838
	Mean value (kN)	801
Pitch angle	Range (deg)	4.5–5.3
	Mean value (deg)	4.9
Surge motion	Range (m)	7.9–9.8
	Mean value (m)	9.1
Heave motion	Range (m)	−0.7–0.7
	Mean value (m)	0.1

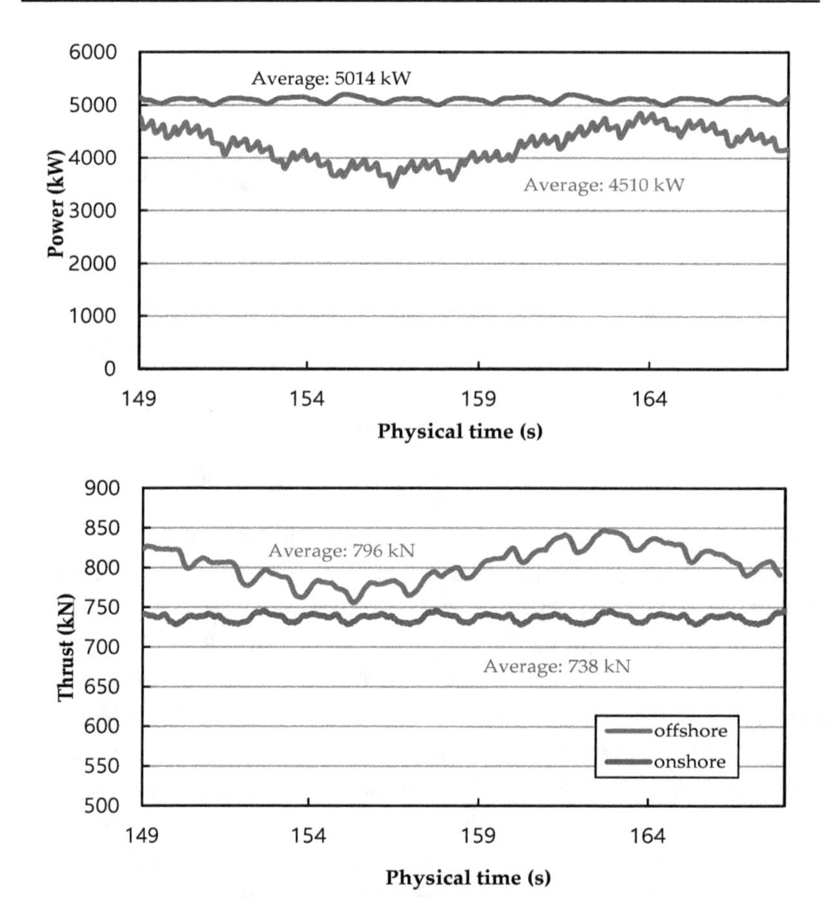

Figure 16. Comparison between onshore and offshore.

7. Conclusions

This investigation performed a CFD numerical analysis for a semi-submersible-type FOWT used in Phase II of the OC5 project, by advanced DFEI method and overlap mesh techno ledge. The full-configuration FOWT in a wind–wave excitation condition has been successfully performed, with simultaneous consideration of the wind turbine movement due to 6-DOF platform dynamics. The RAOs of the surge, heave, and pitch were compared to MARIN test data and data reported in previous studies when only the wave condition was considered. A slight discrepancy was observed between the CFD studies with regard to the pitch, possibly because of the different physical properties of the platform and mooring lines in the OC4 and OC5 projects. There was a relatively large discrepancy in the hydrodynamic response, which can induce large deviations in the prediction procedure. Particularly, pitch natural period showed a 21% discrepancy between OC4 and OC5 projects, as indicated by the results of the free decay test of the pitch and the numerical discrepancy in the RAOs in the regular wave test.

Besides, unsteady blade-tip vortices and strong flow interactions with the turbulent wakes of the tower due to the surge motion of the platform were successfully simulated and visualized using the advanced DFBI and VOF methods. The power and thrust force of the FOWT increased when the floating platform moved in the upwind direction, while the aerodynamic loads decreased as the pitch motion reversed direction. This can be explained by the variation in the angle of attack for each blade section when the FOWT system experiences pitch motion. All the 3-DOFs, including heave, surge, and pitch had smaller amplitudes compared with the results in the regular wave test without wind conditions. Incoming wind from the x-direction obviously has a large effect on the restoring force in the mooring line, and, as a result, the whole FOWT system cannot be restored back to the equilibrium position as in the regular wave test.

In addition, a relatively heavy load on the hub and blades was observed for the FOWT compared with the onshore wind turbine. This is because of the thrust force acting on the top of the tower, due to which the platform moves in the upwind direction to offset the capsizing moment induced by the thrust force. With regard to the power curve, a 10% decrease was observed for the floating offshore wind turbine, which is likely due to the smaller project area and relative income wind speed when the platform experiences pitch motion. Overall, there is a greater variation in the power than in the thrust force, that is, the power output is more sensitive than the thrust force to platform motions.

Until now, all published papers based on an OC4 project which was carried out in 2013 (this was code-to-code comparison project 5 years ago), which found a large discrepancy from the experimental test data of the OC5 project. This study could be a good insight for future studies, as there has not been any specific CFD research based on OC5 test data until now. Examination of the OC5 Phase II project, with a computational fluid dynamics code (which has a higher-fidelity model of the underlying physics), could help determine if there are some deficiencies in the hydrodynamic models being employed by participants in an OC5 code-to-test project [18].

Author Contributions: This research was supervised by B.K. All laboratory work was down by Y.Z.

References

1. Jonkman, J.M. Dynamics Modeling and Loads Analysis of an Offshore Floating Wind Turbine. Ph.D. Thesis, University of Colorado, Denver, CO, USA, 2007.
2. Sebastian, T.; Lackner, M.A. Development of a free vortex wake method code for offshore floating wind turbines. *Renew. Energy* **2012**, *46*, 269–275. [CrossRef]
3. Fernandez-Gamiz, U.; Zulueta, E.; Boyano, A.; Ansoategui, I.; Uriarte, I. Five Megawatt Wind Turbine Power Output Improvements by Passive Flow Control Devices. *Energies* **2017**, *10*, 742. [CrossRef]
4. Fernandez-Gamiz, U.; Errasti, I.; Gutierrez-Amo, R.; Boyano, A.; Barambones, O. Computational Modelling of Rectangular Sub-Boundary Layer Vortex Generators. *Appl. Sci.* **2018**, *8*, 138. [CrossRef]
5. Nematbakhsh, A.; Bachynski, E.E.; Gao, Z.; Moan, T. Comparison of wave load effects on a TLP wind turbine by using computational fluid dynamics and potential flow theory approaches. *Appl. Ocean Res.* **2015**, *53*, 142–154. [CrossRef]
6. De Vaal, J.B.; Hansen, M.O.L.; Moan, T. Effect of wind turbine surge motion on rotor thrust and induced velocity. *Wind Energy* **2014**, *17*, 105–121. [CrossRef]

7. Zhao, W.; Wan, D. Numerical study of interactions between phase II of OC4 wind turbine and its semi-submersible floating support system. *J. Ocean Wind Energy* **2015**, *2*, 45–53.
8. Tran, T.T.; Kim, D.H.; Song, J. Computational fluid dynamic analysis of a floating offshore wind turbine experiencing platform pitching motion. *Energies* **2014**, *7*, 5011–5026. [CrossRef]
9. Tran, T.T.; Kim, D.H. The coupled dynamic response computation for a semi-submersible platform of floating offshore wind turbine. *J. Wind Eng. Ind. Aerodyn.* **2015**, *147*, 104–119. [CrossRef]
10. Liu, Y.; Xiao, Q.; Incecik, A.; Wan, D.C. Investigation of the effects of platform motion on the aerodynamics of a floating offshore wind turbine. *J. Hydrodyn. Ser. B* **2016**, *28*, 95–101. [CrossRef]
11. Ren, N.; Li, Y.; Ou, J. Coupled wind-wave time domain analysis of floating offshore wind turbine based on Computational Fluid Dynamics method. *J. Renew. Sustain. Energy* **2014**, *6*, 1–13. [CrossRef]
12. Quallen, S.; Xing, T.; Carrica, P.; Li, Y.; Xu, J. CFD simulation of a floating offshore wind turbine system using a quasi-static crowfoot mooring-line model. In Proceedings of the Twenty-third International Offshore and Polar Engineering Conference, Anchorage, AK, USA, 30 June–5 July 2013; International Society of Offshore and Polar Engineers: Mountain View, CA, USA, 2013.
13. Tran, T.T.; Kim, D.H. Fully coupled aero-hydrodynamic analysis of a semisubmersible FOWT using a dynamic fluid body interaction approach. *Renew. Energy* **2016**, *92*, 244–261. [CrossRef]
14. Gueydon, S. Aerodynamic damping on a semisubmersible floating foundation for wind turbines. *Energy Procedia* **2016**, *94*, 367–378. [CrossRef]
15. Chen, J.; Hu, Z.; Wan, D.; Xiao, Q. Comparisons of the dynamical charateristics of a semi-submersible floating offshore wind turbine based on two different blade concepts. *Ocean Eng.* **2018**, *153*, 305–318. [CrossRef]
16. Dunbar, A.J.; Craven, B.A.; Paterson, E.G. Development and validation of a tightly coupled CFD/6-DOF solver for simulating floating offshore wind turbine platforms. *Ocean Eng.* **2015**, *110*, 98–105. [CrossRef]
17. Coulling, A.J.; Goupee, A.J.; Robertson, A.N.; Jonkman, J.M.; Dagher, D.J. Validation of a FAST semi-submersible floating wind turbine numerical model with DeepCwind test data. *J. Renew. Sustain. Energy* **2013**, *5*, 16–23. [CrossRef]
18. Robertson, A. OC5 Project Phase II: Validation of Global Loads of the DeepCwind Floating Semisubmersible Wind Turbine. *Energy Procedia* **2017**, *137*, 38–57. [CrossRef]
19. Menter, F.R. Two-Equation Eddy-Viscosity Turbulence Models for Engineering Applications. *AIAA J.* **1994**, *32*, 1598–1605. [CrossRef]
20. Tran, T.T.; Kim, D.H. A CFD study into the influence of unsteady aerodynamic interference on wind turbine surge motion. *Renew. Energy* **2016**, *90*, 204–228. [CrossRef]
21. Sivalingam, K.; Narasimalu, S. Floating Offshore Wind Turbine Rotor Operating State-Modified Tip Loss Factor in BEM and Comparison with CFD. *Int. J. Tech. Res. Appl.* **2015**, *3*, 179–189.
22. Hsu, M.C.; Bazilevs, Y. Fluid-Structure Interaction Modeling of Wind Turbines: Simulating the Full Machine. Ph.D. Thesis, Iowa State University, Ames, IA, USA, 2012.
23. Siddiqui, M.S.; Rasheed, A.; Kvamsdal, T. Quasi-Static & Dynamic Numerical Modeling of Full Scale NREL 5MW Wind Turbine. *Energy Procedia* **2017**, *137*, 460–467.
24. Liu, Y.; Xiao, Q.; Incecik, A.; Wan, D. Establishing a fully coupled CFD analysis tool for floating offshore wind turbines. *Renew. Energy* **2017**, *112*, 280–301. [CrossRef]
25. Tezdogan, T.; Demirel, Y.K.; Kellett, P.; Khorasanchi, M.; Incecik, A.; Turan, O. Full-scale unsteady RANS CFD simulations of ship behavior and performance in head seas due to slow steaming. *Ocean Eng.* **2015**, *97*, 186–206. [CrossRef]

12

Evaluation of Tip Loss Corrections to AD/NS Simulations of Wind Turbine Aerodynamic Performance

Wei Zhong [1], Tong Guang Wang [1], Wei Jun Zhu [2,*] and Wen Zhong Shen [3]

1 Jiangsu Key Laboratory of Hi-Tech Research for Wind Turbine Design, Nanjing University of Aeronautics and Astronautics, Nanjing 210016, China; zhongwei@nuaa.edu.cn (W.Z.); tgwang@nuaa.edu.cn (T.G.W.)
2 College of Electrical, Energy and Power Engineering, Yangzhou University, Yangzhou 225009, China
3 Department of Wind Energy, Technical University of Denmark, 2800 Lyngby, Denmark; wzsh@dtu.dk
* Correspondence: wjzhu@yzu.edu.cn

Abstract: The Actuator Disc/Navier-Stokes (AD/NS) method has played a significant role in wind farm simulations. It is based on the assumption that the flow is azimuthally uniform in the rotor plane, and thus, requires a tip loss correction to take into account the effect of a finite number of blades. All existing tip loss corrections were originally proposed for the Blade-Element Momentum Theory (BEMT), and their implementations have to be changed when transplanted into the AD/NS method. The special focus of the present study is to investigate the performance of tip loss corrections combined in the AD/NS method. The study is conducted by using an axisymmetric AD/NS solver to simulate the flow past the experimental *NREL Phase VI* wind turbine and the virtual *NREL 5MW* wind turbine. Three different implementations of the widely used Glauert tip loss function F are discussed and evaluated. In addition, a newly developed tip loss correction is applied and compared with the above implementations. For both the small and large rotors under investigation, the three different implementations show a certain degree of difference to each other, although the relative difference in blade loads is generally no more than 4%. Their performance is roughly consistent with the standard Glauert correction employed in the BEMT, but they all tend to make the blade tip loads over-predicted. As an alternative method, the new tip loss correction shows superior performance in various flow conditions. A further investigation into the flow around and behind the rotors indicates that tip loss correction has a significant influence on the velocity development in the wake.

Keywords: wind turbine aerodynamics; actuator disc; AD/NS; tip loss correction; blade element momentum

---

1. Introduction

Wind energy is nowadays an important and increasing source of electric power. It has been the biggest contributor of renewable electricity except for hydropower, sharing about 5.5% of the global electricity production in 2018 [1]. As a primary subject of wind turbine technology, aerodynamics [2] largely determines the efficiency of wind energy extraction of an individual wind turbine or a wind farm. Along with the extensive development of high quality wind resources onshore, there are several trends in wind power industry: A large number of newly installed wind turbines have to be installed in areas with lower wind speeds and complex terrain; the layout optimization of wind turbine array becomes very important for wind farm design; offshore wind power development is accelerating, and the rotor size is continuously increasing. These trends need to be supported by more advanced and refined aerodynamic tools. More accurate aerodynamic load prediction is required for the design of a new generation of wind turbines with high efficiency and relatively low weight. Furthermore, a wind farm with dozens of wind turbines needs to be studied as a whole in order to take into account the complex interference between wind turbines.

Blade Element Momentum Theory (BEMT) [3,4] and Computational Fluid Dynamics (CFD) [5,6] are two essential methods of wind turbine aerodynamic computation. BEMT is no doubt the key method for rotor design [7], while CFD gives the most refined data of aerodynamic loads and flow parameters. A full CFD simulation with resolved rotor geometry is usually employed for an individual wind turbine [8,9]. However, full CFD is not suitable for a wind turbine array, due to the huge computational cost. The Actuator Disc/Navier-Stokes (AD/NS) method [10–13] was developed for this situation, in which the rotor geometry is not resolved, and thus, the number of mesh cells is greatly reduced. AD/NS is based on a combination of the blade-element theory and CFD. The flow field is still solved by CFD, while the rotor entity is replaced by a virtual actuator disc on which an external body force is acted. If the blade entities are represented by virtual actuator lines, it is called the Actuator Lines/Navier-Stokes (AL/NS) method [14–17]. As compared to AD/NS, the flow around the rotor solved by AL/NS is closer to the reality, since the rotor vortices are simulated. However, AL/NS requires more grid cells to describe the actuator lines, and thus, is seldom employed in wind farm simulations.

The AD/NS method has played a significant role in wind farm simulations involving wake interaction [18,19], complex terrain [20,21], atmospheric boundary layer [22,23], noise propagation [24,25], etc. Nevertheless, the AD model assumes that the number of blades is infinite, and thus, no tip loss is simulated, causing an over-prediction of the blade tip loads and power output. In order to take tip loss into account, a reliable engineering model has to be embedded into the numerical solver, which is usually called tip loss correction. However, all tip loss corrections were originally proposed for BEMT, and there is no tip loss correction specially developed for AD/NS. Most of the literature about AD/NS simulations either did not mention tip loss or declared that the Glauert tip loss correction [26] was employed. It is worth mentioning that AD/NS and BEMT solve the momentum of the flow past the rotor by using two completely different approaches. The former employs CFD, while the latter applies the momentum theory, though they share the actuator disc assumption and the blade-element theory. That leads to different implementations of tip loss correction for the two methods. In the BEMT, tip loss correction is realized by applying a correction factor F into the induced velocity through the rotor. In the AD/NS, the velocity is naturally found by the NS solver, and factor F can only be used to modify the external body force. The question is whether a tip loss correction has a good global performance when it is transplanted into AD/NS. Even for the BEMT itself, evaluations indicate that accurate tip loss correction is still not well achieved [27], and the development of new correction models is still going on [28–33]. In contrast with the massive study in the BEMT, tip loss corrections applied to AD/NS lack a comprehensive evaluation.

In the present work, the 2-Dimensional (2D) axisymmetric AD model is employed, and steady-state simulations are performed. The code solves the incompressible axisymmetric NS equations for the experimental *NREL Phase VI* wind turbine [34] and the virtual *NREL 5MW* [35] wind turbine under various axial inflow conditions. The main purpose of the numerical study is to evaluate the performance of the tip loss corrections applied to AD/NS. Three different implementations of the Glauert tip loss correction are discussed. In addition, a new tip loss correction recently proposed by Zhong et al. [33] is introduced and compared. The normal and tangential forces of blade cross-sections are chosen as the key parameters for the present evaluation. Because of the long arm of force, computational errors of the forces acted on the tip region are most likely to cause non-ignorable errors of the blade bending moment and the rotor torque (power generation). A BEMT study on the *NREL Phase VI* wind turbine by Branlard [27] showed that the power generation would be overestimated by 15% if no tip loss correction was made and a deviation of about 5% exists when various existing tip loss corrections were applied. That highlights the significance of the present study on tip loss correction.

The innovation of the present study lies in the following items. (a) Different implementations of the Glauert tip loss factor F are gathered, discussed and compared, and finally, one of them is recommended according to its best performance. Such kind of work has never been reported in the existing literature. (b) The Glauert correction is found to have a similar performance when it is transplanted from BEMT to AD/NS. (c) The new tip loss correction of Zhong et al. [33] is for the first time employed in AD/NS simulations. It is found to be generally superior to the Glauert-type corrections. That provides an alternative choice for a more accurate prediction of the blade tip loads. (d) Tip loss correction is found to have a significant influence on the velocity field, which highlights the importance of an accurate tip loss correction for not only the blade loads, but also the wake development.

The paper is organized as follows: In Section 2, the axisymmetric AD/NS method is introduced, including the governing equations of the NS approach and the formulae of the AD model; In Section 3, the Glauert and the new tip loss corrections are introduced; In Section 4, the implementations of the tip loss corrections used in AD/NS are described; In Section 5, the numerical setup, as well as the involved simulation cases, are introduced; In Section 6, all simulation results are presented and discussed; Final conclusions are made in the last section.

2. Axisymmetric AD/NS Method

2.1. Governing Equations

The governing equations of the present AD/NS simulation are the incompressible axisymmetric NS equations. In a cylindrical coordinate system as shown in Figure 1, the axial direction is defined along the z-axis, the radial direction is represented by r, and the tangential direction is represented by θ, the continuity equation is written as

$$\nabla \cdot \vec{u} = \frac{\partial u_z}{\partial z} + \frac{\partial u_r}{\partial r} + \frac{u_r}{r} = 0, \tag{1}$$

the axial and radial momentum equations read

$$\frac{\partial u_z}{\partial t} + \frac{1}{r}\frac{\partial}{\partial z}(ru_z^2) + \frac{1}{r}\frac{\partial}{\partial r}(ru_r u_z) = -\frac{1}{\rho}\frac{\partial p}{\partial z} + \frac{1}{r}\frac{\partial}{\partial z}\left[r\nu\left(2\frac{\partial u_z}{\partial z} - \frac{2}{3}(\nabla \cdot \vec{u})\right)\right] + \frac{1}{r}\frac{\partial}{\partial r}\left[r\nu\left(\frac{\partial u_z}{\partial r} + \frac{\partial u_r}{\partial z}\right)\right] + f_z, \tag{2}$$

$$\begin{aligned} &\tfrac{\partial u_r}{\partial t} + \tfrac{1}{r}\tfrac{\partial}{\partial z}(ru_z u_r) + \tfrac{1}{r}\tfrac{\partial}{\partial r}(ru_r^2) = \\ &-\tfrac{1}{\rho}\tfrac{\partial p}{\partial r} + \tfrac{1}{r}\tfrac{\partial}{\partial z}\left[r\nu\left(\tfrac{\partial u_r}{\partial z} + \tfrac{\partial u_z}{\partial r}\right)\right] + \tfrac{1}{r}\tfrac{\partial}{\partial r}\left[r\nu\left(2\tfrac{\partial u_r}{\partial r} - \tfrac{2}{3}(\nabla \cdot \vec{u})\right)\right] - 2\nu\tfrac{u_r}{r^2} + \tfrac{2}{3}\tfrac{\nu}{r}(\nabla \cdot \vec{u}) + \tfrac{u_\theta^2}{r} + f_r \end{aligned} \tag{3}$$

and the tangential momentum equation is

$$\frac{\partial u_\theta}{\partial t} + \frac{1}{r}\frac{\partial}{\partial z}(ru_z u_\theta) + \frac{1}{r}\frac{\partial}{\partial r}(ru_r u_\theta) = \frac{1}{r}\frac{\partial}{\partial z}\left(r\nu\frac{\partial u_\theta}{\partial z}\right) + \frac{1}{r^2}\frac{\partial}{\partial r}\left[r^3\nu\frac{\partial}{\partial r}\left(\frac{u_\theta}{r}\right)\right] - \frac{u_r u_\theta}{r} + f_\theta. \tag{4}$$

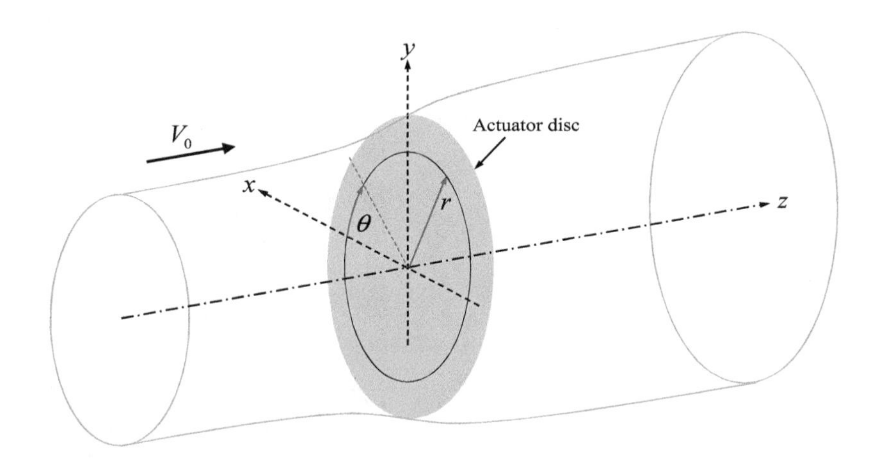

Figure 1. Definition of coordinates and the stream tube through an actuator disc.

In the above equations, $u_z/u_r/u_\theta$ is the axial/radial/tangential velocity, p is the static pressure, ρ is the air density, ν is the kinematic viscosity coefficient of air, f_z, f_r, f_θ are the source term of the axial, radial, tangential external body forces, respectively.

2.2. Force on Actuator Disc

The conceptual idea of the AD/NS method is to solve the aerodynamic force of rotor blades by using the blade-element theory and then applying its counterforce as an external body force into the momentum equations.

At each time step of solving the NS equations, the velocity passing through the actuator disc is detected and used to determine the axial induced velocity W_a and the tangential induced velocity W_t,

$$W_a = V_0 - u_z, \tag{5}$$

$$W_t = -u_\theta. \tag{6}$$

The axial interference factor a and tangential interference factor a' are then determined by,

$$a = \frac{W_a}{V_0}, \tag{7}$$

$$a' = \frac{W_t}{\Omega r}, \tag{8}$$

where V_0 is the wind speed, and Ω is the rotating speed of the wind turbine rotor.

The above interference factors are azimuthally unchanged according to the axisymmetric condition, which implies an infinite number of blades and zero tip loss. In order to estimate the tip loss of the realistic rotor with a finite number of blades, the interference factors need to be corrected as

$$\tilde{a} = f_{corr}(a), \tag{9}$$

$$\tilde{a}' = f'_{corr}(a'), \tag{10}$$

where $\tilde{a}$ and $\tilde{a}'$ denote the corrected axial and tangential interference factors, f_{corr} and f'_{corr} represent correction functions.

The axial and tangential induced velocities after the correction are written as

$$\tilde{W}_a = V_0\tilde{a}, \tag{11}$$

$$\tilde{W}_t = \Omega r\tilde{a}'. \tag{12}$$

According to the velocity triangle shown in Figure 2, the flow angle ϕ, angle of attack α, and relative velocity V_{rel} are then determined by

$$\phi = \tan^{-1}\left(\frac{V_0 - \tilde{W}_a}{\Omega r + \tilde{W}_t}\right), \tag{13}$$

$$\alpha = \phi - \beta, \tag{14}$$

$$V_{rel}^2 = \left(V_0 - \tilde{W}_a\right)^2 + \left(\Omega r + \tilde{W}_t\right)^2, \tag{15}$$

where β is the local pitch angle of a blade cross-section.

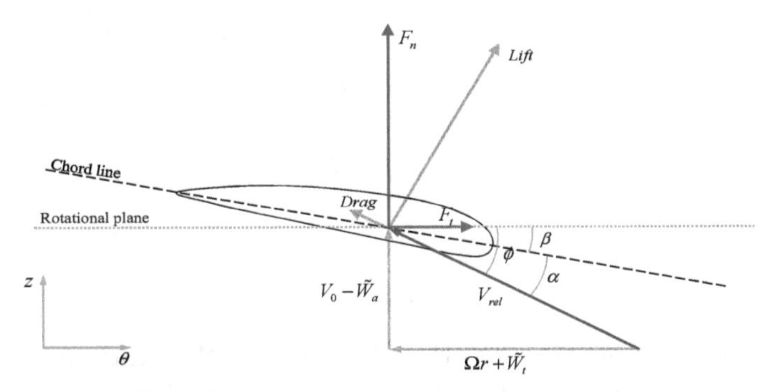

Figure 2. Illustration of the velocity and the aerodynamic force for a blade cross-section.

With the determined angle of attack and relative velocity, the lift coefficient C_l and drag coefficient C_d of each cross-section of the blade can be obtained from tabulated airfoil data. According to the relationship of force projection, the normal and tangential force coefficients are determined by

$$C_n = C_l \cos\phi + C_d \sin\phi, \tag{16}$$

$$C_t = C_l \sin\phi - C_d \cos\phi, \tag{17}$$

The axial force F_z and tangential force F_θ per radial length of the rotor are then given by

$$F_z = \frac{1}{2}\rho V_{rel}^2 cBC_n = \frac{1}{2}\rho V_{rel}^2 cB(C_l \cos\phi + C_d \sin\phi), \tag{18}$$

$$F_\theta = \frac{1}{2}\rho V_{rel}^2 cBC_t = \frac{1}{2}\rho V_{rel}^2 cB(C_l \sin\phi - C_d \cos\phi), \tag{19}$$

where c is the local chord length, and B is the number of blades. Ignoring the effect of the radial flow, the vector form of the counterforce acting on the air is

$$\vec{F} = (-F_z, 0, -F_\theta). \tag{20}$$

Rather than distributing the force only on the disc, the above force is regularized by the following Gaussian distribution along the axial direction in order to avoid a numerical singularity. The Gaussian distribution has been widely used and proven to be proper for AD/NS and AL/NS simulations [10,14–17].

$$\vec{F}_\varepsilon = \eta_\varepsilon \vec{F},\ \eta_\varepsilon = \frac{1}{\varepsilon\sqrt{\pi}} \exp\left[-\left(\frac{z - z_0}{\varepsilon}\right)^2\right], \tag{21}$$

where z_0 is the axial position of the disc. The parameter ε serves to adjust the concentration of the regularized force, which is in the present study set to $\varepsilon = 0.02R$ where R is the rotor radius.

The source terms of the external body force in the momentum Equations (2)–(4) need to be replaced with the above force per unit volume. In the axisymmetric coordinate system, a 2D grid cell with an axial length of Δz and a radial height of Δr represents a volume of $2\pi r \Delta r \Delta z$, and thus, the force per unit volume is

$$\vec{f} = \frac{\vec{F}_\varepsilon \Delta r \Delta z}{2\pi r \Delta r \Delta z} = \frac{\vec{F}_\varepsilon}{2\pi r}, \tag{22}$$

Using Equations (20) and (21), it reads

$$\vec{f} = \frac{\eta_\varepsilon \vec{F}}{2\pi r} = \left(-\frac{\eta_\varepsilon F_z}{2\pi r}, 0, -\frac{\eta_\varepsilon F_\theta}{2\pi r}\right), \tag{23}$$

i.e.,

$$\begin{cases} f_z = -\frac{\eta_\varepsilon F_z}{2\pi r} \\ f_r = 0 \\ f_\theta = -\frac{\eta_\varepsilon F_\theta}{2\pi r} \end{cases}. \tag{24}$$

3. Tip Loss Corrections

3.1. Glauert Correction

The tip loss correction of Glauert [26] was developed from the study of Prandtl [36]. A function F, which was later recognized as the first tip loss factor, was derived by Prandtl making Betz's optimal circulation [37] go to zero at the blade tip,

$$F = \frac{2}{\pi}\arccos\left\{\exp\left[-\frac{B}{2}\left(1-\frac{r}{R}\right)\sqrt{1+\lambda^2}\right]\right\}, \tag{25}$$

where B is the number of blades, r is the local radial location, R is the rotor radius, and λ is the tip speed ratio. The original derivation of Prandtl was written very briefly and was elaborated more in [4,27,38] as for a more detailed derivation.

Later on, Glauert made further contributions. First, he interpreted the physical meaning of factor F as the ratio between the azimuthally averaged induced velocity and the induced velocity at the blade position, leading to

$$F = \frac{\bar{a}}{a_B} = \frac{\overline{a'}}{a'_B}, \tag{26}$$

where $\bar{a} = \frac{1}{2\pi}\int_0^{2\pi} a\mathrm{d}\theta$ and $\overline{a'} = \frac{1}{2\pi}\int_0^{2\pi} a'\mathrm{d}\theta$ are the azimuthally averaged axial and tangential interference factors, and a_B and a'_B are the interference factors at the blade position. Second, the local inflow angle ϕ was introduced into the function in order to make it consistent with the local treatment of the BEMT, leading to the following new formula of factor F,

$$F = \frac{2}{\pi}\arccos\left\{\exp\left[-\frac{B(R-r)}{2r\sin\phi}\right]\right\}. \tag{27}$$

The factor F was then applied into the momentum theory through a straightforward way of multiplying the axial velocity at the rotor plane with factor F, resulting in the following thrust and torque for an annular element,

$$\mathrm{d}T = 4\pi r\rho V_0^2 a(1-a)F\mathrm{d}r, \tag{28}$$

$$\mathrm{d}M = 4\pi r^3\rho V_0\Omega a'(1-a)F\mathrm{d}r. \tag{29}$$

Using another two equations of $\mathrm{d}T$ and $\mathrm{d}M$ derived from the blade-element theory,

$$\mathrm{d}T = \frac{1}{2}\rho V_{rel}^2 cC_n B\mathrm{d}r = \frac{1}{2}\rho V_{rel}^2 c(C_l\cos\phi + C_d\sin\phi)B\mathrm{d}r, \tag{30}$$

$$\mathrm{d}M = \frac{1}{2}\rho V_{rel}^2 cC_t Br\mathrm{d}r = \frac{1}{2}\rho V_{rel}^2 c(C_l\sin\phi - C_d\cos\phi)Br\mathrm{d}r, \tag{31}$$

and the velocity triangle at the rotor plane, the equations for the interference factors in the BEMT approach was derived to be:

$$\frac{a}{1-a} = \frac{\sigma(C_l\cos\phi + C_d\sin\phi)}{4F\sin^2\phi}, \tag{32}$$

$$\frac{a'}{1+a'} = \frac{\sigma(C_l\sin\phi - C_d\cos\phi)}{4F\sin\phi\cos\phi}. \tag{33}$$

The above two equations lead to the final iterative formulae of the BEMT approach with the Glauert tip loss correction. Their difference from the baseline BEMT approach is only the appearance of factor F in the equations. Obviously, the application of the Glauert correction in BEMT is very simple, which is also a great advantage. There are several variations of the Glauert correction in which the way of applying factor F is different [39,40], but the original Glauert tip loss correction is the most commonly used form till today.

3.2. A Newly Developed Correction

A new tip loss correction was recently proposed for BEMT by Zhong et al. [33], based on a novel insight into tip loss. In contrast with the Prandtl/Glauert series corrections that begin with an actuator disc and estimate the effect of the finite number of blades, the new correction begins with a non-rotating blade and estimates the effect of rotation on tip loss. It has been validated in BEMT computations and showed superior performances in the cases involving flow separation or high axial interference factor.

The correction was realized by using two factors of F_R and F_S that treat the rotational effect and the 3D effect, respectively.

$$F_R = 2 - \frac{2}{\pi}\arccos\left\{\exp\left[-2B(1-r/R)\sqrt{1+\lambda^2}\right]\right\}, \tag{34}$$

$$F_S = \frac{2}{\pi}\arccos\left\{\exp\left[-\left(\frac{1-r/R}{\bar{c}/R}\right)^{3/4}\right]\right\},\ \bar{c} = \frac{S_t}{R-r}, \tag{35}$$

where $\bar{c}$ denotes a geometric mean chord length, and S_t is the projected area of the blade between the present cross-section and the tip. The purpose of introducing $\bar{c}$ is to deal with tapered blades and those with sharp tips.

The factor F_R was applied to the BEMT approach by using:

$$\frac{aF_R(1-aF_R)}{(1-a)} = \frac{\sigma\left(\tilde{C}_l\cos\phi + \tilde{C}_d\sin\phi\right)}{4\sin^2\phi}, \tag{36}$$

$$\frac{a'F_R(1-aF_R)}{(1+a')(1-a)} = \frac{\sigma\left(\tilde{C}_l\sin\phi - \tilde{C}_d\cos\phi\right)}{4\sin\phi\cos\phi}, \tag{37}$$

in which the employed lift and drag coefficients were corrected by factor F_S, rather than the direct use of the airfoil data.

$$\tilde{C}_l = \frac{1}{2}\left[C_l(\alpha)F_S + C_l^*\right], \tag{38}$$

$$\tilde{C}_d = \frac{1}{\cos^2\alpha_i}\left[C_d(\alpha_e)\cos\alpha_i + \left(\tilde{C}_l\cos\alpha_i + C_d(\alpha_e)\tan\alpha_i\right)\sin\alpha_i\right]; \tag{39}$$

where

$$C_l^* = \frac{1}{\cos^2\alpha_i}[C_l(\alpha_e)\cos\alpha_i - C_d(\alpha_e)\sin\alpha_i], \tag{40}$$

$$\alpha_i = \frac{C_l(\alpha)}{m}(1-F_S), \tag{41}$$

where m is the slope of the lift-curve of the airfoil before flow separation, α_i is called the downwash angle, and α_e is called the effective angle of attack.

By comparing with the Glauert tip loss correction, the new tip loss correction appears more complicated in application. It is simplified in the present study: First, Equations (36) and (37) will not be employed naturally in the AD/NS method; Second, Equations (38)–(41) are further simplified to

$$\tilde{C}_l = \frac{1}{2}[C_l(\alpha)F_S + C_l(\alpha_e)],\ \alpha_e = \alpha - \alpha_i, \tag{42}$$

$$\tilde{C}_d = C_d(\alpha_e) + \tilde{C}_l \tan \alpha_i, \ \alpha_i = \frac{C_l(\alpha)}{m}(1 - F_S). \quad (43)$$

The simplification is derived by using the fact that α_i and C_d are relatively small. More detailed applications are shown in the next section.

4. Applying Corrections to AD/NS Simulation

4.1. Application of Glauert Tip Loss Factor F

Equations (32) and (33) where the Glauert tip loss factor F is applied are not employed in the AD/NS method because the flow is simulated by the NS solver instead of the momentum theory. As a result, an alternative way has to be employed correctly to apply the tip loss factor F. Nevertheless, among the literature studies, little literature mentions the detail of how the tip loss factor is applied to the AD/NS simulations. After an extensive literature review, we have found three representative documents in which Sørensen et al. [41], Mikkelsen [42], and Shen et al. [43] described their implementations. In order to facilitate the distinction, the implementations adopted by the three studies are denoted as Glauert-A, Glauert-B and Glauert-C in the present paper, respectively.

4.1.1. Glauert-A Correction

Sørensen et al. [41] performed a tip loss correction by applying factor F to modify the aerodynamic force that determines the external body force in the NS equations. They replaced the lift coefficient C_l by C_l/F (there was no need to modify the drag in their study as the drag was assumed not to produce the external body force). In the present Glauert-A correction, C_n and C_t, in Equations (18) and (19), are replaced by:

$$\tilde{C}_n = C_n/F, \quad (44)$$

$$\tilde{C}_t = C_t/F. \quad (45)$$

That is equivalent to replacing C_l by C_l/F and C_d by C_d/F, according to Equations (16) and (17).

Sørensen et al. [41] did not explain the reason why factor F could be directly used to modify the force. They might be inspired by Equations (32) and (33) in which the existence of F can be looked as corrections to C_n and C_t, although from a physical point of view it is a correction to the interference factors. It is important to note that the corrected force is only used for determining the external body force, so as the flowfield, while the force acting on the blade should be regarded as the original one. In addition, the interference factors in Equations (9) and (10) should no longer be corrected, leading to $\tilde{a} = a$ and $\tilde{a}' = a'$.

This implementation involves a division operation with denominator F, which causes unreasonable big values of $\tilde{C}_n$ and $\tilde{C}_t$ at the extreme tip where $F \to 0$. However, there is no exact criterion for defining what a big value is unreasonable because the $\tilde{C}_n$ and $\tilde{C}_t$ themselves are introduced as an engineering correction rather than a physical concept. Sørensen et al. [41] did not mention this problem, and no obvious numerical fluctuation is observed in his result. That is possible because the problem is limited to a very small area at the tip, and thus, the integral effect of the resulted unreasonable body force is also small.

4.1.2. Glauert-B Correction

Mikkelsen [42] compared Equations (32) and (33) to the corresponding baseline equations without factor F. The baseline equations are

$$\frac{a}{1-a} = \frac{\sigma(C_l \cos\phi + C_d \sin\phi)}{4\sin^2\phi}, \quad (46)$$

$$\frac{a'}{1+a'} = \frac{\sigma(C_l \sin\phi - C_d \cos\phi)}{4\sin\phi\cos\phi}. \quad (47)$$

In order to distinguish from the parameters in the baseline equations, we here rewrite Equations (32) and (33) to

$$\frac{F\tilde{a}}{(1-\tilde{a})} = \frac{\sigma\left(\tilde{C}_l \cos\tilde{\phi} + \tilde{C}_d \sin\tilde{\phi}\right)}{4\sin^2\tilde{\phi}}, \tag{48}$$

$$\frac{F\tilde{a}'}{(1+\tilde{a}')} = \frac{\sigma\left(\tilde{C}_l \sin\tilde{\phi} - \tilde{C}_d \cos\tilde{\phi}\right)}{4\sin\tilde{\phi}\cos\tilde{\phi}}. \tag{49}$$

The implementation of Mikkelsen [42] adopts the following equations which implies an assumption of $\phi = \tilde{\phi}$ which in fact does not accurately hold,

$$\frac{a}{1-a} = \frac{\sigma(C_l \cos\phi + C_d \sin\phi)}{4\sin^2\phi} = \frac{F\tilde{a}}{(1-\tilde{a})}, \tag{50}$$

$$\frac{a'}{1+a'} = \frac{\sigma(C_l \sin\phi - C_d \cos\phi)}{4\sin\phi\cos\phi} = \frac{F\tilde{a}'}{(1+\tilde{a}')}. \tag{51}$$

The corrected interference factors were, thus, determined by

$$\tilde{a} = \frac{a}{F(1-a)+a}, \tag{52}$$

$$\tilde{a}' = \frac{a'}{F(1+a')-a'}. \tag{53}$$

The tip loss correction was completed as the above functions for $\tilde{a}$ and $\tilde{a}'$ were used to replace Equations (9) and (10).

4.1.3. Glauert-C Correction

Shen et al. [43] performed their correction by directly using Equation (26) which represents the physical meaning of factor F. Using the azimuthally uniform condition ($\overline{a} = a$), the correction was completed by replacing Equations (9) and (10) with

$$\tilde{a} = a_B = a/F, \tag{54}$$

$$\tilde{a}' = a'_B = a'/F. \tag{55}$$

Similar to the Glauert-A correction, this implementation involves a division operation with denominator F. That causes an unphysical big value of $\tilde{a}$ at the extreme tip where $F \rightarrow 0$. A limiter is set in the present study to force the result to be $\tilde{a} = 1$ when it is greater than 1. The value of $\tilde{a}'$ is usually much less than $\tilde{a}$ and is not limited here.

4.2. Application of New Correction

The application of the new correction consists of two steps: The first is using factor F_R determined by Equation (34) and the second is using factor F_S determined by Equation (35).

In the first step, an implementation similar to the Glauert-C correction is adopted, which is performed by replacing Equations (9) and (10) with

$$\tilde{a} = a/F_R, \tag{56}$$

$$\tilde{a}' = a'/F_R. \tag{57}$$

A limiter of $\tilde{a} = 1$ is used when a is greater than 1. The angle of attack α can then be determined by calculations using Equations (11)–(14).

The second step is to correct the lift and drag coefficients by using Equations (42) and (43), the result of which is used to replace the C_l and C_d in Equations (18) and (19).

5. Computational Setup

5.1. Flow Solver and Mesh Configuration

The 2D axisymmetric NS equations are solved by using an in-house code EllipSys2D [44,45] developed at Technical University of Denmark (DTU). The code is a general incompressible flow solver with multi-block and multi-grid strategy. The equations are discretized with a second-order finite volume method. In the spatial discretization, a central difference scheme is applied to the diffusive terms and the QUICK (Quadratic Upstream Interpolation for Convective Kinematics) upwind scheme is applied to the convective terms. The SIMPLEC (Semi-Implicit Method for Pressure-Linked Equations) scheme is used for the velocity-pressure decoupling. The turbulence flow is simulated using the method of Reynolds-averaged Navier-Stokes equations (RANS) in which the k-ω turbulence model of Menter [46] is employed with a modification for a better simulation of the turbulence quantities in the free-stream flow [47,48].

The coordinate for the AD/NS simulations is defined in the z-y plane. A computational mesh is generated, as shown in Figure 3 where the inflow, outflow and axisymmetric boundaries are indicated, the length of the computational domain is 30, which is nondimensionalized with the rotor radius. The blade is positioned at $z = 0$ and the grid cells are clustered around $z = 0$ and $y = 1$ to ensure a better resolution near the blade tip. The mesh is composed of four blocks where 64×64 grid points are used for each block. There are 64 grid points on the AD along the radial direction, and 20 points in the range of [-0.02, 0.02] in the axial direction where the Gaussian distribution of the body force plays a significant role. Since the axisymmetric boundary condition is applied, the 2D flow solution is regarded as an azimuthal slice of a full 3D field.

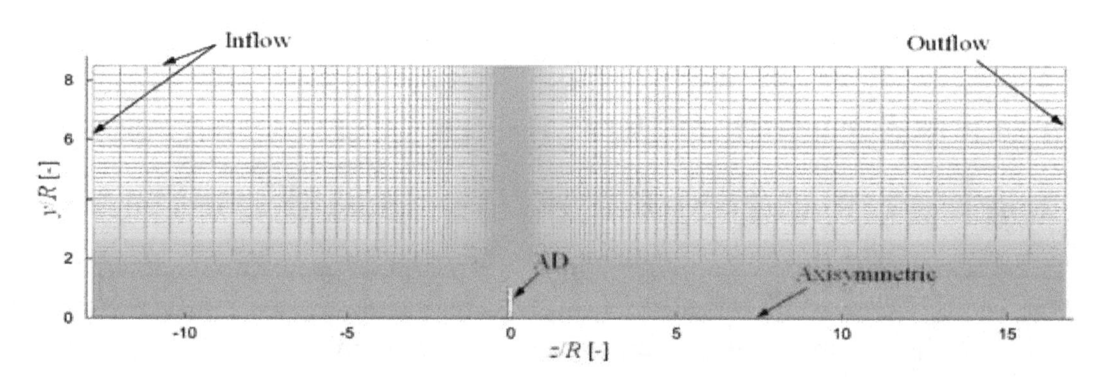

Figure 3. Mesh configuration for the Actuator Disc/Navier-Stokes (AD/NS) simulations.

The above computational setup has been proven to perform well in AD/NS simulations, as shown in the previous work of Cao et al. [49] where both blade loads and wake flows were validated against experiments.

5.2. Simulation Cases

Simulations are performed for two different wind turbines of the *NREL Phase VI* [34] and the *NREL 5MW* [35] that represent a small and a large rotor size, respectively. Additionally, the *NREL Phase VI* rotor has a very blunt blade tip shape as compared with the *NREL 5MW* rotor.

The operational conditions of the two rotors are listed in Table 1. Cases 1, 2 and 3 are set to be consistent with the *NREL UAE* experiment in axial inflow conditions [34]. Various wind speeds of 7 m/s, 10 m/s and 13 m/s are considered, while the rotating speed remains unchanged. The flow is fully attached on the blade surface at a wind speed of 7 m/s, but is partly separated at 10 m/s and 13 m/s (Higher wind speed leads to heavier flow separation) [8]. Measured pressure distributions (from which the force can be obtained by pressure integral) at five blade sections (r/R = 0.30, 0.47, 0.63, 0.80, 0.95) are available for these cases [50]. Cases 4 and 5 are two typical points on the designed power curve of the *NREL 5MW* wind turbine [35]. The wind speed and rotating speed of Case 5 are the rated parameters of this wind turbine. Case 4 represents a condition with a lower wind speed of 8 m/s at which the rotating speed is reduced to 9.22 rpm for tracking the optimum tip speed ratio.

Table 1. Simulation cases for two different wind turbines.

Rotor Name	Number	Wind Speed (V_0, m/s)	Rotating Speed (Ω, rpm)	Tip Speed Ratio (λ)	Tip Pitch Angle (°)
NREL Phase VI	1	7.0	72	5.4	3
	2	10.0	72	3.8	3
	3	13.0	72	2.9	3
NREL 5MW	4	8.0	9.22	7.60	0
	5	11.4	12.06	6.98	0

These cases cover the following multiple situations: Wind turbines with remarkably different rotor sizes and tip shapes; flow with fully attached and separated conditions; operating conditions with various wind speeds and rotating speeds. It is, therefore, interesting to see the performance of the tip loss corrections in these cases.

Experimental data are preferred as the reference for the computational results of the *NREL Phase VI* rotor, as well as the full CFD data are also displayed in some results. For the cases of the *NREL 5MW* rotor, there is no experimental data available such that the full CFD data are the only reference. The related full CFD simulations were previously conducted by the authors, see Zhong et al. [8,33].

6. Simulation Results

6.1. Results of Glauert-Type Corrections

6.1.1. Glauert-A/B/C Corrections

The blade loads for Cases 1, 2 and 3 are gathered in Figure 4, which represent results obtained from the *NREL Phase VI* rotor. The *NREL Phase VI* blade has a linear change of chord distribution starting from r = 1.3 m with a constant slope of 0.1. The chord length is 0.358 m at the tip, which is about 48% of the largest chord length in the blade inboard part. Therefore, the loads do not converge to zero without a tip loss correction.

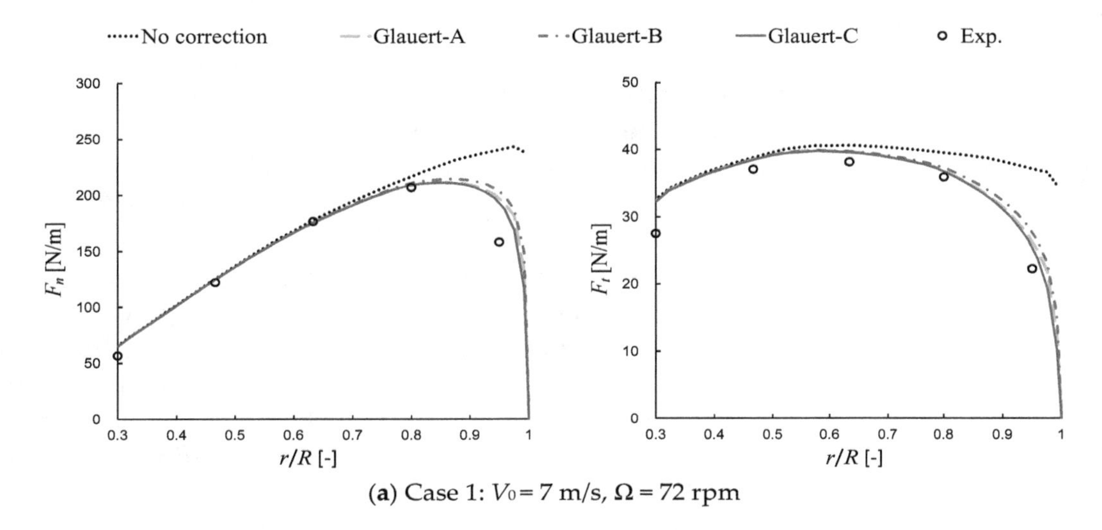

(**a**) Case 1: V_0 = 7 m/s, Ω = 72 rpm

Figure 4. *Cont.*

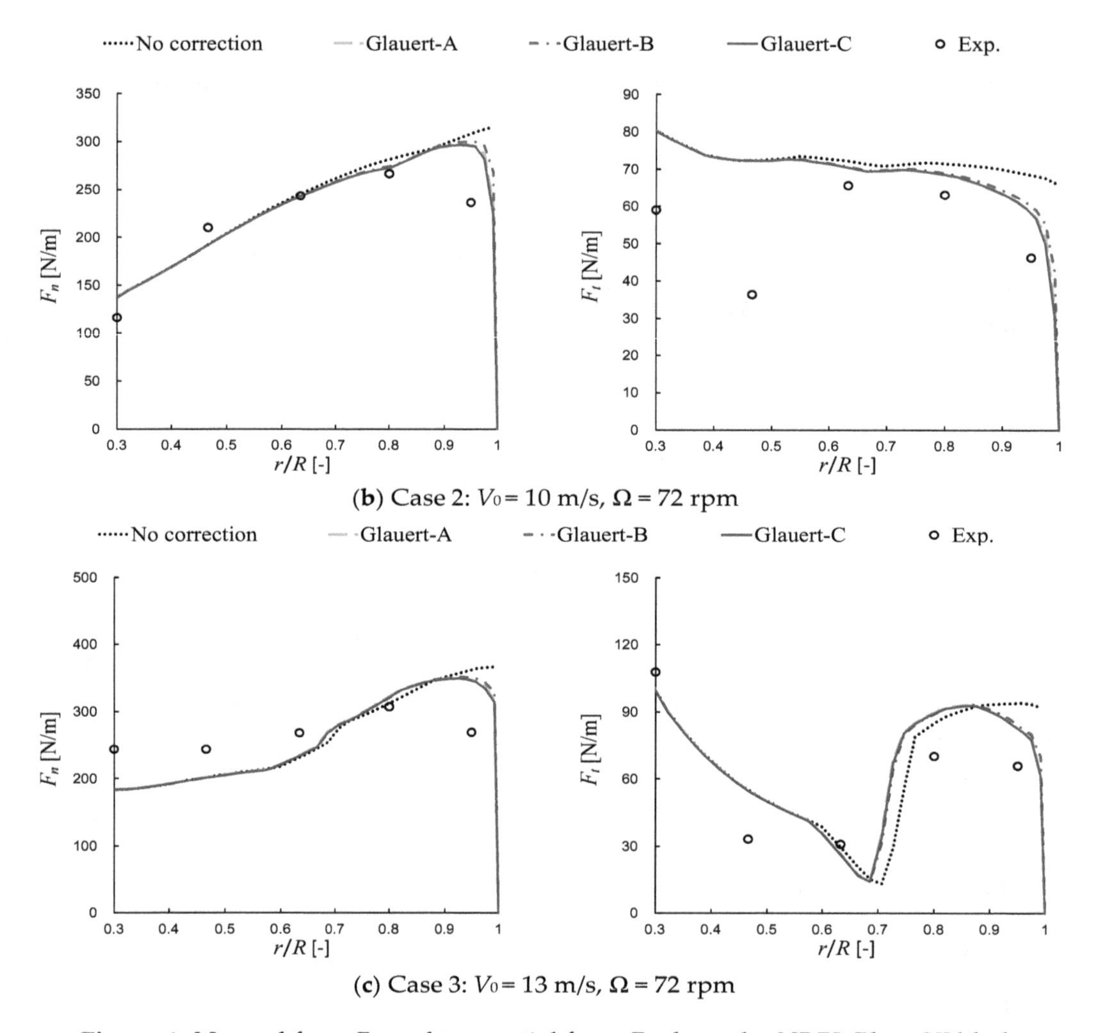

Figure 4. Normal force F_n and tangential force F_t along the *NREL Phase VI* blade.

It is noticed that there is a certain degree of difference between the results of the Glauert-A/B/C corrections in the tip region (e.g., at $r/R > 0.8$). In general, Glauert-C leads to lower loads which are closer to the experimental data, Glauert-A gives results close to, but slightly higher than, those of Glauert-C, while Glauert-B results in relatively higher loads near the tip. Taking the F_n in Figure 4a as an example of quantitative comparison, the relative difference between the results of Glauert-B and Glauert-C is about 4% at $r/R = 0.95$. The difference in other cases is not larger than this value.

At a wind speed of 7 m/s, all the curves of F_n generally agree with the five experimental data except that at $r/R = 0.95$ where an overestimation is observed. This overestimation becomes much more remarkable as the wind speed increases to 10 m/s and 13 m/s. At 10 m/s, there is no much difference around $r/R = 0.9$ between the results with no correction and with the Glauert-type corrections, indicating that almost no effective correction is made here. At 13 m/s, the curves of the Glauert-type corrections even exceed the uncorrected curve in a r/R range of about [0.66, 0.88]. It is clear that the corrections perform much worse at the higher wind speeds as compared with 7 m/s. The overestimation at $r/R = 0.95$ is also observed in the results for F_t, although it looks better to some extent, especially at 7 m/s.

Considering the fact that the Glauert correction was derived based on the potential hypothesis, it is reasonable to believe that the unusually poor performance at the higher wind speed is caused by the flow separation. According to the results of the *NREL UAE Phase VI* experiments [50], the angle of attack at most cross-sections of the blade exceeds the linear range of their aerodynamic polar when the wind speed is increased to be higher than 10m/s. Taking the cross-section of $r/R = 0.63$ as an example, the angles of attack are 5.9°, 11.8°, 16.9° at a wind speed of 7 m/s, 10 m/s, 13 m/s, respectively. The exact operating points for three cases on the lift polar of the S809 airfoil (the airfoil of all cross-sections of the *NREL Phase VI* blade) are depicted in Figure 5. It clearly shows that Case 2 and Case 3 are in the nonlinear lift region corresponding to flow separation, while Case 1 is in the linear lift region corresponding to attached flow, which implies that flow separation occurs at 10 m/s and 13 m/s.

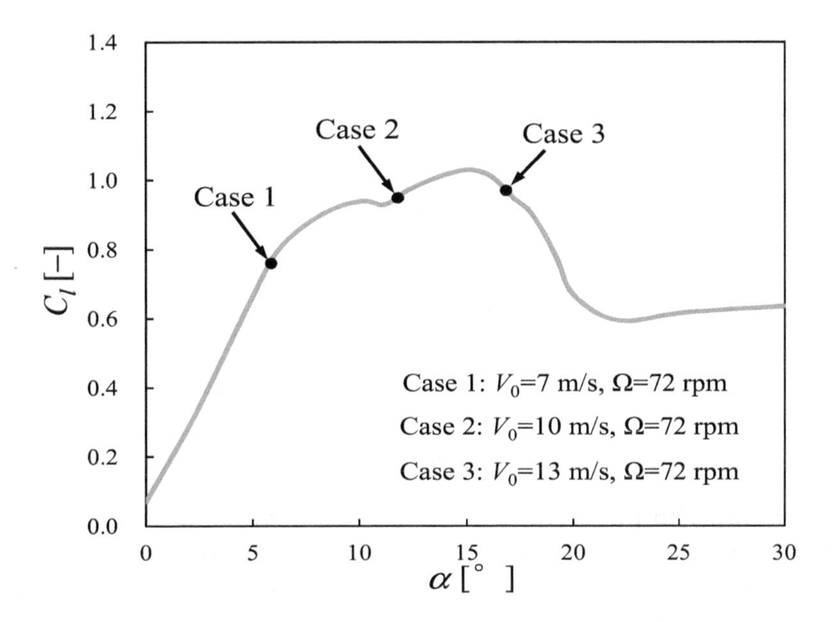

Figure 5. Operating points of the cross-section of r/R = 0.63 on the lift polar of the S809 airfoil. (The Reynolds number of the airfoil is 1×10^6 which is close to that of the cross-section).

The normal and tangential forces along a blade of the *NREL 5MW* rotor are plotted in Figure 6. As a widely studied virtual rotor, full CFD solutions [33] are often used as the reference data. Although the typical rotating speed of the *NREL 5MW* rotor is a few times less than the *NREL Phase VI* rotor, the resulted tip speed ratio is higher, which is closer to the optimal design point. The blade tip of the *NREL 5MW* rotor is smoothly sharpened where the slope at the last 4 m is 0.46, and the chord length at the tip is only 4.3% of the largest chord length in the blade inboard part. It can be seen in Figure 6 that the loads at the tip naturally approach to zero even without a tip loss correction.

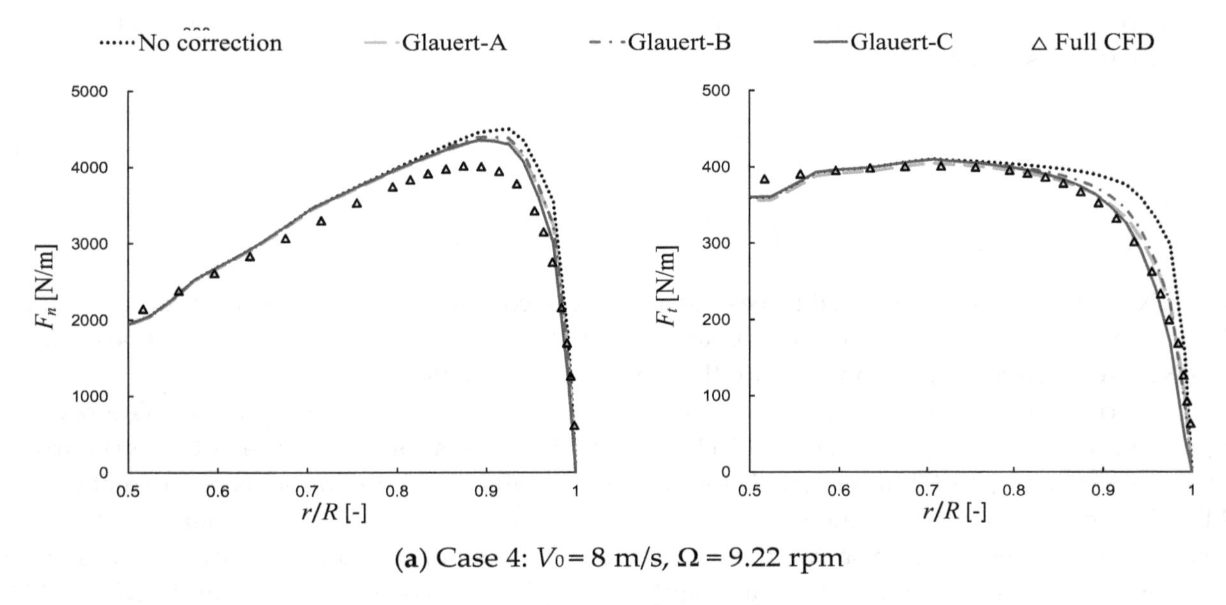

(a) Case 4: V_0 = 8 m/s, Ω = 9.22 rpm

Figure 6. *Cont.*

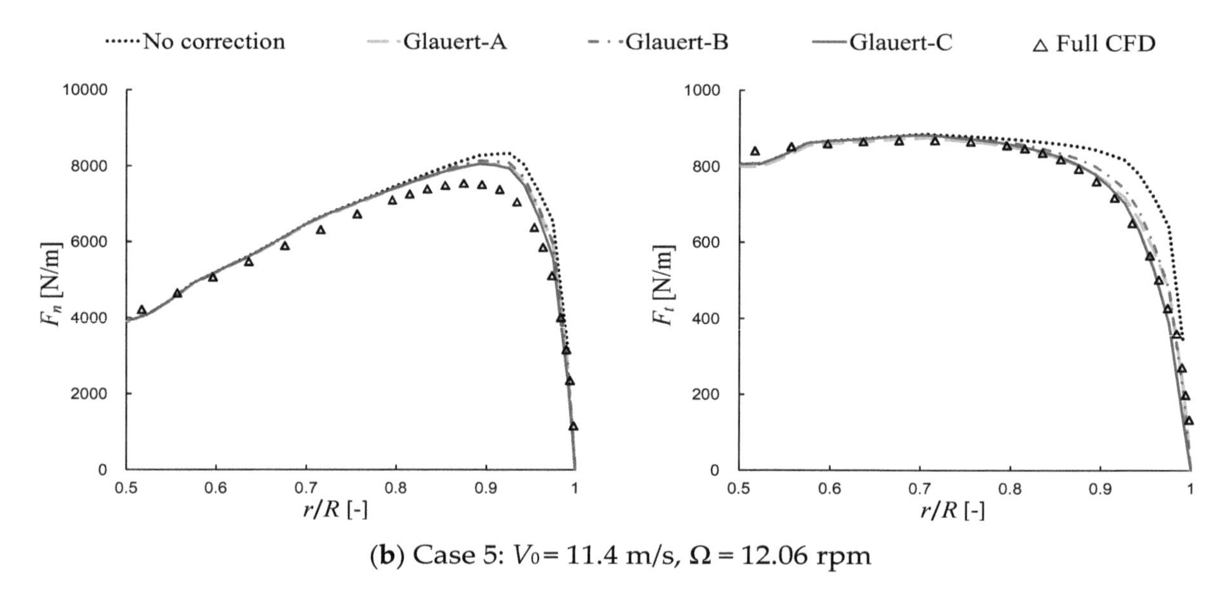

(**b**) Case 5: V_0 = 11.4 m/s, Ω = 12.06 rpm

Figure 6. Normal force F_n and tangential force F_t along the *NREL 5MW* blade.

As a large wind turbine with pitch control, there is no flow separation on the most area of the blade, including the tip. That means the flow pattern ideally keeps the same on the blade as the wind speed increases, which is significantly different from the situation of the *NREL Phase VI* rotor. As seen in Figure 6, by increasing the wind speed from 8 m/s to 11.4 m/s, the forces are greatly increased, whereas, their distributions near the tip region remain similar. All the Glauert-A/B/C corrections over-predict the normal forces near the blade tip, whereas, better agreement with the reference data is found for the tangential forces. Such a result is to some extent similar to that of the *NREL Phase VI* rotor at 7 m/s. Slight discrepancies are also noticed between the three Glauert-type corrections, and the Glauert-C performs better among them.

6.1.2. Comparison with BEMT Results

The Glauert tip loss correction was originally proposed for BEMT. As applied to AD/NS simulations, the correction factor F works in a different way. It is worth to mention that the specific difference in the final results would be caused by the transplantation from BEMT to AD/NS. In order to perform a comparison, BEMT computations under the same conditions of the present study are conducted.

The comparisons were shown by percentages, which represent the relative change of blade loads before and after using a tip loss correction,

$$\delta_n = \frac{F_{n,No} - F_{n,Gl}}{F_{n,No}} \times 100\%, \tag{58}$$

$$\delta_t = \frac{F_{t,No} - F_{t,Gl}}{F_{t,No}} \times 100\%, \tag{59}$$

where $F_{n,No}$ and $F_{n,Gl}$ are the normal forces obtained from computations with no tip loss correction and with the Glauert correction (the standard Glauert correction in BEMT and the Glauert-type corrections in AD/NS), respectively, $F_{t,No}$ and $F_{t,Gl}$ are the corresponding tangential forces.

The above parameters are calculated independently using BEMT and AD/NS. The results are shown in Figure 7 where the Glauert-BEMT denotes the results for the standard Glauert correction used in BEMT, and the Glauert-A/B/C denotes the results for the Glauert-type corrections used in AD/NS. A certain degree of difference between the Glauert-BEMT and the Glauert-A/B/C results is observed. The Glauert-C again shows the best performance, since it is generally most consistent with the Glauert-BEMT, which means the transplantation of the tip loss correction from BEMT to AD/NS does not notably influence its performance. The results for Cases 3 and 4 also agree with the above conclusion, which is not displayed here for simplicity.

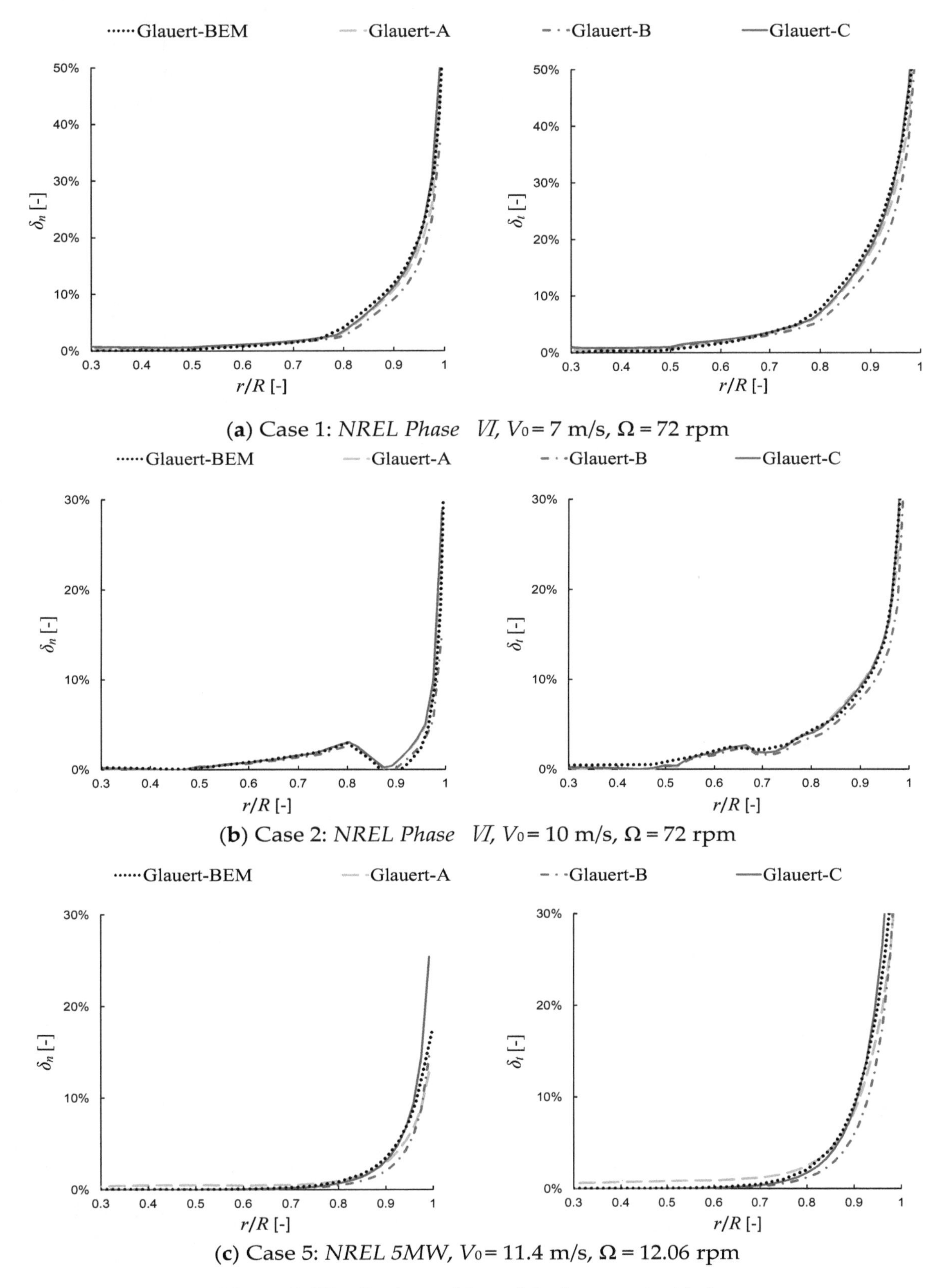

(a) Case 1: *NREL Phase VI*, $V_0 = 7$ m/s, $\Omega = 72$ rpm

(b) Case 2: *NREL Phase VI*, $V_0 = 10$ m/s, $\Omega = 72$ rpm

(c) Case 5: *NREL 5MW*, $V_0 = 11.4$ m/s, $\Omega = 12.06$ rpm

Figure 7. Distributions of δ_n and δ_t along a rotor blade.

6.2. Results of New Tip Loss Correction

The simplified form (see Section 3.2) of the new correction developed by Zhong et al. [33] is used in the present simulations. The results of blade loads for Cases 1, 2 and 5 are displayed in Figure 8a–c, respectively. Case 1 represents a typical condition of the *NREL Phase VI* wind turbine with the fully attached flow on its blades, while Case 2 represents another condition of flow separation (stall). Case 5 represents the rated operating condition of the *NREL 5MW* wind turbine. The results for Cases 3 and 4 are not displayed here for simplicity. (The performance of the correction in Case 3 is similar to that in Case 2, while the performance in Case 4 is similar to that in Case 5.) The Glauert-C correction, which performs best in Glauert-A/B/C, is compared with the new correction. Full CFD results [33] are employed as the reference data for the *NREL 5MW* wind turbine. As compensation for the sparse points of the experimental data, full CFD data [8,33] are also employed for the *NREL Phase VI* wind turbine.

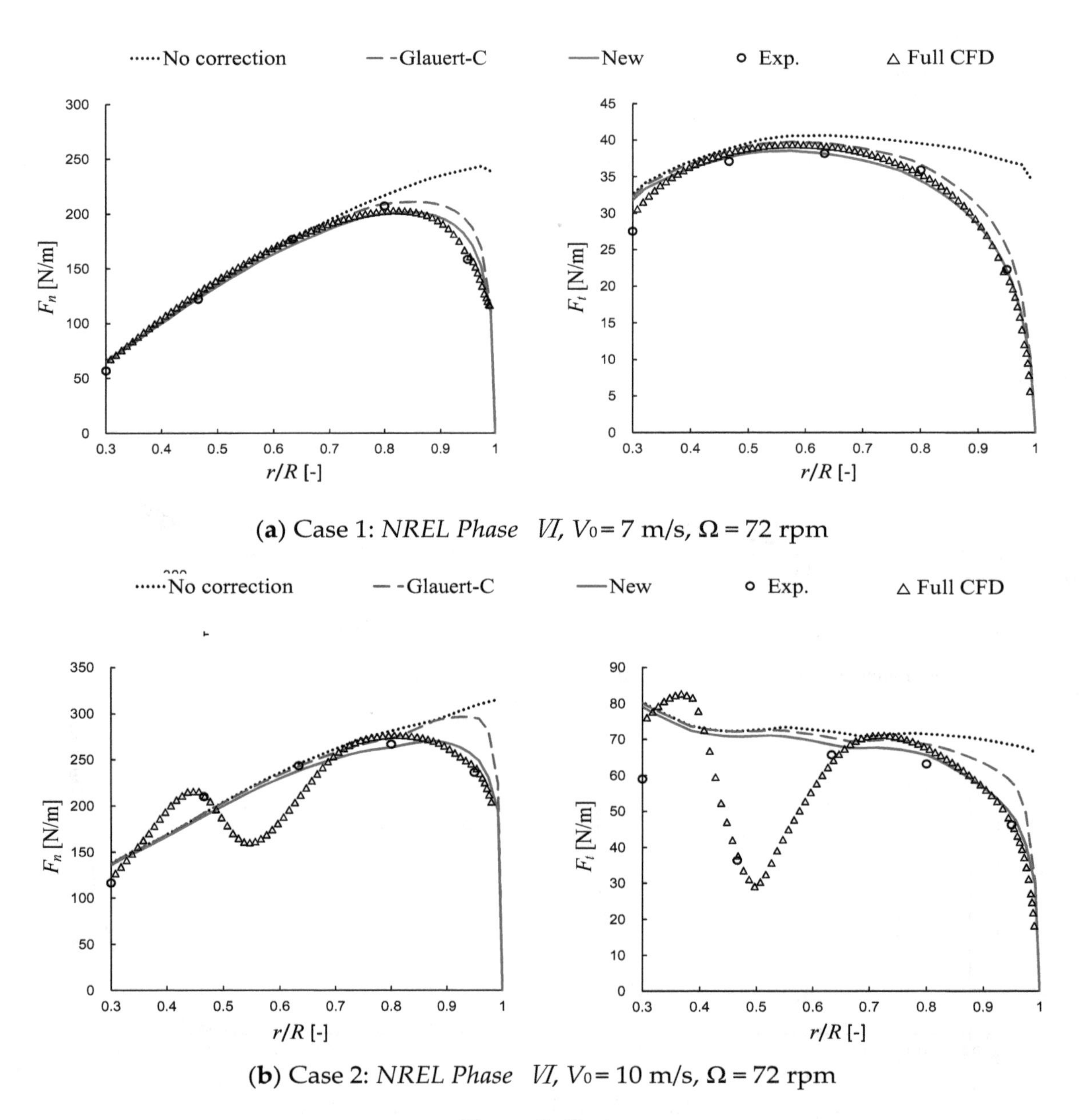

Figure 8. *Cont.*

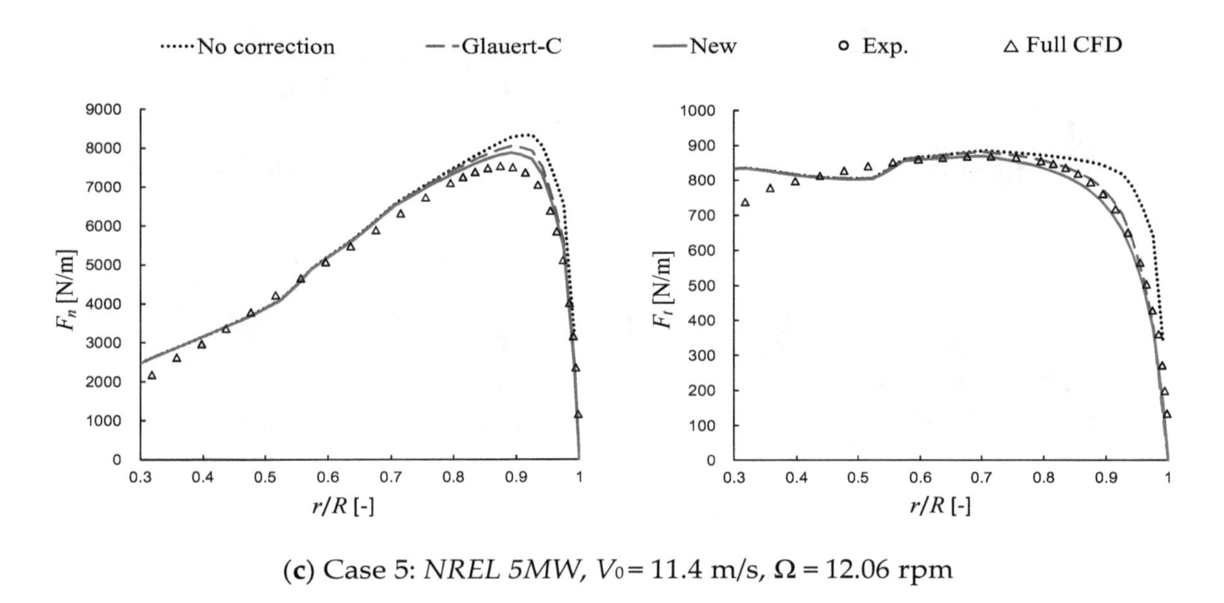

(c) Case 5: *NREL 5MW*, $V_0 = 11.4$ m/s, $\Omega = 12.06$ rpm

Figure 8. Normal force F_n and tangential force F_t along the *NREL 5MW* blade.

It is seen in the result of Case 1, as shown in Figure 8a, that the full CFD result achieves the best agreement with the experimental data and the new correction is superior to the Glauert-C correction. As compared to the Glauert-C correction, the overestimation of F_n in the tip region is properly corrected towards the reference data by the new correction. A similar tendency is seen from the tangential force distribution. In Case 2, as shown in Figure 8b, the new correction maintains its accuracy, while the Glauert-C correction produces larger positive error, due to the occurrence of flow separation. The performance of the two corrections is similar to those in BEMT computations [33].

For the *NREL 5MW* rotor, as shown in Figure 8c, the new correction again makes a better correction to the blade loads in the tip region. The corrected curve of F_n for the new correction is closer to the reference data compared to that for the Glauert-C correction. The difference between the two corrections in tangential force (F_t) is small, and both the corrections can be considered to be doing well.

6.3. Comparison of Velocity Field

The blade loads are affected by the local flow around the blade, and the forces consequently modify the flow around the blade and further in the wake. The flowfield simulated by the AD/NS demonstrates how the tip loss corrections affect the velocity field. In order to clearly show the influence of using different tip loss corrections, instead of directly showing the velocity field, the following velocity difference is defined,

$$\delta u_z = \frac{|u_{z,new} - u_{z,Gl}|}{u_{z,new}} \times 100\%, \tag{60}$$

where $u_{z,new}$ and $u_{z,Gl}$ are the axial velocity obtained from the simulations using the new tip loss correction and the Glauert-C tip loss correction, respectively.

Figure 9 displays the contours of δu_z in the flow field of the studied two rotors, which highlights the influence of the tip loss corrections in terms of changing the velocity field not only around the blade tip, but also in the wake. The influence of δu_z is more concentrated near the blade tip in the rotor plane, whereas, the inner part is hardly affected. In the wake, the influence range of δu_z tends to cover a wider radial space downstream. The magnitude is even increased in the wake of the *NREL 5MW* rotor.

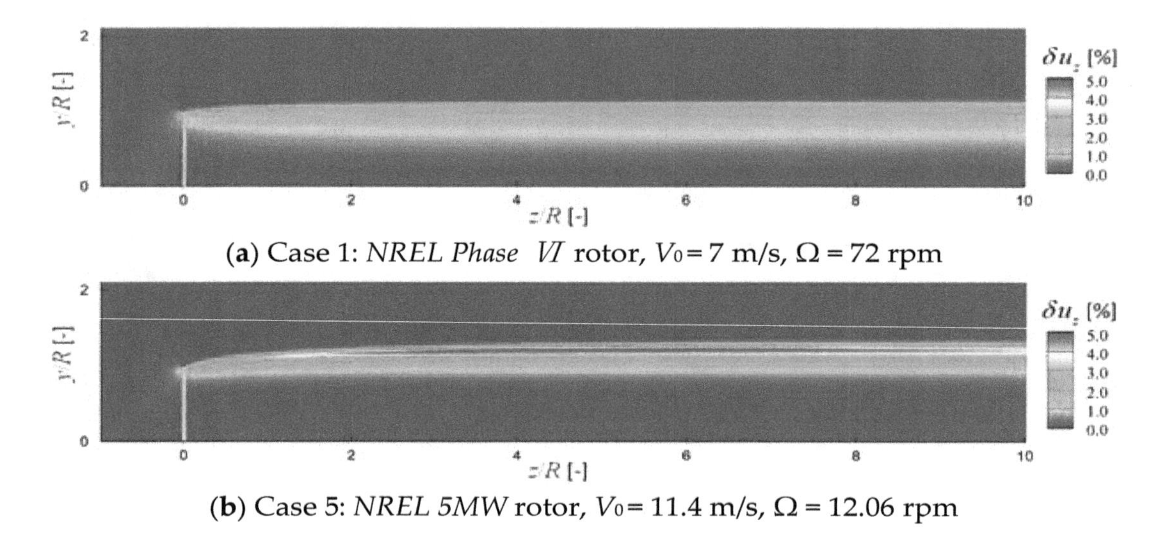

(**a**) Case 1: *NREL Phase VI* rotor, $V_0 = 7$ m/s, $\Omega = 72$ rpm

(**b**) Case 5: *NREL 5MW* rotor, $V_0 = 11.4$ m/s, $\Omega = 12.06$ rpm

Figure 9. Contours of the axial velocity difference between the results of the new and the Glauert-C tip loss corrections.

7. Conclusions

This work presented AD/NS simulations for the *NREL Phase VI* and the *NREL 5MW* rotors, which represent the typical small size and modern size rotors. The primary purpose of the numerical study is to evaluate the performance of tip loss corrections employed in the AD/NS simulations. Three different implementations of the widely used Glauert tip loss factor F, denoted as Glauert-A/B/C, are discussed and evaluated in the study. In addition, a newly developed tip loss correction is also evaluated. As an extension, the influence of employing different tip loss corrections to the velocity field is studied as well. The following conclusions are drawn from the present study:

- The three different implementations of the Glauert tip loss factor showed a certain degree of difference to each other, although the relative difference in blade loads is generally no more than 4%. The Glauert-C correction, in which the tip loss factor F is directly used to be divided by the interference factors, is recommended for AD/NS simulations, since it gives results closest to the reference data in all the studied cases.
- The performance of the Glauert-C correction in AD/NS was found to be almost equal to that of the standard Glauert correction in BEMT. That provides evidence for the reasonableness of the transplantation of the tip loss correction from BEMT into AD/NS.
- The Glauert-type corrections tend to make an over-prediction of the blade tip loads. In contrast, the new correction showed superior performance in various conditions of the present study.
- A difference in tip loss correction leads to the influence of not only on the blade tip loads, but also on the velocity field around and after the rotor. The range of this influence is observed expanding in the rotor wake. In addition, the magnitude of this influence is found to be increased after the *NREL 5 MW* rotor.

The above observations help to understand the performance of different tip loss corrections employed in AD/NS simulations. It is indicated that the blade loads, as well as the wake velocity, can be simulated better by choosing the best implementation of the Glauert tip loss factor or employing the new tip loss correction. That is the major contribution of this study.

Author Contributions: Conceptualization, W.Z.; Formal analysis, W.Z. and W.J.Z.; Funding acquisition, T.G.W.; Investigation, W.Z. and W.J.Z.; Methodology, W.Z.; Supervision, T.G.W. and W.Z.S.; Writing—Original draft, W.Z.; Writing—Review and editing, W.J.Z. and W.Z.S.

References

1. *Renewables 2019 Global Status Report*; REN21: Paris, France, 2019.
2. Shen, W.Z. Special issue on wind turbine aerodynamics. *Appl. Sci.* **2019**, *9*, 1725. [CrossRef]
3. Hansen, M.O.L. The classical blade element momentum method. In *Aerodynamics of Wind Turbines*, 3rd ed.; Routledge: Abingdon, UK, 2015; pp. 38–53.
4. Sørensen, J.N. *General Momentum Theory for Horizontal Axis Wind Turbines;* Springer International Publishing Switzerland: Cham, Switzerland, 2016; pp. 123–132. [CrossRef]
5. Sumner, J.; Watters, C.S.; Masson, C. CFD in wind energy: The virtual, multiscale wind tunnel. *Energies* **2010**, *3*, 989–1013. [CrossRef]
6. Sanderse, B.; van er Pijl, S.P.; Koren, B. Review of computational fluid dynamics for wind turbine wake aerodynamics. *Wind Energy* **2011**, *14*, 799–819. [CrossRef]
7. Rehman, S.; Alam, M.M.; Alhems, L.M.; Rafique, M.M. Horizontal axis wind turbine blade design methodologies for efficiency enhancement—A review. *Energies* **2018**, *11*, 506. [CrossRef]
8. Zhong, W.; Tang, H.W.; Wang, T.G.; Zhu, C. Accurate RANS simulation of wind turbines stall by turbulence coefficient calibration. *Appl. Sci.* **2018**, *8*, 1444. [CrossRef]
9. Rodrigues, R.V.; Lengsfeld, C. Development of a computational system to improve wind farm layout, part I: Medel validation and near wake analysis. *Energies* **2019**, *12*, 940. [CrossRef]
10. Mikkelsen, R.; Sørensen, J.N.; Shen, W.Z. Modelling and analysis of the flow field around a coned rotor. *Wind Energy* **2001**, *4*, 121–135. [CrossRef]
11. Sharpe, D.J. A general momentum theory applied to an energy-extracting actuator disc. *Wind Energy* **2004**, *7*, 177–188. [CrossRef]
12. Moens, M.; Duponcheel, M.; Winckelmans, G.; Chatelain, P. An actuator disk method with tip-loss correction based on local effective upstream velocities. *Wind Energy* **2018**, *21*, 766–782. [CrossRef]
13. Simisiroglou, N.; Polatidis, H.; Ivanell, S. Wind farm power production assessment: Introduction of a new actuator disc method and comparison with existing models in the context of a case study. *Appl. Sci.* **2019**, *9*, 431. [CrossRef]
14. Sørensen, J.N.; Shen, W.Z. Numerical modeling of wind turbine wakes. *J. Fluids Eng.* **2002**, *124*, 393–399. [CrossRef]
15. Shen, W.Z.; Zhu, W.J.; Sørensen, J.N. Actuator line/Navier-Stokes computations for the MEXICO rotor: Comparison with detailed measurements. *Wind Energy* **2012**, *15*, 811–825. [CrossRef]
16. Martínez-Tossas, L.A.; Churchfield, M.J.; Meneveau, C. A highly resolved large-eddy simulation of a wind turbine using an actuator line model with optimal body force projection. *J. Phys. Conf. Ser.* **2016**, *753*, 082014. [CrossRef]
17. Martínez-Tossas, L.A.; Meneveau, C. Filtered lifting line theory and application to the actuator line model. *J. Fluid Mech.* **2019**, *863*, 269–292. [CrossRef]
18. Castellani, F.; Vignaroli, A. An application of the actuator disc model for wind turbine wakes calculations. *Appl. Energy* **2013**, *101*, 432–440. [CrossRef]
19. Wu, Y.T.; Porte-Agel, F. Modeling turbine wakes and power losses within a wind farm using LES: An application to the Horns Rev offshore wind farm. *Renew. Energy* **2015**, *75*, 945–955. [CrossRef]
20. Sessarego, M.; Shen, W.Z.; van der Laan, M.P.; Hansen, K.S.; Zhu, W.J. CFD simulations of flows in a wind farm in complex terrain and comparisons to measurements. *Appl. Sci.* **2018**, *8*, 788. [CrossRef]
21. Diaz, G.P.N.; Saulo, A.C.; Otero, A.D. Wind farm interference and terrain interaction simulation by means of an adaptive actuator disc. *J. Wind Eng. Ind. Aerodyn.* **2019**, *186*, 58–67. [CrossRef]
22. Gargallo-Peiró, A.; Avila, M.; Owen, H.; Prieto-Godino, L.; Folcha, L. Mesh generation, sizing and convergence for onshore and offshore wind farm Atmospheric Boundary Layer flow simulation with actuator discs. *J. Comput. Phys.* **2018**, *375*, 209–227. [CrossRef]
23. Mao, X.; Sørensen, J.N. Far-wake meandering induced by atmospheric eddies in flow past a wind turbine. *J. Fluid Mech.* **2018**, *846*, 190–209. [CrossRef]
24. Zhu, W.J.; Shen, W.Z.; Barlas, E.; Bertagnolio, F.; Sørensen, J.N. Wind turbine noise generation and propagation modeling at DTU Wind Energy: A review. *Renew. Sustain. Energy Rev.* **2018**, *88*, 133–150. [CrossRef]
25. Shen, W.Z.; Zhu, W.J.; Barlas, E; Barlas, E.; Li, Y. Advanced flow and noise simulation method for wind farm assessment in complex terrain. *Renew. Energy* **2019**, *143*, 1812–1825. [CrossRef]

26. Glauert, H. Airplane propellers. In *Aerodynamic Theory*; Durand, W.F., Ed.; Dover: New York, NY, USA, 1963; pp. 169–360.
27. Branlard, E. Wind Turbine Tip-Loss Corrections. Master's Thesis, Technical University of Denmark, Lyngby, Denmark, 2011.
28. Shen, W.Z.; Mikkelsen, R.; Sørensen, J.N.; Bak, C. Tip loss corrections for wind turbine computations. *Wind Energy* **2005**, *8*, 457–475. [CrossRef]
29. Sørensen, J.N.; Dag, K.O.; Ramos-Garía, N. A new tip correction based on the decambering approach. *J. Phys. Conf. Ser.* **2015**, *524*, 012097. [CrossRef]
30. Wood, D.H.; Okulov, V.L.; Bhattacharjee, D. Direct calculation of wind turbine tip loss. *Renew. Energy* **2016**, *95*, 269–276. [CrossRef]
31. Schmitz, S.; Maniaci, D.C. Methodology to determine a tip-loss factor for highly loaded wind turbines. *AIAA J.* **2017**, *55*, 341–351. [CrossRef]
32. Wimshurst, A.; Willden, R.H.J. Computatioanl observations of the tip loss mechanism experienced by horizontal axis rotors. *Wind Energy* **2018**, *21*, 544–557. [CrossRef]
33. Zhong, W.; Shen, W.Z.; Wang, T.; Li, Y. A tip loss correction model for wind turbine aerodynamic performance prediction. *Renew. Energy* **2020**, *147*, 223–238. [CrossRef]
34. Hand, M.M.; Simms, D.A.; Fingersh, L.J.; Jager, D.W.; Cotrell, J.R.; Schreck, S.; Larwood, S.M. *Unsteady Aerodynamics Experiment Phase Vi: Wind Tunnel Test Configurations and Available Data Campaigns*; NREL/TP-500-29955; National Renewable Energy Laboratory: Golden, CO, USA, 2001.
35. Jonkman, J.; Butterfield, S.; Musial, W.; Scott, G. *Definition of A 5-Mw Reference Wind Turbine for Offshore System Development*; Technical Report NREL/TP-500-38060; National Renewable Energy Laboratory: Golden, CO, USA, 2009.
36. Prandtl, L. *Applications of Modern Hydrodynamics to Aeronautics*; NACA: Washington DC, USA, 1921; No. 116.
37. Prandtl, L.; Betz, A. *Vier Abhandlungen zur Hydrodynamik und Aerodynamik*; Göttingen Nachrichten: Göttingen, Germany, 1927. (In German)
38. Ramdin, S.F. Prandtl Tip Loss Factor Assessed. Master's Thesis, Delft University of Technology, Delft, The Netherlands, 2017.
39. Wilson, R.E.; Lissaman, P.B.S. *Applied Aerodynamics of Wind Power Machines*; NSF/RA/N-74113; Oregon State University: Corvallis, OR, USA, 1974.
40. De Vries, O. *Fluid Dynamic Aspects of Wind Energy Conversion*; AGARD: Brussels, Belgium, 1979; AG-243, Chapter 4; pp. 1–50.
41. Sørensen, J.N.; Kock, C.W. A model for unsteady rotor aerodynamics. *J. Wind Eng. Ind. Aerodyn.* **1995**, *58*, 259–275. [CrossRef]
42. Mikkelsen, R. Actuator Disc Methods Applied to Wind Turbines. Ph.D. Thesis, Technical University of Denmark, Lyngby, Denmark, 2003.
43. Shen, W.Z.; Sørensen, J.N.; Mikkelsen, R. Tip loss correction for Actuator/Navier-Stokes computations. *J. Sol. Energy Eng.* **2005**, *127*, 209–213. [CrossRef]
44. Michelsen, J.A. *Basis3d—A Platform for Development of Multiblock PDE Solvers*; Technical Report AFM; Technical University of Denmark: Lyngby, Denmark, 1992.
45. Sørensen, N.N. *General Purpose Flow Solver Applied Over Hills*; RISØ-R-827-(EN); Risø National Laboratory: Roskilde, Denmark, 1995.
46. Menter, F.R. Two-equation eddy-viscosity turbulence models for engineering applications. *AIAA* **1994**, *32*, 1598–1605. [CrossRef]
47. Tian, L.L.; Zhu, W.J.; Shen, W.Z.; Sørensen, J.N.; Zhao, N. Investigation of modified AD/RANS models for wind turbine wake predictions in large wind farm. *J. Phys. Conf. Ser.* **2014**, *524*, 012151. [CrossRef]
48. Prospathopoulos, J.M.; Politis, E.S. Evaluation of the effects of turbulence model enhancements on wind turbine wake predictions. *Wind Energy* **2011**, *14*, 285–300. [CrossRef]
49. Cao, J.; Zhu, W.; Shen, W; Sørensen, J.N.; Wang, T. Development of a CFD-based wind turbine rotor optimization tool in considering wake effects. *Appl. Sci.* **2018**, *8*, 1056. [CrossRef]
50. Jonkman, J.M. *Modeling of the UAE Wind Turbine for Refinement of FAST_AD*; Technical Report NREL/TP-500-34755; National Renewable Energy Lab.: Golden, CO, USA, 2003.

13

Assessment of Turbulence Modelling in the Wake of an Actuator Disk with a Decaying Turbulence Inflow

Hugo Olivares-Espinosa [1,2,*], Simon-Philippe Breton [2,3], Karl Nilsson [2], Christian Masson [1], Louis Dufresne [1] and Stefan Ivanell [2]

[1] Department of Mechanical Engineering, École de Technologie Supérieure, 1100 Notre-Dame Ouest, Montréal, QC H3C 1K3, Canada; Christian.Masson@etsmtl.ca (C.M.); Louis.Dufresne@etsmtl.ca (L.D.)
[2] Wind Energy Section, Department of Earth Sciences, Uppsala University Campus Gotland, Cramérgatan 3, 62165 Visby, Sweden; Simon-Philippe.Breton@geo.uu.se (S.-P.B.); karl.nilsson@geo.uu.se (K.N.); stefan.ivanell@geo.uu.se (S.I.)
[3] Nergica, 70 rue Bolduc, Gaspé, QC G4X 1G2, Canada
* Correspondence: hugo.olivares@geo.uu.se

Abstract: The characteristics of the turbulence field in the wake produced by a wind turbine model are studied. To this aim, a methodology is developed and applied to replicate wake measurements obtained in a decaying homogeneous turbulence inflow produced by a wind tunnel. In this method, a synthetic turbulence field is generated to be employed as an inflow of Large-Eddy Simulations performed to model the flow development of the decaying turbulence as well as the wake flow behind an actuator disk. The implementation is carried out on the OpenFOAM platform, resembling a well-documented procedure used for wake flow simulations. The proposed methodology is validated by comparing with experimental results, for two levels of turbulence at inflow and disks with two different porosities. It is found that mean velocities and turbulent kinetic energy behind the disk are well estimated. The development of turbulence lengthscales behind the disk resembles what is observed in the free flow, predicting the ambient turbulence lengthscales to dominate across the wake, with little effect of shear from the wake envelope. However, observations of the power spectra confirm that shear yields a boost to the turbulence energy within the wake noticeable only in the low turbulence case. The results obtained show that the present implementation can successfully be used in the modelling and analysis of turbulence in wake flows.

Keywords: wind turbine wakes; turbulence; actuator disk; LES; wind tunnel; OpenFOAM

---

1. Introduction

Studies of turbulence in the wake of wind turbines are often made with the aim of analyzing the influence of the flow and rotor models in the reproduction of wake characteristics. Investigations with various rotor models in the Atmospheric Boundary Layer (ABL) have been made either with the goal of improving the production efficiency of a cluster of turbines (e.g., [1–4]) or aimed at comparing the characteristics of wakes with respect to measurements of real or downscaled turbines (e.g., [5–8]). When numerical simulations of large wind parks are made, computational limitations oblige employing simpler rotor models that, while numerically less expensive, are requested to produce a minimum level of detail in wake features that yields an acceptable reproduction of the interaction of the ensuing wakes and the downstream wind turbines, as well as other wakes. Amongst the simplest formulations of the rotor model is the Actuator Disk (AD) [9,10] which reproduces the effect of the rotor in the incoming flow by means of a permeable surface in the shape of a disk that acts as a momentum sink. In its most basic conception, the AD constitutes a one-dimensional force opposite to the flow, perpendicular to the rotor plane, with no wake rotation or airfoil properties considered. Different studies [8,11,12] have shown the capability of the AD technique to approximate the characteristics of a real turbine wake within its far region. An experimental investigation of the wakes produced by porous disks has been performed by [13,14], where the disks are made of metallic meshes representing different solidities. In recent work, similar research has been made of wakes produced by a decaying turbulence inflow [15,16]. Unlike the ABL, no shear-produced turbulence occurs, greatly simplifying the modelling of the flow physics and permitting to isolate the features of the wake from those deriving from the inflow (e.g., the vertical momentum flux in the ABL). With data collected in a related measurement campaign, the authors of [17] employed Reynolds-Averaged Navier–Stokes (RANS) models to reproduce measurements in the wakes produced by the porous disks. Although good results are obtained, the experimental study represents an opportunity to perform comparisons with numerical simulations that allow a greater detail in the reproduction of the turbulence characteristics. Therefore, the reproduction of these experiments employing Large-Eddy Simulations (LES) seems appealing.

To reproduce the inflow properties measured experimentally, first it is necessary to model the flow of decaying turbulence produced in a wind tunnel. In this regard, different works have been dedicated to investigate various methodologies to produce adequate inlet conditions [18–20]. Amongst the different techniques, the method developed by Mann [21–23] to create a synthetic turbulence field has often been used to create inflow conditions for simulations in ABL [5,6,24,25] as well as in homogeneous turbulence [26–28]. In these works, it has been shown that the Mann method is capable of producing realistic turbulence fields, resembling the second order statistics of real turbulence [5,29]. This algorithm permits creating synthetic fields of homogeneous turbulence by prescribing two parameters, the turbulence lengthscale and (albeit indirectly) the turbulence intensity. An anisotropy factor is also used in the algorithm to create boundary layer fields. The transition and evolution of turbulence characteristics when synthetic fields are introduced in LES domains, especially of integral lengthscales, has been previously studied by [27] for homogeneous turbulence and recently by [24,30] (using the turbulence generation method of [20]) in sheared flows.

The objective of this work is the study of turbulence characteristics in the wake flow of a wind turbine model. To fulfill this purpose, a methodology is developed to replicate: (1) the turbulence characteristics of a homogeneous wind tunnel flow and (2) the wake field arising from the introduction of porous disks representing the wind turbine. This procedure is employed to reproduce the inflow and wake characteristics measured in the experimental campaign carried out by G. Espana and S. Aubrun [17,31], so the comparison with wind tunnel data serves as method of validation. The study of diverse features of the turbulence field both in the free decaying flow and in the wake is presented next to such comparison. The computational platform employed is OpenFOAM® v.2.1 [32,33] (the OpenFOAM Foundation Ltd., London, UK) an open-source code amply used in flow simulations, chosen for its availability and access to apply ad-hoc modifications to existing solvers and utilities. While the synthetic turbulence is created with the Mann method, the flow is reproduced using LES computations. The methodology implemented follows—in part—a procedure developed on the CFD software EllipSys3D [34–36] (Department of Mechanical Engineering and RisøCampus, DTU, Lyngby, Denmark) to simulate wake flows with turbulent inflows [5]. It has been shown that this method provides good results to introduce ABL as well as homogeneous turbulence conditions [5,6,24–28]. By using this approach, we expect to assess: (a) how well the main turbulence characteristics can be reproduced (e.g., turbulence intensity and integral lengthscale) in the inflow and in the wake with LES, (b) how the main turbulence characteristics in the wind tunnel change due to the presence of the wind turbine model and (c) how the LES modelling change along the wake compared to the undisturbed flow (resolved/modelled velocity fluctuations). It should be noted that these questions are formulated in the context of a *limited mesh resolution*, which makes it more relevant for the wind energy field since it is often desired to minimize the computational requirements while successfully reproducing the requested flow features, which in this case consists mainly of the integral lengthscale.

The work presented here is organized as follows: a brief theoretical background is provided about the homogenenous isotropic turbulence and its experimental approximation, the decaying homogeneous turbulence, next to a description of the measurement campaign and data to be reproduced. Later, the methodology is described in detail, with special attention to the generation of turbulence inflow and the reproduction of the adequate characteristics in the absence of disk. This also comprises the description of the methodology employed to introduce the synthetic turbulence in the LES domain. Next, the results are presented and discussed for the different quantities computed in the turbulence field of the wakes. A final section presents a summary and the conclusions of the work.

2. Homogeneous Turbulence

For a given flow, the instantaneous velocity vector is referred to as u_i whose components in the streamwise, vertical and spanwise directions (x, y, z) are $u_i = (u, v, w)$. The Reynolds decomposition in the streamwise direction is defined as $u = \langle U \rangle + u'$, where "$\langle \cdot \rangle$" denotes time-average (capital letters are also used to emphasize averaged magnitudes). The characteristic size of the largest eddies is identified as the distance L required to nullify the correlation function $\mathcal{R}_{ij}(\mathbf{r}, t)$. With this assumption, the integral lengthscale

$$L_{ij}^{(d)} = \int_0^\infty \mathcal{R}_{ij}(\mathbf{e}_d r, t) dr \tag{1}$$

in the direction d (set by the unitary vector $\mathbf{e}$) is defined [37]. From all scales defined by this expression, those most commonly used are the longitudinal integral lengthscale $L_1 = L_{11}^{(1)}$ as well as the transversal one $L_2 = L_{22}^{(1)}$. Similarly, the longitudinal Taylor lengthscale (or micro-scale) $\lambda_1 = \lambda_{11}^{(1)}$ is defined by the osculating parabola to the correlation function $\mathcal{R}_{ij}(\mathbf{r}, t)$. If isotropy is assumed (or at least between the 1 and 2 directions) the equivalences $L_{22}^{(1)} = L_{11}^{(2)}$ and $\lambda_{22}^{(1)} = \lambda_{11}^{(2)}$ are also valid.

In the absence of shear, the Taylor hypothesis of frozen turbulence can be adopted more comfortably. This is, it is assumed that the turbulence field does not change as it is convected by the mean wind at $\langle U \rangle$, which yields the equivalence between the spatial and temporal correlations. In this way, correlations can be made from the time series of each velocity component. In particular, the autocorrelation will provide the integral time scales $\mathcal{T}_{11}$ and $\mathcal{T}_{22}$ from where the integral lengthscales can be computed by means of $L_1 = \langle U \rangle \mathcal{T}_{11}$ and $L_2 = \langle U \rangle \mathcal{T}_{22}$. Likewise, the longitudinal Taylor lengthscale can be calculated from the expression

$$\frac{1}{\lambda_1^2} = \frac{\langle U \rangle^{-2}}{2\left\langle u_1'^2 \right\rangle} \left\langle \left(\frac{\partial u_1'}{\partial t} \right)^2 \right\rangle \tag{2}$$

as seen in [38] (note that Equation (2) differs from the one presented in that report by a factor of $\sqrt{2}$). Considering the turbulent kinetic energy of a field $k = \frac{1}{2}\left(\langle u_1'^2 \rangle + \langle v_2'^2 \rangle + \langle w_3'^2 \rangle\right)$ and the assumption of isotropy, λ_2 is related to the amount of dissipation of k by

$$\varepsilon = \frac{15\nu \left\langle u_1'^2 \right\rangle}{\lambda_2^2} = \frac{30\nu \left\langle u_1'^2 \right\rangle}{\lambda_1^2} \ . \tag{3}$$

When grid turbulence is used to approximate the theoretical case of decaying isotropic turbulence, the characteristics observed at different positions downstream from the grid correspond to the time evolution of isotropic turbulence with zero mean velocity. Thus, a decay during the interval Δt is approximated by that occurring within Δx in a wind tunnel. In this manner, it has been observed that the decay of k follows the expression

$$\frac{k}{\langle U \rangle^2} = c_A \left(\frac{x - x_0}{M} \right)^{-n} , \tag{4}$$

where M is the turbulence grid size, c_A (also written as $1/A$) is a fitting parameter, n is the decay exponent and x_0 a virtual origin. Equation (4) is commonly employed to track the streamwise turbulence intensity decay, replacing k for $\left\langle u_1'^2 \right\rangle$. While Ref. [37] mentions $1.1 \leq n \leq 1.3$ and $c_A \simeq 1/30$, Ref. [39] reports to have observed $n = 1.25$ with $x_0 = 0$, whereas Ref. [40] mentions $1.15 \leq n \leq 1.45$, remarking that c_A varies greatly depending on the geometry of the grid and Re_λ. Ref. [37] indicates that the integral lengthscale evolves downstream according to

$$L_2 \simeq c_{B_1} M \left[\frac{x - x_0}{M} \right]^{n_1} \tag{5}$$

with $c_{B_1} = 0.06$ and $n_1 = 0.35$.

3. Experimental Setup and Measurement Campaigns

The measurements used in this work come from experiments performed in the Eiffel-type wind tunnel of the Prisme laboratory of the University of Orléans. The test section has a width and a height of 0.5 m and a length of 2 m. Two different grids were used to generate turbulence at the entrance of the test section, resulting in two different turbulence intensities. At a distance of $x = 0.5$ m from that grid, the reported values of turbulence intensity and integral lengthscale (measured at the centreline) were TI $= 3\%$ and $L_1 = 0.01$ m as well as TI $= 12\%$ and $L_1 = 0.03$ m. These two inflow cases are henceforward identified as Ti3 and Ti12, respectively. The streamwise position where these values are reported is referred to as the *target position* x_D.

Later, disks made of a metallic mesh were located at x_D to simulate the effect of the AD model (a porous surface) on the flow. Two disks were used, each with a diameter of $D = 0.1$ m but made with a different wire to produce different induction factors. The thrust coefficient C_T of each disk is calculated following the procedure presented by [13] and revisited by [17], based on the measurement of the velocity deficit in the wake. In total, six experimental cases are considered for this work as shown in Table 1. Complete details about the experimental setup, the measurement techniques as well as the characteristics of the flows generated by this wind tunnel can be found in [15,31].

Table 1. Reference parameters of flow and disks used in the experiments.

TI [%]	L_1 [m]	Case
3	0.01	No-disk
		$C_T = 0.42$
		$C_T = 0.62$
12	0.03	No-disk
		$C_T = 0.45$
		$C_T = 0.71$

The data used in this work was obtained using two different techniques. Firstly, with the aim of obtaining time-series of the flow velocity, a Hot-Wire Anemometry (HWA) probe was located along vertical lines at $x = 3D$, $4D$ and $6D$ from the disk centre (the origin of the reference system $x, y, z = 0, 0, 0$ is set there). The probe moved along each vertical line between $0 \leq y/D \leq 1.5$, registering data at different steps (data recorded every $0.1D$ is employed). A scheme of the measuring locations with respect to the experimental arrangement is shown in Figure 1. At each probe position, data was acquired with a sampling frequency of $f_{acq} = 2$ kHz during about 1 min. A low-pass filter was also used, with a cut-off frequency fixed at $f_c = 1$ kHz. The reference velocity during these measurements was $U_\infty = 3$ m/s. Of the measurements made with this technique, only the database corresponding to the cases of Ti12 is used in the comparisons shown here since the sampling rate was assessed to be too low for the turbulence scales of the Ti3 case. Due to this, HWA measurements from [16] are used to complement the experimental data for the comparison of the Ti3 case. These were made using the same experimental setup as the other HWA measurements, with TI $\simeq 3\%$ also at the target position but with $U_\infty = 20$ m/s and in consequence a higher Reynolds number.

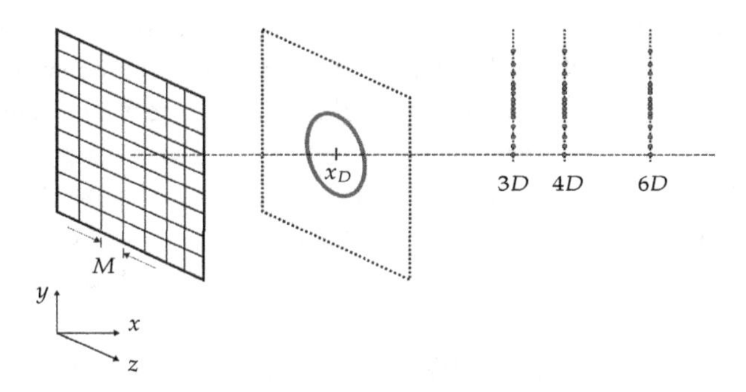

Figure 1. Representation of the measurement positions of the hot-wire. The turbulence is generated by a grid of spacing M. The reported values of TI are measured at $x = 0.5$ m from such grid, where the ADs are subsequently located. This position is referred to as x_D. Time-series of the velocity are recorded at various positions along vertical lines at $3D$, $4D$ and $6D$.

Secondly, a Laser-Doppler-Anemometer (LDA) was used to simultaneously measure two components of velocity (u, v) with the main purpose of obtaining time-averaged information of the wake. Measurements behind the disks were made along the vertical directions at $x = 2D, 4D, 6D, 8D$ and $10D$ from the disk centre. The recording positions were aligned in the vertical direction, performed in steps (generally) of $0.1D$ between $-1.5 \leq y/D \leq 1.5$. For the Ti3 cases, the positions in the vertical direction where data is available vary slightly, but most of them are made in steps of $0.1D$ between $-1.0 \leq y/D \leq 1.0$. Measurements were made using a non-uniform sampling frequency, with an average of 1 KHz during 90 s. The reference velocity was $U_\infty = 10$ m/s for the cases Ti3 and $U_\infty = 6$ m/s for Ti12. As it was shown by [41] and later work, various estimations in grid-generated turbulence can be considered Reynolds independent (but not for observations such as the scaling region of the spectrum, as shown by [42]. Therefore, non-dimensional results of mean velocities and root-mean-square (RMS) statistics obtained with the LDA technique will be used despite the differences in reference velocity, set in the simulations to $U_\infty = 3$ m/s. LDA data are used to compare with the obtained results of velocity deficit and k, while the velocity time-series from HWA are used to compute the other quantities examined in this work.

4. Model Description

4.1. Numerical Model

In order to reproduce the measured characteristics of the wake field, an LES model is employed. This allows for resolving the large (energy-containing) motions, whereas the effects of the smaller eddies are modelled. To achieve this, the Navier–Stokes equations are decomposed into a filtered (or resolved) component and a residual (or subgrid scale, SGS) component. The classic Smagorinsky model [43] is used for the parametrization of the residual scales. There, the subgrid viscosity ν_{SGS} is assumed to be proportional to the Smagorinsky coefficient C_s and the local cell length $\Delta = (\Delta_x \Delta_y \Delta_z)^{1/3}$ and the resolved strain-rate $\overline{S}_{ij}$ as $\nu_{SGS} = (C_s\Delta)^2 \sqrt{2\overline{S}_{ij}\overline{S}_{ij}}$. The value of C_S is set to 0.168, which comes from adjustments made to reproduce decaying homogeneous turbulence [44]. This model is chosen for its ubiquity and because the absence of its best known disadvantage, associated to bounded flows: overdissipation close to walls [40,45]. The interpolation scheme for the convective term consists of a dynamic blend where, depending on the velocity flux and the magnitude of the velocity gradients at the cell faces, an amount of up to 20% upwind is used in combination with the linear interpolation. This scheme is called *filteredLinear* within OpenFOAM v2.1 . In this way, the upwind part is employed only in regions of steep velocity gradients while the flow maintains the second-order accuracy of the linear scheme elsewhere. The PISO algorithm is employed for the solution of the pressure–velocity equations.

4.2. Computational Domain and Grid Resolution

The dimensions of the computational domain are set to imitate those of the measuring region in the wind tunnel. The domain and grid sizes of the LES computations as well as of the synthetic velocity field, identified as *turbulence box* (use as inflow) are listed in Table 2. As in the experiments, x_D corresponds to the origin of the coordinate system for the computations, at the centre of the crosswise y–z plane and at $5D$ from the inlet. The reason to imitate the dimensions of the experiment, in particular in the crosswise directions, is to reproduce the potential effects of blockage on the wake development. A small blockage of 1.3% in average has been reported for measurements in this wind tunnel [17].

The grid size is determined by the optimum number of cells per L, or in fact L_1. Unlike the ABL, where L_1 is typically two to three times the diameter of the rotor, the turbulence grids used in the wind tunnel produce turbulence with an eddy size approximately ten to three times smaller—at x_D—than the diameter of the AD. Evidently, this imposes a strict demand for the cell resolution, particularly for the turbulence box as the turbulence scale there ($L_{1,\mathrm{B}}$) should be even smaller to account for its increase along the flow direction once turbulence is introduced in the LES domain.

On account of these limitations, the determination of the cell resolution of the turbulence boxes is based on what is physically realizable, due to the constraints represented by the total number of cells. For this, it should be considered that a box with twice the length of the desired lateral dimensions must be generated, as suggested by [22] since the simulated velocity field is periodic in all directions, so it is recommended to create a turbulence box with cross-flow dimensions twice as big as the desired size and only use one quarter of the simulated field. The box length is determined by the recycling period of the box into the computational domain (having assumed the equivalence between the streamwise direction of the box and the time for its convection in the domain), which is wished to be kept to a minimum. It is chosen to create boxes with length equivalent to at least two longitudinal flow-times, abbreviated as LFT (1 LFT is defined as $L_x / \langle U_\infty \rangle$). Considering these arguments, two turbulence boxes were created using the Mann implementation for each TI case. The parameters of these boxes are listed in Table 2. Note that the dimensions of the boxes are set to multiples of 2^n ($n \in \mathbb{N}^+$) due to the Fourier techniques used in the generation algorithm. As the grid size limit of the OpenFOAM installation in the cluster where computations were submitted has been found to be $\sim$180 $\times$ 10^6 cells (perhaps due to the floating-point precision used to store the grid data) it is easy to see that a larger mesh than the one used for the Ti3 case (e.g., 2048 $\times$ 512 $\times$ 512) would have surpassed this ceiling. The turbulence generator has been implemented outside the OpenFOAM framework so when the turbulence is originally generated, with twice as many points in the lateral directions, the cell number is not restricted by this limit. Since the mesh of the Ti12 case is coarser, it was possible to increase the length of the domain, allowing for a smaller recycling rate of the turbulence box.

In each TI case, the mesh of the computational domain is set according to the resolution used for the corresponding turbulence box, so that the cell sizes in the domain and box are equal. Unlike the turbulence box, the grid used in the LES is not completely uniform across the domain. Instead, only a central region of $20D \times 3.6D \times 3.6D$ of uniform (cubic) cells is defined. This region is needed to assure a consistent filtering for the SGS scales, as implicit filtering is used in the LES. In addition, the uniformly distributed cells should comprise all the positions of measurement, which includes those made up to $y = 1.5D$ from the centreline. Outside the uniform grid region, the cells are stretched towards the lateral boundaries with an aspect ratio $\Delta z_{max} / \Delta z_{min} = \Delta y_{max} / \Delta y_{min} = 4$, where $\Delta z_{min} = \Delta y_{min} = \Delta x = \Delta$ is the cell side length in the uniform region. This central region has approximately the same cell size as in the corresponding turbulence boxes. The slight differences arise from the purpose of accommodating an integer number of cells along the diameter of the AD. This way, the central region in each case consists of uniform cells with a side length Δ of 0.002 m (Ti3) and 0.004 m (Ti12). Hence, taking L_1 at x_D as a reference, the cell resolution of the integral scales L_1/Δ corresponds to 5 and 7.5 cells, respectively.

Table 2. Main parameters of the computational domains of LES and synthetic field (turbulence box). Dimensions of computational domains are given as $L_x \times L_y \times L_z$ with grids containing $N_x \times N_y \times N_z$ cells. Synthetic field domains are given as $L_{B,x} \times L_{B,y} \times L_{B,z}$ containing $N_{B,x} \times N_{B,y} \times N_{B,z}$ cells. Lengths measured in $D = 0.1$ m.

LES Domain Size		**$20D \times 5D \times 5D$**
	Layout	Uniform region $20D \times 3.6D \times 3.6D$
Case Ti3	LES domain grid	$1000 \times 208 \times 208$ cells
	Turbulence box	$40D \times 5D \times 5D$
	Box grid	$2048 \times 256 \times 256$ cells
Case Ti12	LES domain grid	$500 \times 104 \times 104$ cells
	Turbulence box	$80D \times 5D \times 5D$
	Box grid	$2048 \times 128 \times 128$ cells

4.3. Generation of Turbulent Inflow, Introduction into the Computational Domain and Boundary Conditions

In the homogeneous case, the use of the Mann method requires adjusting two parameters to produce the turbulence with the demanded characteristics: the lengthscale L and TI. The latter is normally controlled by means of varying the coefficients $\alpha\varepsilon^{2/3}$ (α is the Kolmogorov constant) of the von Kármán energy spectrum [22] until the desired TI is achieved. As it is not straightforward to give an exact relationship between ε and the generated TI, the procedure suggested by [46] is used. Instead of changing ε, a scaling factor $SF = \sqrt{\sigma^2_{target}/\sigma^2}$, is used, where σ^2_{target} is the desired average variance of the turbulence and σ^2 is the variance of the turbulence field in each direction. In this way, the desired TI can be obtained by multiplying SF by each velocity component of the turbulence box. It is expected that when the HIT field is convected at a uniform velocity, the TI will decay monotonically in the streamwise direction. To estimate the turbulence intensity value that the box (TI_B) should have in order to attain the desired TI at the given position, empirical relations obtained from fits to experimental results can be used [37,39,40]. However, such relations depend on fitting parameters that are reported to vary within a certain margin. This fact and the expected numerical dissipation cause that TI_B cannot be estimated beforehand with precision. Furthermore, the averaged value of TI at the position where turbulence is introduced in the computational domain does not correspond to TI_B (due to the interpolations between turbulence planes and the adjustment of the synthetic velocity field to the LES conditions, to be discussed later on). In consequence, some testing was necessary to find the right value. Likewise, empirical relations for the development of L_1 [37] are not sufficient to predict the value of $L_{1,B}$, so tests were necessary to determine its magnitude. The values found for TI_B and $L_{1,B}$ are presented in the results section.

Boundary conditions are set to replicate the conditions in the wind tunnel. Thus, slip conditions are used for the lateral boundaries, whereas the outlet is set to zero gradient. For the time derivative, a second-order backward scheme is employed. When the disks are introduced, these are located at x_D, this is $x = 0.5$ m from the inlet, at the centre of the y–z plane. Assuming the Taylor hypothesis of equivalent spatial and temporal correlations, the streamwise axis of the turbulence box is assumed equivalent to time. To introduce the synthetic turbulence into the computational domain, cross-sectional planes of the turbulence box are taken for every available longitudinal position and their velocity values are mapped onto the inlet of the computational domain. The procedure is illustrated in Figure 2, where the planes extracted from the synthetic turbulence field are separated by Δx_B. As the crosswise locations of the cell centres of the synthetic turbulence do not exactly correspond to those of the computational domain, linear interpolations are used to evaluate the velocity values at the required positions. Different Courant–Friedrichs–Lewy (CFL) conditions are used in each case. For the Ti3, CFL ≈ 0.8 while for the Ti12, CFL ≈ 0.5. These are the maximum CFL values over the whole computational domain, which are attained next to the inlet, where the velocity fluctuation is the highest. In comparison, their domain-averaged values are $\approx$0.3 and $\approx$0.1, respectively. The time-steps used in the computations are $\Delta t = 2 \times 10^{-4}$ s and $\Delta t = 1.2 \times 10^{-4}$ s for each case. Linear interpolations are used in the streamwise direction (i.e., between planes of the turbulence box) to compute the required velocity values at the given time.

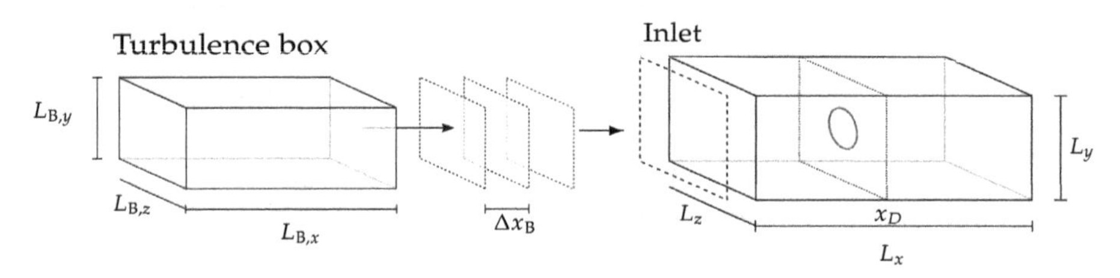

Figure 2. Introduction of synthetic turbulence field into the LES.

Simulations are allowed to run initially for 5 LFT to allow the stabilization of the flow. After this, measurements are made during a time equivalent to 20 LFT, which is equal to approximately 13.33 s in real time. Since the turbulence boxes defined in Table 2 are only enough to supply an inflow during 2 LFT (Ti3) and 4 LFT (Ti12), the boxes are recycled for the duration of the computations. Velocity data are sampled at every time-step, resulting in a higher sampling frequency than the one used in the experiments, although it is made during a shorter period (13.33 s compared to $\sim$60 s). The length of the computations is chosen as to maximize the bandwidth of the data employed for the calculations of spectra and correlations and to avoid fringe patterns in the average fields.

4.4. Estimation of Integral Lengthscales

The integral lengthscales are calculated from the autocorrelation curves of u and v in the longitudinal direction in the synthetic turbulence or from their time-series in the LES. In this way, L_1 and L_2 are obtained by making use of Equation (1). The method used to compute L_i consists of approximating the autocorrelation curve by a sum of six decaying exponentials. A similar procedure has been also used by [16,31], based on a technique first suggested by [47]. This technique is used as it avoids the uncertainty of determining the crossing of the oscillating function $\mathcal{R}$ around zero as well as approximating better the expected asymptotic behaviour of a theoretical autocorrelation sampled to infinity, $\lim_{x\to\infty} \mathcal{R}_{ii}(x,t) = 0$, yielded by the exponentials. While this method provides L_i for the synthetic turbulence, in the LES, the autocorrelations provide a time-scale that can be translated into a spatial one only under the assumption of the Taylor hypothesis (here, the integral time-scale obtained from the autocorrelations is multiplied by the average streamwise velocity at the point where the data are registered, which can be slightly different from U_∞).

4.5. Actuator Disk Model

In line with the experiments, the rotor of a horizontal-axis wind turbine is modelled in the computations as an actuator disk [9], where the effect of the blades on the wind flow is reproduced by forces distributed over a disk. As the actual geometry of the blades is not reproduced, the load of the turbine is distributed over the area swept by the rotor. The simplest conception of the model is employed here, where it is assumed that the forces over the AD point only in the axial direction, opposite to the flow and are distributed uniformly over the disk. If U_∞ is the inflow velocity, the force is calculated as

$$F_x = -\frac{1}{2}\rho U_\infty^2 C_T A, \tag{6}$$

where A is the area of the disk. The diameter of the disk is $D = 0.1$ m and C_T is equal to the values shown in Table 1. Since the introduction of the forces represents an abrupt discontinuity in the flow field, large velocity gradients occur in the vicinity of the AD and spatial oscillations (wiggles) on the velocity field may appear due to pressure-velocity decoupling inherent to collocated grids. To avoid this effect, the forces that comprise the AD are distributed in the axisymmetric direction. This is done by taking the convolution of the forces with a Gaussian distribution

$$g(x) = \frac{1}{\sigma\sqrt{2\pi}} \exp\left(-\frac{x^2}{2\sigma^2}\right). \quad (7)$$

In this manner, the value of the standard deviation σ (i.e., the distribution width) will define the thickness of the disk. The force distribution is defined between the limits $[-3\sigma, 3\sigma]$ so that it contains 99.7% of magnitude of the forces computed for the original—one cell thick—disk. A value of $\sigma = 2\Delta x$ is used, yielding a disk thickness equal to $12\Delta x$. Therefore, the absolute magnitude of the thickness changes according to the cell length.

4.6. About RANS Results

In a study by [17], RANS computations were performed to reproduce the same LDA measurements used in our study. In their work, a RANS turbulence model, identified as "Sumner and Masson", based on modifications to the k-ε model of [48] is proposed. While the latter model attempts to correct the well known overestimation of turbulent stresses [49] by introducing a dissipative term proportional to the turbulence production in the ε-equation, Sumner and Masson pursue the same objective by neglecting some terms of turbulence production also in the vicinity of the disk (the cylindrical volume centred at the AD, extending $\pm 0.25D$ in the axial direction), obtaining a good comparison for the velocity deficit and k along the wake of the disks. The results obtained with this model are also included in the comparisons as they serve as a reference element of the capabilities of an industry standard to reproduce the evolution of turbulence features in the wake. Note that since the simulations of [17] were made for only half the wake, their results (velocity deficit, k and ε) are shown mirrored in the vertical direction.

5. Results and Discussion

5.1. Turbulence Decay and Integral Lengthscale Development without Disk

The first step of the investigation consists of the calibration of the parameters of the synthetic turbulence. This is finding $\mathrm{TI_B}$ and $L_{1,\mathrm{B}}$ so that when a turbulence box is introduced in the computational domain, the desired target values are attained after a distance of $5D$, at x_D. The parameters of the synthetic turbulence for all boxes are shown in Table 3. These were computed longitudinally and averaged over the whole volume. It is immediately noticed that high TI values were necessary to reproduce the evolution of the turbulence intensities reported by the experiments. Consequently, the approach followed could be seen as rather crude, on account of the Taylor approximation. However, the results reproduce, for the most part, the longitudinal evolution of turbulence predicted also by the empirical relations found in the literature. This can be seen in Figure 3, which shows the free (no disk) homogeneous turbulence decay in each TI case obtained with LES. There, every value represents the average of the TIs computed from time-series stored in nine probes distributed in a crosswise plane, in turn located at every longitudinal position indicated by the marks in the curves. LES results are compared to the values measured in the wind tunnel and to the least-squares fit with Equation (4). As mentioned in Section 3, experimental values from [16] are used in the Ti3 case. To complement HWA data (which in the Ti12 case has only 3 points), LDA measurements are used, obtained in the experiment with the low thrust disks but outside the wake envelope ($y = \pm 1$D). It can be seen that for both Ti3 and Ti12 the decay predicted by the LES follows fairly well the experimental values. The subframes in Figure 3 represent the same data but plotted in a log-log scale, so the power law decay of the TI (of slope

$-n$ in Equation (4) can be better appreciated. This permits seeing that, after some distance, the decay rate is approximately equal for both TIs. Furthermore, it is also seen that the LDA (measurements outer wake) in the Ti3 case deviates from a constant decay rate and therefore from the LES. Conversely, the LDA data in the Ti12 compares very well with the LES prediction.

Table 3. Characteristics of the different boxes of synthetic turbulence used as inflow for the LES.

	$\mathbf{TI_B}$ [%]	$L_{1,B} \times 10^{-3}$ [m]	$k/\frac{1}{2}U_\infty^2$ [-]
Case Ti3	35.0	5.82	0.37
Case Ti12	60.2	15.3	1.08

Equation (4) is an empirical relation first proposed by [41] to describe the decay in homogeneous turbulence seen in a wind tunnel. The applicability of this relation has been proven in a wide range of Re flows in later work [39,42,50,51]. In most of the results reported in the literature, a fit is produced setting the virtual origin x_0 to zero in the equation, which neglects the agreement close to the grid or the place where turbulence originates as the stations where measurements or calculations are reported are generally far from such region. However, as the complete evolution of TI is monitored, a better fit is obtained by setting x_0 to a position different from where the turbulence is introduced (in particular, to an upstream location). The fit of Equation (4) for the curves shown in Figure 3 yields the results shown in Table 4. If the fit is made using $x_0 = 0$, the parameters are closer to those reported in the literature (see Section 2), although the curve would display a much higher TI at the inlet than the one given with $x_0 \neq 0$, this is, $\sim$60% for Ti3 and $\sim$100% for Ti12. The mesh spacings used for the fits are $M = 0.0225$ m for Ti3 and $M = 0.20$ m for Ti12. The problem of setting x_0 has been discussed by [52], which show that when x_0 is properly determined, its value and the exponent of the power law decay n is Re-independent, while A is indeed a function of initial conditions (including Re), so the turbulence decay takes a universal self-similar behaviour.

Table 4. Parameters of the fit of TI decay of LES to Equation (4). x_0 denotes the virtual origin of the curve with respect to the inlet.

	$x_0 = 0$		$x_0 \neq 0$		
	$1/A$	n	$1/A$	n	x_0 [m]
Case Ti3	24.11	1.281	9.85	1.519	−0.021
Case Ti12	28.49	1.15	11.43	1.661	−0.0845

Table 3 also shows the magnitude of the $L_{1,B}$ in the synthetic turbulence. According to these results, it is seen that the requirement of two cells per L_1 (assuming the applicability of the minimum condition to represent a wavelength [53]) is sufficiently fulfilled. In effect, for the case Ti3, $L_1/\Delta \simeq 3$ is obtained. The resolution of eddies is somewhat improved in the case of Ti12, where $L_1/\Delta \simeq 4$. These resolutions that seem a priori rather coarse are the result of a series of compromises that have been previously explained. In [30] a similar resolution was used for the synthetic inflow in LES computations of the wake of a rectangular channel, obtaining a good comparison with experimental results related to the flow structures. The results of the turbulence decay in the absence of the disk show that these values are enough to supply an integral lengthscale to procure the desired magnitudes $L_1 = 0.01$ m (Ti3) and $L_1 = 0.03$ m (Ti12) at x_D. Figure 4 shows the development of the longitudinal and transversal integral lengthscales, L_1 and L_2, along the flow. These are computed using velocity time-series (recorded at the same positions for TI above) and employing the method described in Section 4.4. The measurements from [16] are also used for comparison in the Ti3 case. There, the comparison of the measurements with the LES shows a small overestimation of L_1, although it should be considered that the difference is increased in the figure as the experimental value is below the target $L_1 = 0.01$ m. A fit of Equation (5) is also made for L_2 obtained with LES. In the Ti3 case, the least-squares fit method applied yields

$B_1 = 0.089$, $n_1 = 0.392$ and an origin set at $x_0 = -0.188$ m (upstream) from the inlet. Therefore, according to reference values provided along Equation (5), the LES slightly overestimates L_2 for Ti3. In the Ti12 case, the comparison with measurements is made with the only three points available so it is less decisive, although they suggest a higher growth rate. The fit of Equation (5) to the LES predicted L_2 yields $B_1 = 0.064$ and $n_1 = 0.254$ with $x_0 = -0.068$ m, which, compared to the values in the literature, also indicates a slight overestimation by the computations. For the purposes of this work and considering the mesh restrictions, the development of the integral lengthscales in the flow predicted by the LES is deemed satisfactory.

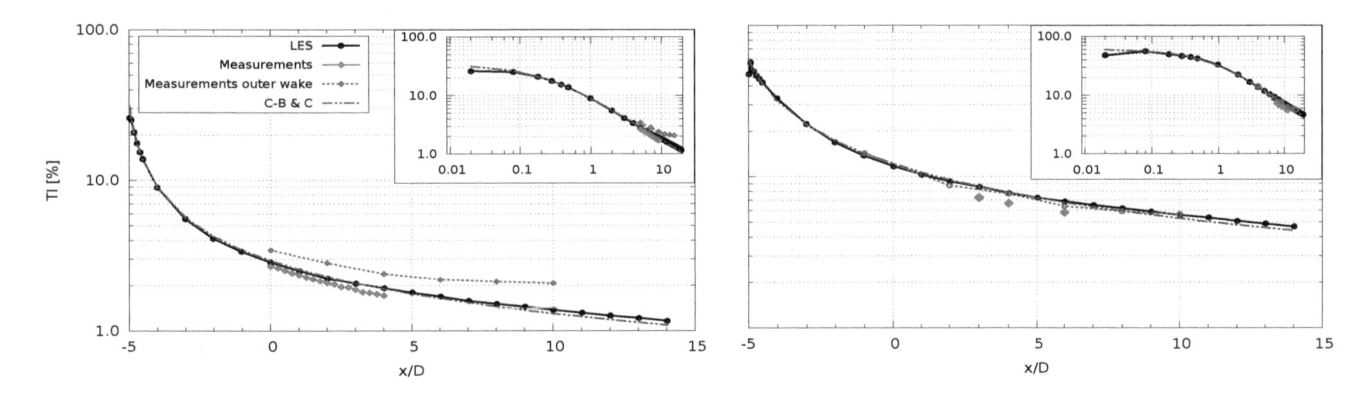

Figure 3. Comparison of the TI decay for the Ti3 (**left**) and Ti12 (**right**) cases without disk. C-B & C corresponds to Equation (5) as found in [41]. The inserted frames contain the same data but with a logarithmic scale also in both axes and denoting the distance from the inlet in x/D units.

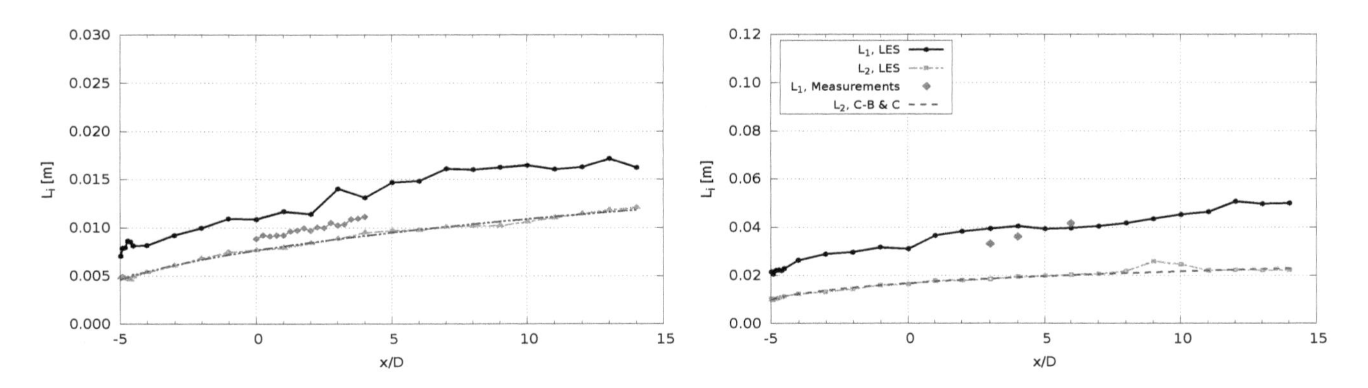

Figure 4. Longitudinal evolution of L_1 and L_2 for the Ti3 (**left**) and Ti12 (**right**) cases without disk.

In both TI and L_i results, we can observe small oscillations for the first positions next to the inlet, which are likely the result of the adjustment of the synthetic velocity field to the LES. Specifically, the incompressibility conditions imposed by the solver as the original formulation of the Mann algorithm does not produce divergence free fields, as pointed out by [29].

5.2. Velocity Deficit

The results for the wake simulations obtained with the AD are now shown. The first comparison is made from the results of the streamwise velocity deficit along the vertical direction at different longitudinal positions, normalized by the freestream velocity at $y = 1.5D$. For these and other quantities extracted across the wake field shown in Sections 5.2–5.5, the values are sampled at every cell centre along a line in the y-direction (at the mid-plane $z = 0$) at different x/D positions. These quantities correspond to time averages made during 20 LFT (13.33 s). No further spatial averaging is made. Note that LES results at $x = 3D$ have been added to the available x-positions of the LDA data as this is a position where values computed from HWA are later shown.

Figure 5 shows the results for the high and low solidity disks under the inflow Ti3. The agreement to the experimental results is very good, with the larger difference observed around the shear layer (i.e., the wake envelope at $y \simeq \pm 0.5D$) from the disk edges, especially for the disk with higher thrust.

In the case of the Ti12 inflow, Figure 6 shows a minor reduction in the agreement of the LES with the measurements. The predictions commence to differ when moving further into the far wake, around the centreline and shear layer regions. Remarkably, the results of RANS are almost identical to those from LES in both TI cases.

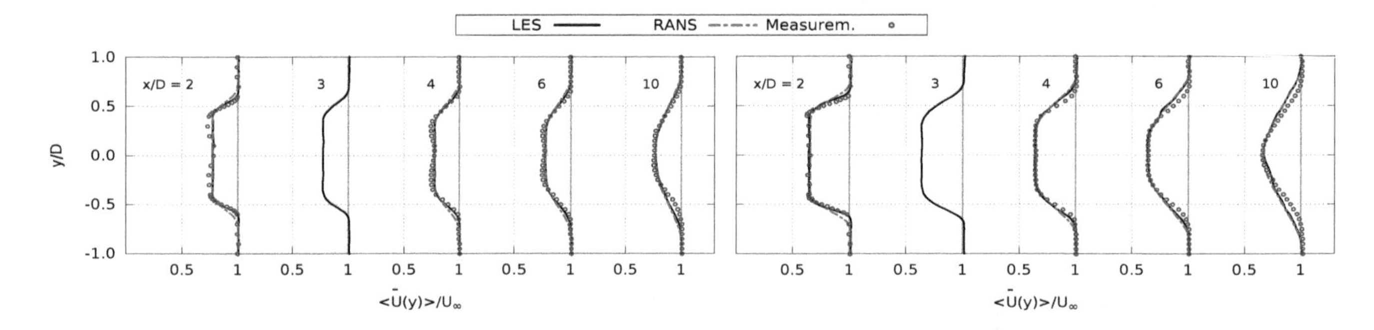

Figure 5. Vertical profiles of velocity deficit behind the disk $C_T = 0.42$ (**left**) and $C_T = 0.62$ (**right**), Ti3 case. RANS results in all figures are by Sumner et al. [17].

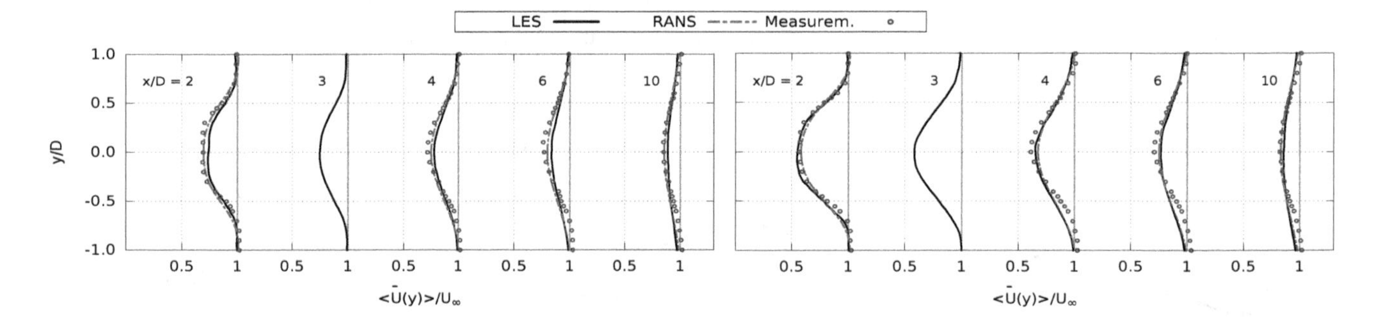

Figure 6. Vertical profiles of velocity deficit behind the disk $C_T = 0.45$ (**left**) and $C_T = 0.71$ (**right**), Ti12 case.

5.3. Turbulence Kinetic Energy in the Wake

It is expected that the wake created by the disks augments the turbulence level with respect to the ambient value. It is now investigated how the computations of the added turbulence compare to the experimental results within the wake. Figure 7 shows the profiles of k (this is, $k_{tot} = k_{SGS} + k_{res}$ for the LES) at different downstream positions along the wake, when the inflow of the case Ti3 is used. There, it is observed that the LES results match quite well the measured (LDA data) turbulence levels behind both disks. However, it is noticed that, except for the nearest position to the disk, the LES predicts a higher diffusion of shear turbulence in the crosswise direction, an effect that is increased with the disk thrust.

For the disks in the Ti12 case in Figure 8, the LES compare mostly well with the experimental data, although a small overestimation of k can be seen just behind the disk ($x = 2D$). It is also observed that the shear layer originating at the edges of the disk is mixing faster with the ambient turbulence compared to the Ti3 inflow. Indeed, the effect of shear prevails deeper into the wake in the LES with the highest thrust disk, whereas it is mixed faster into the ambient turbulence when the thrust is lower. Results from the RANS computations are discussed in the next section.

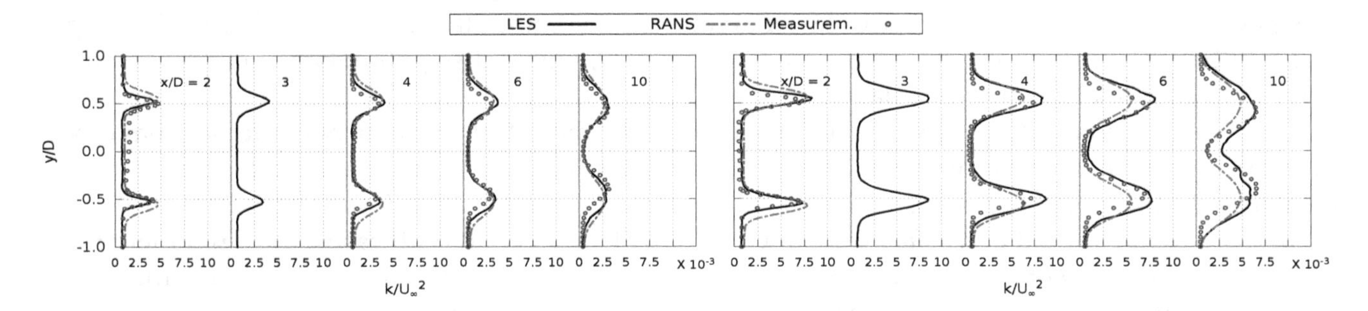

Figure 7. Vertical profiles of k behind the disk $C_T = 0.42$ (**left**) and $C_T = 0.62$ (**right**), Ti3 case.

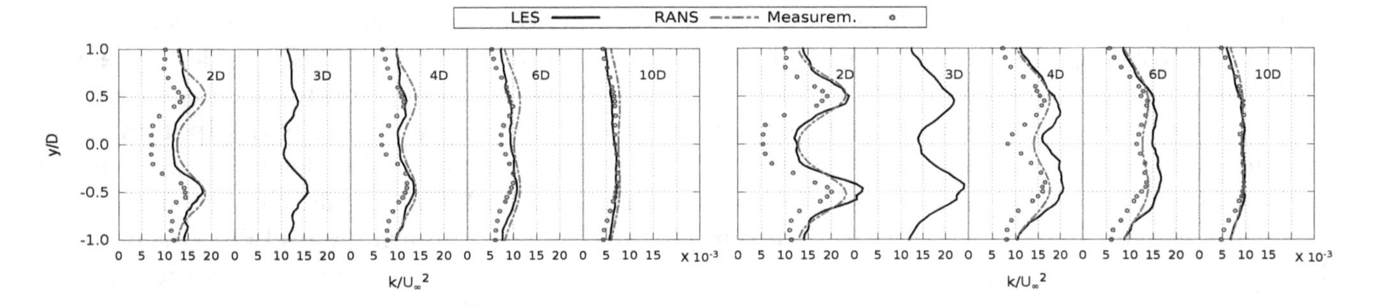

Figure 8. Vertical profiles of k behind the disk $C_T = 0.45$ (**left**) and $C_T = 0.71$ (**right**), Ti12 case.

It is also noticed that some inhomogeneities appear along the curves of LES, very noticeable in the simulations with the Ti12 inflow. These seem to indicate a footprint of the turbulence structures of the inflow turbulence. Although this feature could provide evidence of the need of performing averages in the azimuthal direction or creating synthetic turbulence that would cover longer simulation periods, it is thought that the results shown in the figures are sufficient for the purpose of these comparisons.

5.4. Turbulence Dissipation in the Wake

In the LES computations, the dissipation shown corresponds to $\varepsilon_{tot} = \varepsilon_{res} + \varepsilon_{SGS}$. The dissipation is calculated from the HWA data using Equations (2) and (3), which was only available in the Ti12 case. Therefore, results of Ti3 are shown only for completeness in Figure 9, where it can be seen that, as before, differences between LES and RANS are small, as the curves differ only at $x = 2D$ where RANS predicts a higher dissipation within the shear layer.

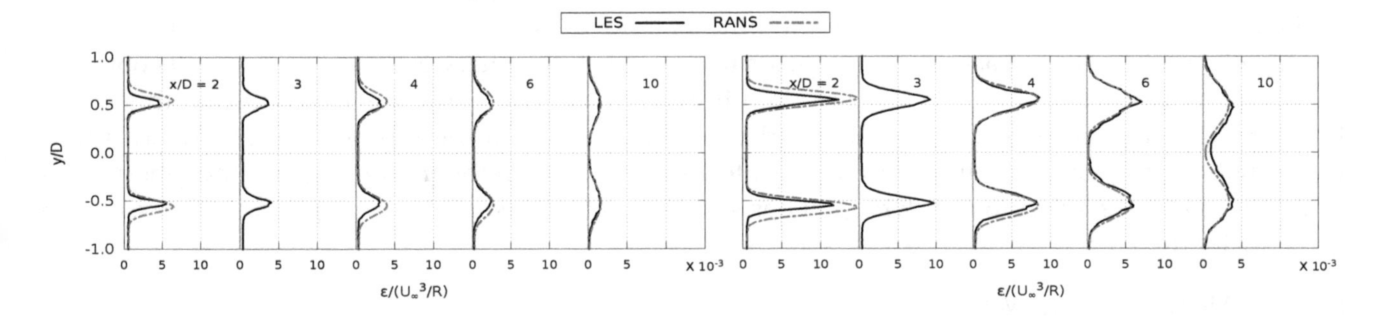

Figure 9. Vertical profiles of ε behind the disk $C_T = 0.42$ (**left**) and $C_T = 0.62$ (**right**), Ti3 case.

For the results with the Ti12 inflow in Figure 10, it is seen that, for the disk $C_T = 0.45$, the LES predictions compare well with measured values. For the disk $C_T = 0.71$, the measurements reveal a large increase of dissipation within the shear layer, compared to the data of computations with the lower thrust AD. Furthermore, at least within the three longitudinal positions available, measured dissipation in the shear layer is more or less maintained. The computations with $C_T = 0.71$ also display a somewhat stronger mixing of turbulence from $x = 4D$, where dissipation becomes more uniform and less predominant in the shear layer.

The RANS computations with the modified $k - \varepsilon$ model of Sumner and Masson have been previously shown capable of reproducing the turbulence level in the wake [17]. In the present comparison, it is seen that for the Ti3 inflow the agreement is very good for the disk $C_T = 0.42$ while it falls somewhat behind in the far wake of $C_T = 0.62$. However, in both cases, the agreement in the computed dissipation of RANS and LES is very good except for $x = 2D$. Interestingly, it is the vicinity of the disk where the k-ε is often corrected by adding dissipative terms to the ε equation to overcome the miscalculated turbulence stresses [49]. The results with the Ti12 show the opposite picture with regard to the estimation of k, as the agreement with measurements becomes better only for farther distances from the disk. For the closest position, the turbulence level is overestimated (as it is in the LES) despite the drop of the turbulence production terms near the disk ($x = 2D$ is outside this region). Dissipation seems overestimated in the case of $C_T = 0.45$ when comparing to the measurements. This is less certain for the higher thrust disk, where at $x = 4D$ the peak value of dissipation seems equal to the measured one, but much smaller in the case of $x = 6D$. Notably, ε from RANS is always higher than any LES in the wakes of the Ti12 inflow. Previous work [49] has shown that, in the ABL, the k-ε model overestimates the dissipation around the disk when comparing with LES. This has been observed to occur even upstream of the disk, where ε has been seen to increase unlike computations of LES, where this value does not grow until $0.5D$ downstream from the rotor.

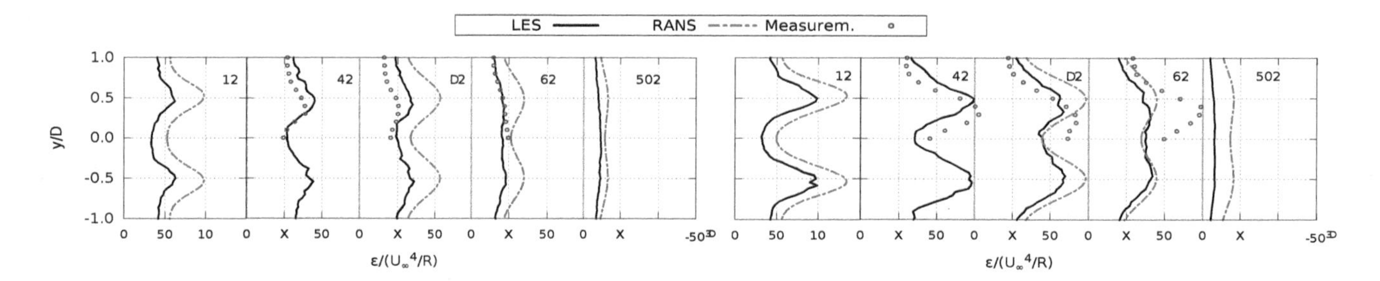

Figure 10. Vertical profiles of ε behind the disk $C_T = 0.45$ (**left**) and $C_T = 0.71$ (**right**), Ti12 case. The scale for the curves at $x = 2D$ has been doubled to accommodate the larger values.

It should be remarked that Sumner and Masson [17] showed that results with various turbulence closures (standard k-ε, RNG, El Kasmi and Masson model [48] and their own model) compare, in essence, equally well to the measurements, with no apparent advantage of their proposed correction to the k-ε model (although ε yielded by the different closures was not compared). The fact that all models compare well to measurements appears to contradict the otherwise inadequate results obtained in simulations of wakes in the ABL flow cited in that work. There, it is also argued that this is due to the relative decrease of the modelled turbulent viscosity ν_t in the reproduction of wind-tunnel wakes with homogeneous inflow with respect to its proportion in the modelling of atmospheric flow. In those conditions, previous work by [49] has successfully proved the advantages of LES to estimate the velocity deficit and turbulence levels in the wake.

5.5. LES Modelling in the Wake

The previous results for k and ε indicate that the LES running in OpenFOAM is able to predict with relative accuracy not only the velocity deficit in the wake, but also the level of turbulence and its dissipation in the case where the TI in the inflow is low (~3%). For the high TI inflow (~12%), the prediction becomes more imprecise, according to the comparison with the experimental data. In the absence of disks, it is observed that the fraction of the turbulence kinetic energy that is resolved by the LES with respect to the total k_{res}/k_{tot} occurs for the most part in the resolved scales, at around 90%

in both TI cases throughout the domain and only somewhat smaller close to the inlet [54]. With the introduction of the disks, this does not appreciably change except for the shear layer nearby the AD at around $x = 1D$ in the Ti3 case, where the resolved part is slightly reduced [54]. In the Ti12 case, this difference is not noticeable and the value of k_{res}/k_{tot} remains throughout.

Figure 11 shows the ratio of subgrid dissipation with respect to the total value $\varepsilon_{SGS}/\varepsilon_{tot}$ along the wake for the Ti3 case. Note that, as the positions of comparison are no longer restricted to those of the available experimental data, profiles are shown at different longitudinal positions from other figures. In Figure 11, it is noticed that the LES has an appreciable increment in subgrid dissipation within the shear layer. Furthermore, this increase persists longitudinally even as far as when the wake appears to reach a state of transversally-uniform dissipation, i.e. at $x = 12D$ with disk $C_T = 0.62$. This is consistent with the hypothesis that turbulence is created at smaller lengthscales than the ambient turbulence at the disk edge. Due to the limited resolution of turbulence lengthscales in the Ti3 flow (missing in the synthetic flow as well), the increase in subgrid dissipation is produced at scales that seem absent in the incoming flow. These results also show that the wake envelope becomes the main carrier of dissipation. The subgrid dissipation part is also larger with higher thrust, yet by a small margin. It is observed that, in the absence of disks, most of the dissipation comes from the resolved fluctuations, in a proportion very similar to that shown outside the wake, at $y \pm 1.0D$ [54]. It is seen in Figure 12 that when the inflow turbulence raises (which comprises better resolved lengthscales), the increment of subgrid dissipation in the region of the wake envelope is greatly reduced. As a result, the modelling ratio seen outside the wake is essentially conserved. From these results, it can be deduced that the LES modelling across the wake is largely determined by the ambient turbulence.

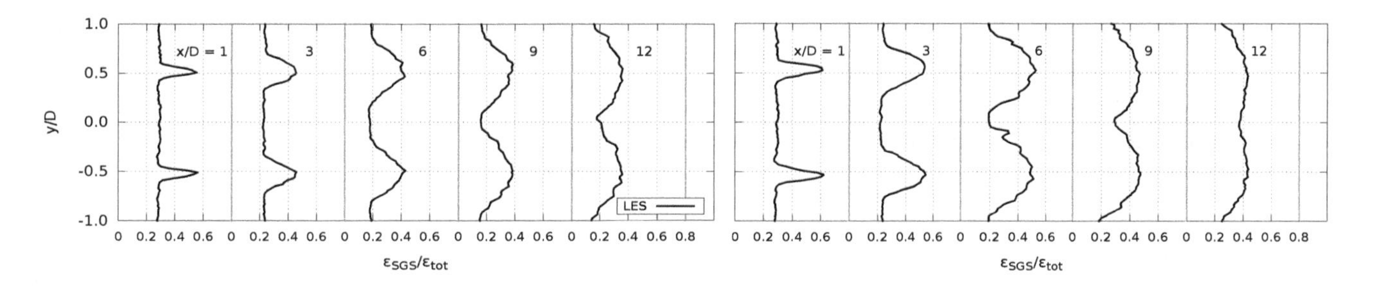

Figure 11. Vertical profiles of $\varepsilon_{SGS}/\varepsilon_{tot}$ behind the disk $C_T = 0.42$ (**left**) and $C_T = 0.62$ (**right**), Ti3 case.

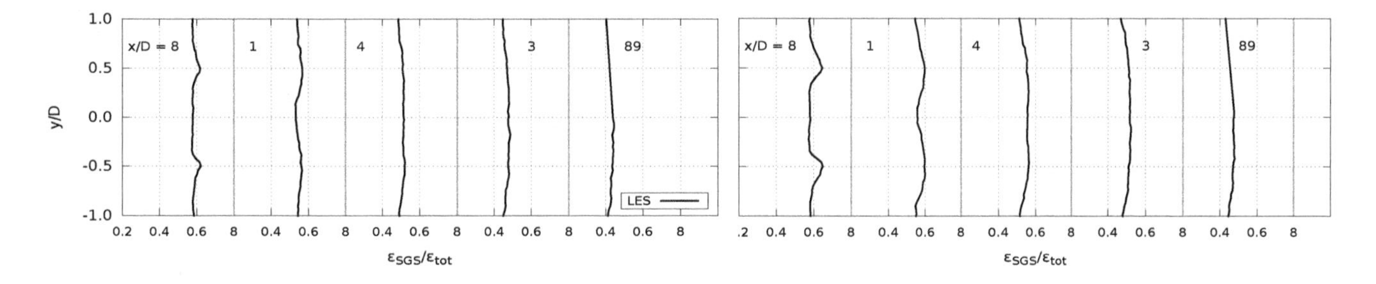

Figure 12. Vertical profiles of $\varepsilon_{SGS}/\varepsilon_{tot}$ behind the disk $C_T = 0.45$ (**left**) and $C_T = 0.71$ (**right**), Ti12 case.

5.6. Integral Lengthscale in the Wake

The changes in L_1 caused by the presence of the AD and the wake are now investigated. The computation of L_1 is performed as described in Section 4.4, which involves the assumption of the Taylor hypothesis to transform the computed time-scales into lengthscales. Evidently, this supposition becomes more difficult to accept when shear is present in the flow. However, previous work has reported satisfactory results in wake studies that support the continuing applicability of the hypothesis. For instance, Ref. [16] has compared the lateral distribution of L_1 behind the wake produced by a porous disk (in a setting similar to the experiments used in this work) computed from

HWA with the one obtained from PIV. They did not find a difference in the results obtained from either technique, despite the fact that HWA uses the local mean velocity to calculate the lengthscale, compared to the direct spatial measurement offered by PIV. Making the same assumption, the evolution of the integral lengthscale behind the AD computations is studied.

Figure 13 displays the values of L_1 computed from the LES in each code with the Ti3 inflow, from $y = 0$ to $y = 1.5$, at $3D$, $4D$ and $6D$, which correspond to the positions where HWA data for the Ti12 inflow is available. It is first noticed that there is not a clear influence of the shear layer in the size of the turbulence scales. However, for a region about $0.5 \leq y/D \leq 1.0$, next to the the shear layer, larger lengthscales can be discerned amongst the variations in the profile. Indeed, the maximum values of L_1 at each $x-$position are at close $y = 0.5D$ in the wake of the disk $C_T = 0.42$. This is consistent with the previous results with regard to the location of the shear layer along the wake (e.g. k and ε). Conversely, for the other disk the maxima of L_1 would suggest a wake that expands to about $y = 0.75D$ at $x = 6D$, which is larger than what the previous computations indicate.

Results for the Ti12 inflow are shown in Figure 14. Notably, the computed values from the experimental time-series do not reveal a variation of the lengthscale values at the shear layer. In fact, there is no evident change in L_1 within the wake. This trait is similarly observed in the LES results. The only variations in computations are observed at the upper part of the curves or, in the case of the disk $C_T = 0.45$, towards the bottom part where L_1 is larger (but this effect is reduced further downstream).

Previous experimental work by [16] showed that in the wake of a porous disk with a solidity of 45%, L_1 is approximately 1.5 times larger within the shear layer with respect to the values within the wake or outside the envelope. However, these measurements were obtained using an inflow with very low turbulence (TI $<$ 0.4%), which clearly sets a different scenario in comparison to this study. Precisely, the absence of a variation of L_1 in the shear layer can be explained considering the previous results, which point at a dominance of the ambient turbulence characteristics over the wake in the case of the inflow Ti12. Although the turbulence production is visibly higher when the disk thrust is larger (Figure 8), its effect does not appear to have an impact on the turbulence lengthscales. Similarly, the use of a lower turbulence inflow (Ti3) does not seem to decidedly increase the magnitude of the lengthscales in the area of turbulence production, or at least not in the computations performed for this work. In this regard, the fact that the characteristic lengthscales of the Ti12 inflow are better resolved by the mesh and the LES compared to the Ti3 cases can be a factor to consider. This is, if resolution is not adequate within the shear layer, it is to be expected that a sizeable part of the turbulence being produced would fall into the modelled part instead of being resolved, therefore affecting the magnitude of the computed scales. This has been studied in Section 5.5, where it is shown that the LES modelling does not vary within the wake with respect to the external flow aside from very close to the disk ($x = 1D$) in both codes. Nevertheless, it has been seen that despite the limited resolution, the LES computations have been able to reproduce other principal features along the wake, such as the turbulence levels.

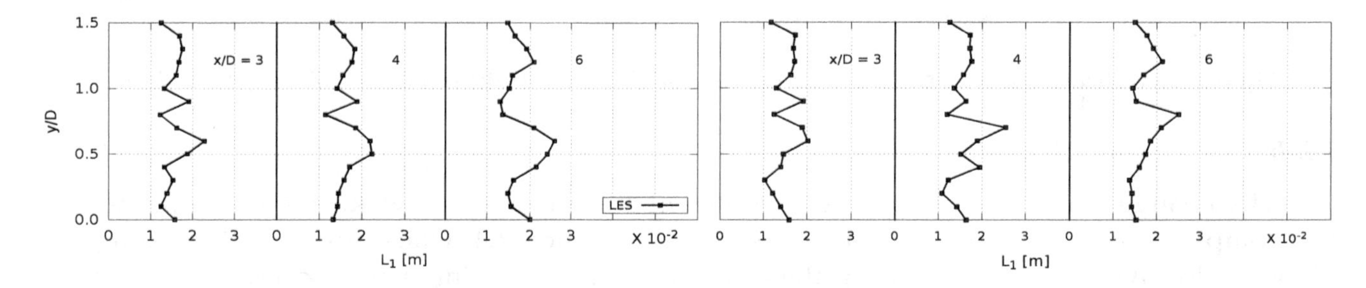

Figure 13. Vertical profiles of L_1 behind the AD with inflow Ti3, disks: $C_T = 0.42$ (**left**) and $C_T = 0.62$ (**right**).

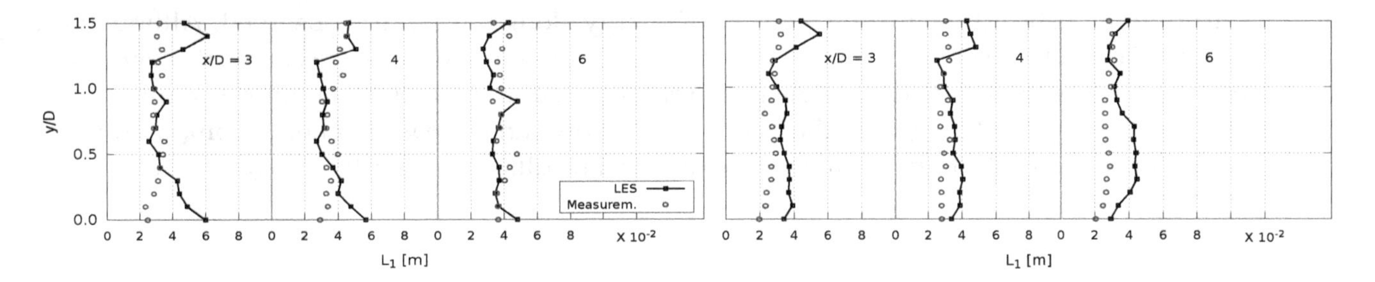

Figure 14. Vertical profiles of L_1 behind the AD with inflow Ti12, disk $C_T = 0.45$ (**left**) and $C_T = 0.71$ (**right**).

5.7. Spectra behind Disks

To study the redistribution of turbulence energy along the wake, the spectra obtained at different longitudinal positions for every disk are compared with the spectrum of the free decaying turbulence. Power Spectral Density (PSD) curves are calculated from only one measuring position at centreline, so to reduce the noise in the spectral curves, the time-series of each register are divided into eight non-overlapping blocks with an equal number of samples. Then, the PSD of all blocks are averaged to produce the curve at each longitudinal position. However, noise remains along the curves that make the comparison very difficult, so a smoothing procedure is needed to be performed. To this aim, an exponential moving average is used to filter the spectra computed at each longitudinal position (a rational transfer function is employed, see [55]). Hence, the spectra shown in the following figures have been processed with this technique, with the sole exception of that obtained from measurements without a disk, which was spatially averaged with results obtained at the other eight locations distributed crosswisely. As the spectra are calculated from data at a fixed location (sampled in time), the Taylor hypothesis is applied to transform the frequency spectra into a wavenumber spectra using $\kappa_1 = 2\pi f / \langle U \rangle$, where f is f_{acq} for measurements or $f = 1/\Delta t$ for the LES. In this way, it is possible to also compare with the PSD from the synthetic turbulence, which is calculated as the volume average of the spectra computed in the longitudinal direction.

The results for the inflow Ti3 are shown in Figure 15. In the results without the AD, a constant decay of energy is observed as the flow moves downstream. The spectra from the synthetic box serve to mark the extension of the resolved wavenumbers ($\kappa_{max} = \pi/\Delta = 1571$ m^{-1}) since the spatial resolution in the box is the same as in the LES. Note that the abrupt drop in the turbulence energy spectra is attributed to a combination of numerical diffusion and the limited spatial resolution [5]. In case of the disk $C_T = 0.42$, a gain in turbulence energy is seen immediately behind the disk, as the curves at $-1D$ and $1D$ are almost identical. A small decay of energy is observed at $4D$ and, from there, an increase in turbulence energy around the highest levels (lowest κ). For the disk $C_T = 0.62$, the effects are accentuated, and the curves at $4D$ are the only ones displaying a decay and yet only around the inertial range. The energy of the next two longitudinal positions, $10D$ and $14D$, increases for all wavenumbers, which represents an increment of about one order of magnitude at the lowest wavenumber, with respect to the levels displayed by the decaying turbulence without disks. Notably, the spectra of the last two positions seemingly exhibit an inertial range, characterized by the slope of $-5/3$ in the decay rate.

The results for the Ti12 inflow are shown in Figure 16. In this case, the spectra computed from experimental results are also included. The spectra obtained from measurements with disks extend to larger wavenumbers than in the cases without disk, due to the use of a different frequency in the low-pass filter. For the case without disks, the energy at the lowest wavenumbers obtained from the LES proves to decay less as it has been shown before. However, they display a steady decay that adjusts well to the characteristic slope of the intertial range, also discernable in the experimental results. These observations are analogous for the results with the disk $C_T = 0.45$. In contrast to the Ti3 inflow where energy is seen to increase beyond $x = 4D$ for the disk with the same porosity, here a reduction in the contribution of shear towards the increase of energy along the wake is observed. Although the overall levels of turbulence energy in the wake are higher than in the free decaying turbulence, they maintain more or less the same relative decay from one to another. This behaviour is similar in the case of the disk $C_T = 0.71$. There, only the curve at $4D$ shows an increase in energy compared to the previous disk (also matching fairly well the experimental results in the inertial range).

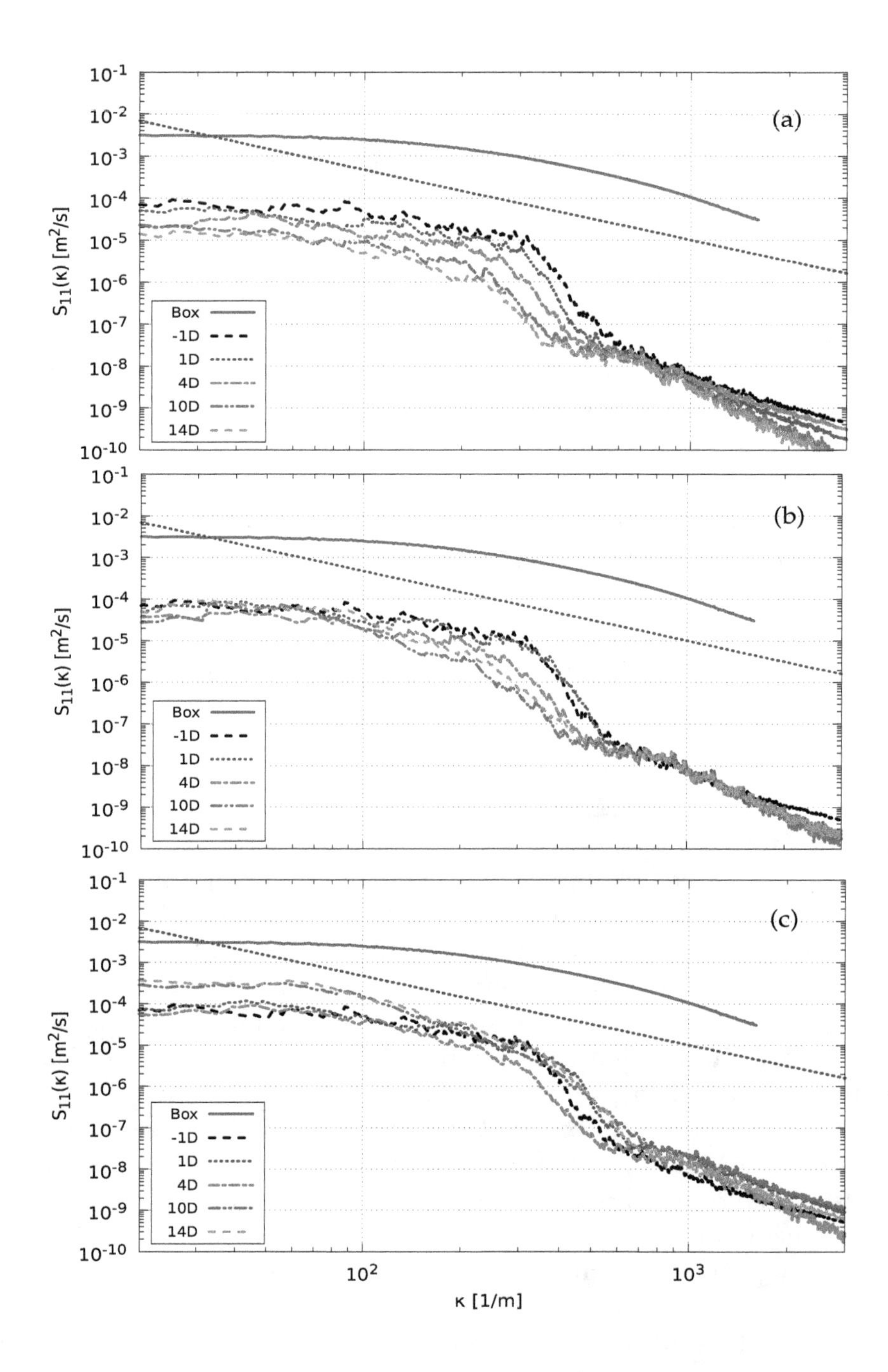

Figure 15. Longitudinal evolution of spectra at centreline using the Ti3 inflow. The results for the free decaying turbulence (without AD) are shown in the top row (**a**), results with disk $C_T = 0.42$ are shown in the middle row (**b**) and results with disk $C_T = 0.62$ are shown in the bottom row (**c**). The straight dotted line marks the $-5/3$ slope that characterizes the inertial range.

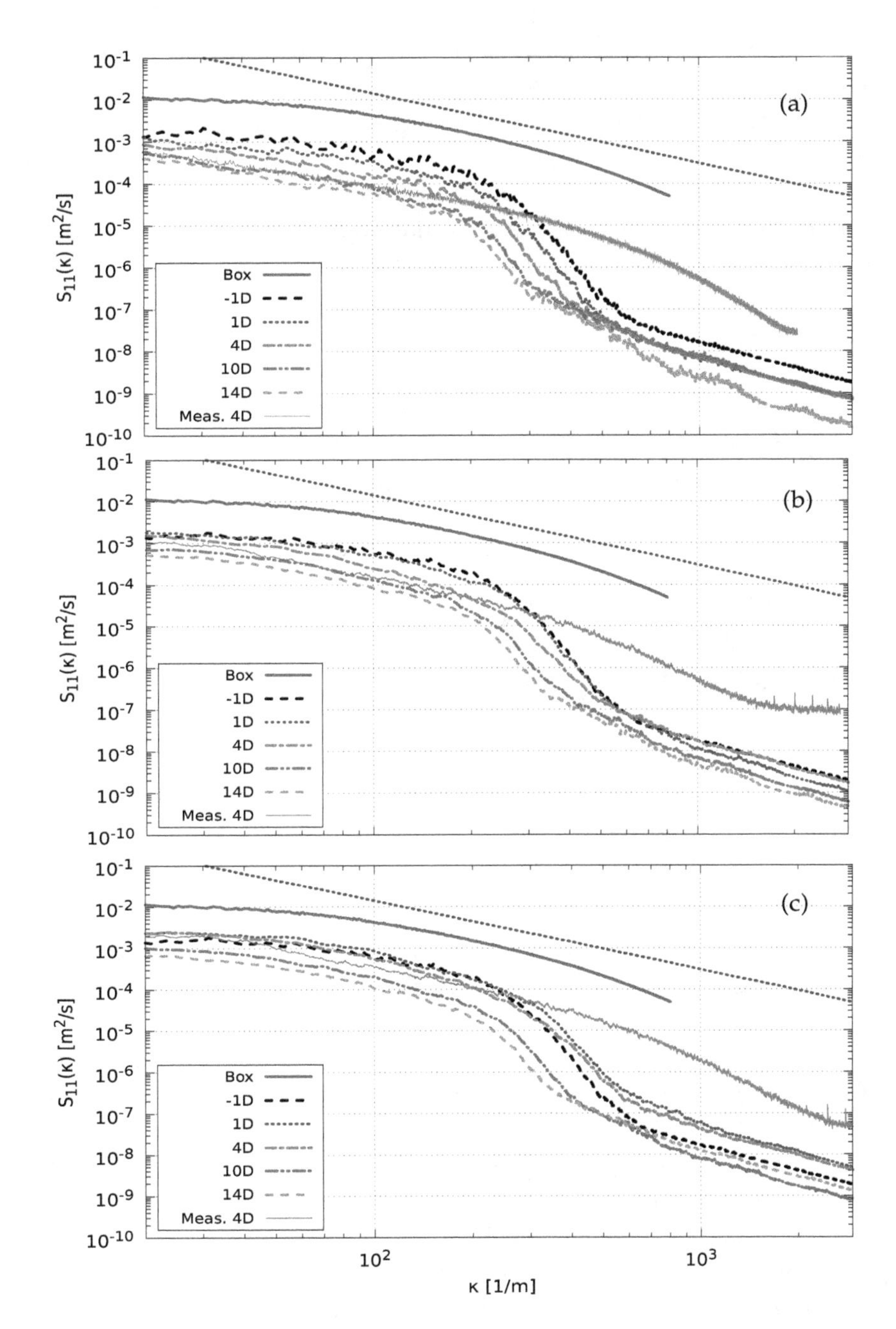

Figure 16. Longitudinal evolution of spectra at centreline using the Ti12 inflow. The results for the free decaying turbulence (without AD) are shown in the top row (**a**), results with disk $C_T = 0.45$ are shown in the middle row (**b**) and results with disk $C_T = 0.71$ are shown in the bottom row (**c**). Spectra computed from measurements are included only for the position $x = 4D$. The straight dotted line marks the $-5/3$ slope that characterizes the inertial range.

5.8. Vorticity Contours

Lastly, to complement all previous results, Figure 17 shows a comparison between the the contours of vorticity obtained in the Ti3 and Ti12 cases and using the highest porosity disk, $C_T = 0.62$ and $C_T = 0.71$, respectively. The images are taken at the x–y plane, at $z = 0$ and correspond to the vorticity field computed at the last time step of the LES runs. Make note that black bars are used to represent the disk position but do not portray the complete longitudinal region where the forces modelling the AD act, distributed using Equation (7). This figure permits visualizing the dominant effect of the turbulence structures from the inflow of the Ti12 case, which, unlike the Ti3 inflow, prevail along the wake. With the Ti3 inflow, there is a clear shear region arising from the edges of the disk with distinct turbulence structures. Conversely, with the use of the Ti12 inflow, the shear region is substantially less noticeable in the higher ambient turbulence. Indeed, the vorticity contours from the inflow appear to dominate in the vicinity of the AD. This is consistent with the comparison of k in Figures 5 and 6, where its increase within the shear layer is less prominent with the increase of ambient TI. In Figure 17, it can also be seen that, for the Ti12 inflow, structures that appear to arise from within the wake become more apparent than those outside from approximately $x \simeq 6D$.

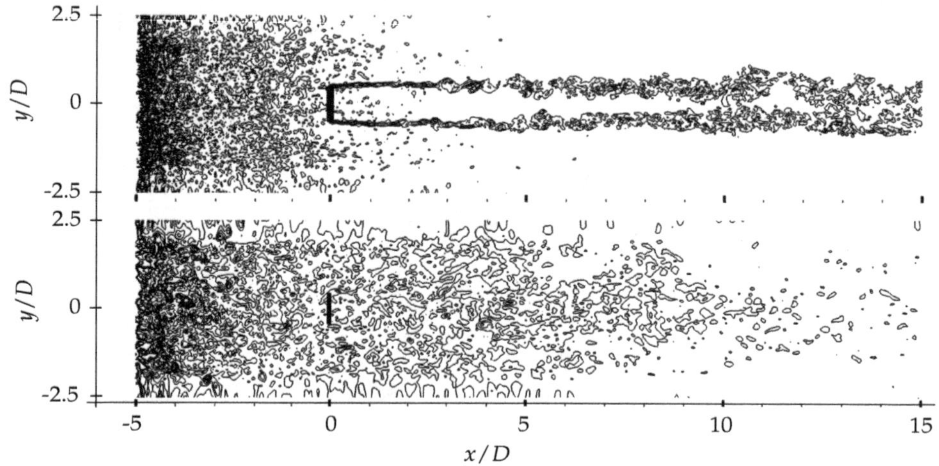

Figure 17. Contours of the vorticity field obtained with the Ti3 inflow and disk $C_T = 0.62$ (**top**) and the Ti12 inflow and disk $C_T = 0.71$ (**bottom**).

6. Summary and Conclusions

This work has been dedicated to study the flow characteristics as well as the modelling of turbulence in wakes produced by a porous disk. For this, disks of two porosities have been used as well as two instances of decaying and homogeneous turbulence inflows, identified by streamwise turbulence intensities (TI) of approximately 3% and 12% and corresponding longitudinal integral lengthscales (L_1) of 0.01 m and 0.03 m. To perform this study, a methodology has been developed and implemented in OpenFOAM that uses Large-Eddy Simulations (LES) to reproduce the main turbulence features of a turbulent flow generated in a wind tunnel. The Actuator Disk (AD) technique was used to model the porous disk and replicate the turbulence properties measured in the wake flow. The employed methodology is based on a procedure previously shown to provide good results in the simulation of wind turbine wakes in homogeneous turbulence and atmospheric flows. A comparison with prior work made with RANS was made wherever possible.

To reproduce the grid-generated turbulence in the wind tunnel, a synthetic field of homogeneous isotropic turbulence was introduced at the inlet of an LES. An analysis of the decaying turbulence simulated by the LES demonstrated that the simulations reproduce the TI decay and the evolution of L_1 predicted by the parametrizations predominant in the literature, in addition to replicating fairly well the experimental values at the measured locations. The wake field results obtained with the introduction of the AD showed a good agreement with the wind tunnel experiments for the velocity deficit, turbulence kinetic energy k and its dissipation ε. It was also found that L_1 evolves, for the most part, as in the decaying homogeneous turbulence. Remarkably, there was no apparent variation of L_1 within the shear layer compared to the center of the wake or outside the wake envelope.

A study of how the LES performs in the wake was also carried out, with the main interest of observing how the modelling ratio (the contribution of the resolved or SGS parts to the total value) is affected by the regions of high shear behind the AD in comparison with the outflow and the center of the wake. It was found that while the modelling ratio of k in the wake is largely maintained with respect to the outside flow, the subgrid part of ε increases within the shear layer, but only when the low TI inflow is used. From the results, it is seen that modelling in the freestream flow prevails in the wake just as the level of inflow turbulence increases. On the other hand, the wake results compare well with the velocity and k obtained with RANS. Likewise, LES and RANS results of ε match also in the low TI case, whereas the values yielded by RANS seem overestimated—in comparison with both measurements and LES—in the high TI case, particularly in regions of strong shear.

Spectra computed at different axial positions in the wake reveal that shear causes a noticeable boost in the energy content of turbulence, but only in the low TI case. This causes that for the two furthermost positions ($x = 10D$ and $14D$), the energy levels are higher or at least as energetic as in the upstream region near the disks. Moreover, the turbulence at those positions shows a clear inertial range that was absent in the decaying turbulence at low TI. Conversely, for the high TI inflow, it is seen that, despite the fact that turbulence energy levels rise in the wake with respect to the decaying homogeneous flow, the relative decay is maintained from one position to the other.

Considering the above, the central conclusions of this investigation are:

- The evolution of L_1 was not noticeably different in the wake in comparison to what was observed in the decaying homogeneous turbulence.
- Turbulence scales in the wake appear to be dominated by the inflow characteristics and this effect increases with the level of TI in the inflow.
- It is seen that the resolved/subgrid modelling ratio in the freestream flow prevails in the wake in relation to the increasing level of ambient turbulence.

The comparison of the LES with experimental results has permitted validating the methodology presented in this work as a suitable approach to investigate the turbulence features in the wake flow of wind turbine models. With this method, it can be possible to study the spatial and transient development of wake characteristics and assess the influence of, for example, a variety of inflow turbulence conditions or wind turbine models of varying sophistication. These aspects are considered for future work. In the same manner, an analogous study can be performed for an array of wind turbine models with the additional interest of observing the influence of downstream turbines in the wake turbulence field, for different layouts. Clearly, a CFD tool like the one presented in this work bears the advantage of changing the simulation setup without needing to perform new experiments each time, while also removing the spatial or temporal limitations associated with the measuring techniques.

Author Contributions: Conceptualization, H.O.-E. and S.-P.B.; Methodology, H.O.-E. and S.-P.B.; Software, H.O.-E., S.-P.B. and K.N.; Validation, H.O.-E.; Formal Analysis, H.O.-E.; Investigation, H.O.-E.; Writing—Original Draft Preparation, H.O.-E. with contributions of all authors to its review and editing; Supervision, C.M. and L.D.; Project Administration, C.M., L.D. and S.I.; Funding Acquisition, H.O.-E., C.M., L.D. and S.I.

Acknowledgments: We thank Sandrine Aubrun and Yann-Aël Muller of Université d'Orléans for facilitating the wind tunnel measurements used in this work. Likewise, we thank Jonathon Sumner from Dawson College (Montreal, Canada) for the RANS data. The computations were performed on resources provided by the Calcul Québec-Compute Canada consortium. Complementary calculations were performed on resources provided by the Swedish National Infrastructure for Computing (SNIC) at the National Supercomputer Centre (NSC).

References

1. Crespo, A.; Hernandez, J.; Frandsen, S. Survey of modelling methods for wind turbine wakes and wind farms. *Wind Energy* **1999**, *2*, 1–24. [CrossRef]
2. Calaf, M.; Meneveau, C.; Meyers, J. Large eddy simulation study of fully developed wind-turbine array boundary layers. *Phys. Fluids* **2010**, *22*, 015110. [CrossRef]
3. Churchfield, M.J.; Lee, S.; Moriarty, P.J.; Martinez, L.A.; Leonardi, S.; Vijayakumar, G.; Brasseur, J.G. A large-eddy simulation of wind-plant aerodynamics. In Proceedings of the 50th AIAA Aerospace Sciences Meeting Including the New Horizons Forum and Aerospace Exposition, Nashville, TN, USA, 9–12 January 2012; p. 537.

4. Nilsson, K.; Ivanell, S.; Hansen, K.S.; Mikkelsen, R.; Sørensen, J.N.; Breton, S.P.; Henningson, D. Large-eddy simulations of the Lillgrund wind farm. *Wind Energy* **2015**, *18*, 449–467. [CrossRef]
5. Troldborg, N. Actuator Line Modeling of Wind Turbine Wakes. Ph.D. Thesis, Technical University of Denmark, Lyngby, Denmark, 2008.
6. Ivanell, S.A. Numerical Computations of Wind Turbine Wakes. Ph.D. Thesis, Royal Institute of Technology, Stockholm, Sweden, 2009.
7. Chamorro, L.P.; Porté-Agel, F. A wind-tunnel investigation of wind-turbine wakes: Bundary-layer turbulence effects. *Bound.-Layer Meteorol.* **2009**, *132*, 129–149. [CrossRef]
8. Porté-Agel, F.; Wu, Y.T.; Lu, H.; Conzemius, R.J. Large-eddy simulation of atmospheric boundary layer flow through wind turbines and wind farms. *J. Wind Eng. Ind. Aerodyn.* **2011**, *99*, 154–168. [CrossRef]
9. Sørensen, J.N.; Myken, A. Unsteady actuator disc model for horizontal axis wind turbines. *J. Wind Eng. Ind. Aerodyn.* **1992**, *39*, 139–149. [CrossRef]
10. Ammara, I.; Leclerc, C.; Masson, C. A viscous three-dimensional differential/actuator-disk method for the aerodynamic analysis of wind farms. *J. Sol. Energy Eng.* **2002**, *124*, 345–356. [CrossRef]
11. Jiménez, A.; Crespo, A.; Migoya, E.; Garcia, J. Large-eddy simulation of spectral coherence in a wind turbine wake. *Environ. Res. Lett.* **2008**, *3*, 015004. [CrossRef]
12. Aubrun, S.; Loyer, S.; Hancock, P.; Hayden, P. Wind turbine wake properties: Comparison between a non-rotating simplified wind turbine model and a rotating model. *J. Wind Eng. Ind. Aerodyn.* **2013**, *120*, 1–8. [CrossRef]
13. Aubrun, S.; Devinant, P.; Espana, G. Physical modelling of the far wake from wind turbines. Application to wind turbine interactions. In Proceedings of the European Wind Energy Conference, Milan, Italy, 7–10 May 2007; pp. 7–10.
14. Espana, G.; Aubrun, S.; Loyer, S.; Devinant, P. Spatial study of the wake meandering using modelled wind turbines in a wind tunnel. *Wind Energy* **2011**, *14*, 923–937. [CrossRef]
15. Espana, G.; Aubrun, S.; Loyer, S.; Devinant, P. Wind tunnel study of the wake meandering downstream of a modelled wind turbine as an effect of large scale turbulent eddies. *J. Wind Eng. Ind. Aerodyn.* **2012**, *101*, 24–33. [CrossRef]
16. Thacker, A.; Loyer, S.; Aubrun, S. Comparison of turbulence length scales assessed with three measurement systems in increasingly complex turbulent flows. *Exp. Therm. Fluid Sci.* **2010**, *34*, 638–645. [CrossRef]
17. Sumner, J.; Espana, G.; Masson, C.; Aubrun, S. Evaluation of RANS/actuator disk modelling of wind turbine wake flow using wind tunnel measurements. *Int. J. Eng. Syst. Model. Simul.* **2013**, *5*, 147–158. [CrossRef]
18. Tabor, G.R.; Baba-Ahmadi, M. Inlet conditions for large eddy simulation: a review. *Comput. Fluids* **2010**, *39*, 553–567. [CrossRef]
19. Lund, T.S.; Wu, X.; Squires, K.D. Generation of turbulent inflow data for spatially-developing boundary layer simulations. *J. Comput. Phys.* **1998**, *140*, 233–258. [CrossRef]
20. Klein, M.; Sadiki, A.; Janicka, J. A digital filter based generation of inflow data for spatially developing direct numerical or large eddy simulations. *J. Comput. Phys.* **2003**, *186*, 652–665. [CrossRef]
21. Mann, J. The spatial structure of neutral atmospheric surface-layer turbulence. *J. Fluid Mech.* **1994**, *273*, 141–168. [CrossRef]
22. Mann, J. Wind field simulation. *Probab. Eng. Mech.* **1998**, *13*, 269–282. [CrossRef]
23. Peña, A.; Hasager, C.; Lange, J.; Anger, J.; Badger, M.; Bingöl, F.; Bischoff, O.; Cariou, J.P.; Dunne, F.; Emeis, S.; et al. *Remote Sensing for Wind Energy*; Technical Report; DTU Wind Energy: Roskilde, Denmark, 2013.
24. Keck, R.E.; Mikkelsen, R.; Troldborg, N.; de Maré, M.; Hansen, K.S. Synthetic atmospheric turbulence and wind shear in large eddy simulations of wind turbine wakes. *Wind Energy* **2014**, *17*, 1247–1267. [CrossRef]
25. Nilsson, K. Numerical Computations of Wind Turbine Wakes and Wake Interaction. Ph.D. Thesis, Royal Institute of Technology, Stockholm, Sweden, 2015.
26. Bechmann, A. Large-Eddy Simulation of Atmospheric Flow over Complex. Ph.D. Thesis, Risø Technical University of Denmark, Roskilde, Denmark, 2006.
27. Gilling, L.; Sørensen, N.N. Imposing resolved turbulence in CFD simulations. *Wind Energy* **2011**, *14*, 661–676. [CrossRef]
28. Troldborg, N.; Zahle, F.; Réthoré, P.E.; Sørensen, N.N. Comparison of wind turbine wake properties in non-sheared inflow predicted by different computational fluid dynamics rotor models. *Wind Energy* **2015**, *18*, 1239–1250. [CrossRef]

29. Gilling, L. *TuGen: Synthetic Turbulence Generator, Manual and User's Guide*; Technical Report DCE-76; Department of Civil Engineering, Aalborg University: Aalborg, Denmark, 2009.
30. Nilsen, K.M.; Kong, B.; Fox, R.O.; Hill, J.C.; Olsen, M.G. Effect of inlet conditions on the accuracy of large eddy simulations of a turbulent rectangular wake. *Chem. Eng. J.* **2014**, *250*, 175–189. [CrossRef]
31. Espana, G. Étude expérimentale du sillage lointain des éoliennes à axe horizontal au moyen d'une modélisation simplifiée en couche limite atmosphérique. Ph.D. Thesis, Université d'Orléans, Orléans, France, 2009. (In French)
32. Weller, H.G.; Tabor, G.; Jasak, H.; Fureby, C. A tensorial approach to computational continuum mechanics using object-oriented techniques. *Comput. Phys.* **1998**, *12*, 620–631. [CrossRef]
33. The OpenFOAM Foundation. OpenFOAM: The Open Source CFD Toolbox; User Guide. 2016. Available online: https://cfd.direct/openfoam/user-guide (accessed on 29 August 2018).
34. Michelsen, J. *Basis3D—A Platform for Development of Multiblock PDE Solvers*; Technical Report AFM 92-05; Technical University of Denmark: Lyngby, Denmark, 1992.
35. Michelsen, J.A. *Block Structured Multigrid Solution of 2D and 3D Elliptic PDE's*; Technical Report AFM 94-06; Technical University of Denmark: Lyngby, Denmark, 1994.
36. Sørensen, N.N. General Purpose Flow Solver Applied to Flow over Hills. Ph.D. Thesis, Risø Technical University of Denmark, Lyngby, Denmark, 1995.
37. Bailly, C.; Comte-Bellot, G. *Turbulence*; CNRS éditions: Paris, France, 2003. (In French)
38. Jiménez, J. (Ed.) *A Selection of Test Cases for the Validation of Large-Eddy Simulations of Turbulent Flows*; Technical Report AGARD Advisory Report No. 345; Working Group 21 of the Fluid Dynamics Panel, North Atlantic Treaty Organization: Neuilly-sur-Seine, France, 1997.
39. Kang, H.S.; Chester, S.; Meneveau, C. Decaying turbulence in an active-grid-generated flow and comparisons with large-eddy simulation. *J. Fluid Mech.* **2003**, *480*, 129–160. [CrossRef]
40. Pope, S.B. *Turbulent Flows*; Cambridge Univ Press: Cambridge, UK, 2000.
41. Comte-Bellot, G.; Corrsin, S. The use of a contraction to improve the isotropy of grid-generated turbulence. *J. Fluid Mech.* **1966**, *25*, 657–682. [CrossRef]
42. Mydlarski, L.; Warhaft, Z. On the onset of high-Reynolds-number grid-generated wind tunnel turbulence. *J. Fluid Mech.* **1996**, *320*, 331–368. [CrossRef]
43. Smagorinsky, J. General circulation experiments with the primitive equations: I. The basic experiment*. *Mon. Weather Rev.* **1963**, *91*, 99–164. [CrossRef]
44. Muller, Y.A. Étude du méandrement du sillage éolien lointain dans différentes conditions de rugosité. Ph.D. Thesis, Université d'Orléans, Orléans, France, 2014. (In French)
45. Porté-Agel, F.; Meneveau, C.; Parlange, M.B. A scale-dependent dynamic model for large-eddy simulation: Application to a neutral atmospheric boundary layer. *J. Fluid Mech.* **2000**, *415*, 261–284. [CrossRef]
46. Larsen, T.J. *Turbulence for the IEA Annex 30 OC4 Project*; Technical Report I-3206; Risø Technical University of Denmark: Roskilde, Denmark, 2013.
47. Kaimal, J.C.; Finnigan, J.J. *Atmospheric Boundary Layer Flows: Their Structure and Measurement*; Oxford University Press: Oxford, UK, 1994.
48. El Kasmi, A.; Masson, C. An extended k–ε model for turbulent flow through horizontal-axis wind turbines. *J. Wind Eng. Ind. Aerodyn.* **2008**, *96*, 103–122. [CrossRef]
49. Réthoré, P.E.M. Wind Turbine Wake in Atmospheric Turbulence. Ph.D. Thesis, Technical University of Denmark, Risø National Laboratory for Sustainable Energy Risø National laboratoriet for Bæredygtig Energi, Lyngby, Denmark, 2009.
50. Comte-Bellot, G.; Corrsin, S. Simple Eulerian time correlation of full-and narrow-band velocity signals in grid-generated, 'isotropic'turbulence. *J. Fluid Mech.* **1971**, *48*, 273–337. [CrossRef]
51. Mydlarski, L.; Warhaft, Z. Passive scalar statistics in high-Péclet-number grid turbulence. *J. Fluid Mech.* **1998**, *358*, 135–175. [CrossRef]
52. Mohamed, M.S.; Larue, J.C. The decay power law in grid-generated turbulence. *J. Fluid Mech.* **1990**, *219*, 195–214. [CrossRef]
53. Fletcher, C. *Computational Techniques for Fluid Dynamics 1*; Springer Science & Business Media: Berlin, Germany, 1991; Volume 1.

54. Olivares-Espinosa, H. Turbulence Modelling in Wind Turbine Wakes. Ph.D. Thesis, École de Technologie Supérieure, Université du Québec, Montreal, QC, Canada, 2017.
55. Oppenheim, A.V.; Schafer, R.W.; Buck, J.R. *Discrete-Time Signal Processing*; Prentice Hall: Upper Saddle River, NJ, USA, 1999.

Permissions

All chapters in this book were first published in MDPI; hereby published with permission under the Creative Commons Attribution License or equivalent. Every chapter published in this book has been scrutinized by our experts. Their significance has been extensively debated. The topics covered herein carry significant findings which will fuel the growth of the discipline. They may even be implemented as practical applications or may be referred to as a beginning point for another development.

The contributors of this book come from diverse backgrounds, making this book a truly international effort. This book will bring forth new frontiers with its revolutionizing research information and detailed analysis of the nascent developments around the world.

We would like to thank all the contributing authors for lending their expertise to make the book truly unique. They have played a crucial role in the development of this book. Without their invaluable contributions this book wouldn't have been possible. They have made vital efforts to compile up to date information on the varied aspects of this subject to make this book a valuable addition to the collection of many professionals and students.

This book was conceptualized with the vision of imparting up-to-date information and advanced data in this field. To ensure the same, a matchless editorial board was set up. Every individual on the board went through rigorous rounds of assessment to prove their worth. After which they invested a large part of their time researching and compiling the most relevant data for our readers.

The editorial board has been involved in producing this book since its inception. They have spent rigorous hours researching and exploring the diverse topics which have resulted in the successful publishing of this book. They have passed on their knowledge of decades through this book. To expedite this challenging task, the publisher supported the team at every step. A small team of assistant editors was also appointed to further simplify the editing procedure and attain best results for the readers.

Apart from the editorial board, the designing team has also invested a significant amount of their time in understanding the subject and creating the most relevant covers. They scrutinized every image to scout for the most suitable representation of the subject and create an appropriate cover for the book.

The publishing team has been an ardent support to the editorial, designing and production team. Their endless efforts to recruit the best for this project, has resulted in the accomplishment of this book. They are a veteran in the field of academics and their pool of knowledge is as vast as their experience in printing. Their expertise and guidance has proved useful at every step. Their uncompromising quality standards have made this book an exceptional effort. Their encouragement from time to time has been an inspiration for everyone.

The publisher and the editorial board hope that this book will prove to be a valuable piece of knowledge for researchers, students, practitioners and scholars across the globe.

List of Contributors

Bofeng Xu, Yue Yuan, Zhenzhou Zhao and Haoming Liu
College of Energy and Electrical Engineering, Hohai University, Nanjing 211100, China

Hadi Sanati, David Wood and Qiao Sun
Department of Mechanical and Manufacturing Engineering, University of Calgary, 2500
University Drive N.W., Calgary, AB T2N 1N4, Canada

Shoutu Li, Congxin Yang, Xuyao Zhang, Qing Wang and Deshun Li
School of Energy and Power Engineering, Lanzhou University of Technology, Lanzhou 730050, China
Gansu Provincial Technology Centre for Wind Turbines, Lanzhou 730050, China
Key Laboratory of Fluid machinery and Systems, Lanzhou 730050, China

Ye Li
School of Naval Architecture, Ocean and Civil Engineering, Shanghai Jiao Tong University, Shanghai 200240, China
State Key Laboratory of Ocean Engineering, School of Naval Architecture, Ocean and Civil Engineering, Shanghai Jiao Tong University, Shanghai 200240, China
Collaborative Innovation Center for Advanced Ship and Deep-Sea Exploration, Shanghai Jiao Tong University, Shanghai 200240, China
Key Laboratory of Hydrodynamics (Ministry of Education), Shanghai Jiao Tong University, Shanghai 200240, China

Rudi Purwo Wijayanto
Graduate School of Natural Science and Technology, Kanazawa University, Kanazawa 920-1192, Japan
The Agency of the Assessment and Application of Technology (BPPT), Jakarta 10340, Indonesia

Takaaki Kono and Takahiro Kiwata
Institute of Science and Engineering, Kanazawa University, Kanazawa 920-1192, Japan

Jianghai Wu, Long Wang and Ning Zhao
Jiangsu Key Laboratory of Hi-Tech Research for Wind Turbine Design, Nanjing University of Aeronautics and Astronautics, Nanjing 210016, China

Ju Feng and Wen Zhong Shen
Department of Wind Energy, Technical University of Denmark, 2800 Kgs. Lyngby, Denmark

Alois Peter Schaffarczyk and Andreas Jeromin
Mechanical Engineering Department, Kiel University of Applied Sciences, D-24149 Kiel, Germany

Qiang Wang and Kangping Liao
College of Shipbuilding Engineering, Harbin Engineering University, Harbin 150001, China

Qingwei Ma
School of Mathematics, Computer Sciences & Engineering, City, University of London, London EC1V 0HB, UK

Lena Vorspel
ForWind, University of Oldenburg, Ammerländer Heerstr. 114-118, 26129 Oldenburg, Germany

Bernhard Stoevesandt
Fraunhofer IWES, Küpkersweg 70, 26129 Oldenburg, Germany

Joachim Peinke
ForWind, University of Oldenburg, Ammerländer Heerstr. 114-118, 26129 Oldenburg, Germany
Fraunhofer IWES, Küpkersweg 70, 26129 Oldenburg, Germany

Chang Xu, Chenyan Hao, Linmin Li, Xingxing Han and Feifei Xue
College of Energy and Electrical Engineering, Hohai University, Nanjing 211100, China

Mingwei Sun
College of Naval Coast Defence Arm, Naval Aeronautical University, Yantai 264001, China

Wenzhong Shen
Department of Wind Energy, Fluid Mechanics Section, Technical University of Denmark, Nils Koppels Allé, Building 403, 2800 Kgs. Lyngby, Denmark

Yin Zhang and Bumsuk Kim
Faculty of Wind Energy Engineering Graduate School, Jeju National University, Jeju City 63243, Korea

Wei Zhong and Tong Guang Wang
Jiangsu Key Laboratory of Hi-Tech Research for Wind Turbine Design, Nanjing University of Aeronautics and Astronautics, Nanjing 210016, China

Wei Jun Zhu
College of Electrical, Energy and Power Engineering, Yangzhou University, Yangzhou 225009, China

Wen Zhong Shen
Department of Wind Energy, Technical University of Denmark, 2800 Lyngby, Denmark

Hugo Olivares-Espinosa
Department of Mechanical Engineering, École de Technologie Supérieure, 1100 Notre-Dame Ouest, Montréal, QC H3C 1K3, Canada
Wind Energy Section, Department of Earth Sciences, Uppsala University Campus Gotland, Cramérgatan 3, 62165 Visby, Sweden

Christian Masson and Louis Dufresne
Department of Mechanical Engineering, École de Technologie Supérieure, 1100 Notre-Dame Ouest, Montréal, QC H3C 1K3, Canada

Karl Nilsson and Stefan Ivanell
Wind Energy Section, Department of Earth Sciences, Uppsala University Campus Gotland, Cramérgatan 3, 62165 Visby, Sweden

Simon-Philippe Breton
Wind Energy Section, Department of Earth Sciences, Uppsala University Campus Gotland, Cramérgatan 3, 62165 Visby, Sweden
Nergica, 70 rue Bolduc, Gaspé, QC G4X 1G2, Canada

Index